RICE SCIENCE

Biotechnological and Molecular Advancements

RICE SCIENCE

Biotechnological and Molecular Advancements

Edited by
Deepak Kumar Verma
Prem Prakash Srivastav, PhD
Altafhusain B. Nadaf, PhD

Apple Academic Press Inc.
3333 Mistwell Crescent
Oakville, ON L6L 0A2 Canada

Apple Academic Press Inc.
9 Spinnaker Way
Waretown, NJ 08758 USA

© 2019 by Apple Academic Press, Inc.

First issued in paperback 2021

Exclusive worldwide distribution by CRC Press, a member of Taylor & Francis Group

No claim to original U.S. Government works

ISBN 13: 978-1-77463-390-8 (pbk)
ISBN 13: 978-1-77188-692-5 (hbk)

Library and Archives Canada Cataloguing in Publication

Rice science : biotechnological and molecular advancements / edited by
  Deepak Kumar Verma, Prem Prakash Srivastav, PhD, Altafhusain B. Nadaf, PhD.

Includes bibliographical references and index.
Issued in print and electronic formats.
ISBN 978-1-77188-692-5 (hardcover).--ISBN 978-1-351-13658-7 (PDF)

1. Rice--Biotechnology. 2. Rice--Molecular genetics. I. Verma, Deepak Kumar, 1986-, editor
  II. Srivastav, Prem Prakash, editor III. Nadaf, Altafhusain B., editor

SB191.R5R53 2018   633.1'823   C2018-905243-0   C2018-905244-9

Library of Congress Cataloging-in-Publication Data

Names: Verma, Deepak Kumar, 1986- editor. | Srivastav, Prem Prakash, editor. | Nadaf, Altafhusain B., editor.

Title: Rice science : biotechnological and molecular advancements / editors: Deepak Kumar Verma, Prem Prakash Srivastav, Altafhusain B. Nadaf.

Description: Oakville, ON ; Waretown, NJ : Apple Academic Press, 2018. | Includes bibliographical references and index.

Identifiers: LCCN 2018042383 (print) | LCCN 2018043927 (ebook) | ISBN 9781351136587 (ebook) | ISBN 9781771886925 (hardcover : alk. paper)

Subjects: LCSH: Rice--Research. | Rice--Biotechnology. | Rice--Molecular aspects.

Classification: LCC SB191.R5 (ebook) | LCC SB191.R5 R5317 2018 (print) | DDC 633.1/8--dc23

LC record available at https://lccn.loc.gov/2018042383

Apple Academic Press also publishes its books in a variety of electronic formats. Some content that appears in print may not be available in electronic format. For information about Apple Academic Press products, visit our website at **www.appleacademicpress.com** and the CRC Press website at **www.crcpress.com**

ABOUT THE EDITORS

Deepak Kumar Verma is an agricultural science professional and is currently a PhD research scholar with specialization in food processing engineering in the Agricultural and Food Engineering Department, Indian Institute of Technology, Kharagpur (WB), India. In 2012, he received a DST-INSPIRE Fellowship for PhD study by the Department of Science & Technology (DST), Ministry of Science and Technology, Government of India. Mr. Verma is currently assigned for research on "isolation and characterization of aroma volatile and flavoring compounds from aromatic and non-aromatic rice cultivars of India." While pursuing his master's degree, his research was entitled "Physico-chemical and Cooking Characteristics of Azad Basmati (CSAR 839-3)—A Newly Evolved Variety of Basmati Rice (*Oryza sativa* L.)." He earned his BSc degree in agricultural science in 2009 from the Faculty of Agriculture, Gorakhpur University, Gorakhpur, and MSc (Agriculture) in agricultural biochemistry in 2011 with first rank and also received a department topper award from the Department of Agricultural Biochemistry, Chandra Shekhar Azad University of Agricultural and Technology, Kanpur, India. Apart from his area of specialization in plant biochemistry, he has also built a sound background in plant physiology, microbiology, plant pathology, genetics and plant breeding, plant biotechnology and genetic engineering, seed science and technology, food science and technology, etc. In addition, he is member of several professional bodies, and his activities and accomplishments include conferences, seminars, workshops, and training. He has published research articles, books, and book chapters.

Prem Prakash Srivastav, PhD, is Associate Professor of food science and technology in the Agricultural and Food Engineering Department, Indian Institute of Technology, Kharagpur (WB), India. He has graduated from Gorakhpur University, Gorakhpur, and received an MSc degree with a major in food technology and a minor in process engineering from G.B. Pant University of Agriculture and Technology, Pantnagar, India. He was awarded a PhD from the Indian Institute of Technology, Kharagpur. He teaches various UG, PG and PhD level courses and has guided many research projects at the PhD, master's, and undergraduate levels. His research interest includes development of specially designed convenience, functional and therapeutic foods; extraction of nutraceuticals; and development of various low-cost food processing machineries. He has organized many sponsored short-term courses and completed sponsored research projects and consultancies. He has published various research papers in peer-reviewed international and national journals and proceedings and many technical bulletins and monographs as well. Other publications include books and book chapters along with many patents. He has attended, chaired, and presented various papers at international and national conferences and delivered many invited lectures in various summer/winter schools. He received best poster paper awards from ISAE–2009, ICTF–2010, FOQSAT–2011, and IFT (USA)–2014. He is a life member of various professional bodies, such as ISTE, AFST(I), IDA, and AMI, and a member of the American Society of Agricultural and Biological Engineers and the Institute of Food Technologists (USA).

Altafhusain Nadaf, PhD, is working as an Associate Professor at the Department of Botany, Savitribai Phule Pune University, Pune, India, in the area of biochemistry and molecular genetics of scented rice for past 13 years. He was awarded an Erasmus Mundus Action 2 India4EU II scholarship to visit the University of Bolgona, Italy, as a visiting professor (2014) and a DST-BOYSCAST Fellowship to work as a visiting fellow at the Centre for Plant Conservation Genetics, Southern Cross University, Lismore, Australia, for one year in 2010–2011. He has received research grants from several funding agencies, including the DST-Fast Track Scheme for Young Scientists and DBT-Rapid Grant for Young Investigators (RGYI). He has successfully guided several PhD and MPhil students and women scientists (under DST-Women Scientists Scheme A & B). He is working as a reviewer for many international journals. He has to his credit two books and more than 50 research papers published in peer-reviewed national and international journals of repute. He has presented his research at national and international levels. His research interests are biochemistry and molecular genetics of rice aroma volatiles. He has characterized scented rice varieties of India and other plant species containing rice aroma volatiles following the HS-SPME-GC-MS/FID approach. At the molecular level, he has also characterized the betaine aldehyde dehydrogenase2 (badh2) gene in the plant species having 2 acetyl-1-pyroline (2AP) expression. In addition, he has revised Indian Pandanaceae thoroughly with the contribution of three new species and assessed phylogenetic relationship among the taxa using cpDNA regions. Some of the species are now being characterized for salt-tolerant genes and bioactive compounds.

CONTENTS

CONTRIBUTORS

Aziz Ahmad
School of Fundamental Sciences, c/o Centre for Fundamental and Liberal Education, Universiti Malaysia Terengganu, 21030, Kuala Terengganu, Terengganu, Malaysia, Tel: +609 6683176, Mob.: +6 19 3475443, Fax: +609 6683434, E-mail: aaziz@umt.edu.my

Bavita Asthir
Department of Biochemistry, Collage of Basic Sciences and Humanities, Punjab Agriculture University, Ludhiana, Punjab 141004, India, Mobile: +91-9216292388, Tel.: +91-1612562967, E-mail: b.asthir@rediffmail.com, basthir@pau.edu

Shanmugaraj Bala Murugan
Plant Genetic Engineering Laboratory, Department of Biotechnology, Bharathiar University, Coimbatore, Tamil Nadu 641046, India, Mob.: +919003608162, E-mail: balagene3030@gmail.com

Srinivasan Balamurugan
Plant Genetic Engineering Laboratory, Department of Biotechnology, Bharathiar University, Coimbatore, Tamil Nadu 641046, India, Mob.: +919566878007, E-mail: bala.svm@gmail.com

B. K. Das
Nuclear Agriculture and Biotechnology Division (NA&BTD), Bhabha Atomic Research Centre, Mumbai, Maharashtra 400085, India, Tel.: +91-22-25592640, E-mail: bkdas@barc.gov.in

Raj Kumar Joshi
Centre of Biotechnology, Siksha O Anusandhan University, Bhubaneswar, Odisha 751030, India, Mob.: +91-9437684176, E-mail: rajjoshi@soauniversity.ac.in

Tushar Khare
Department of Biotechnology, Modern College of Arts, Science and Commerce, Savitribai Phule Pune University, Ganeshkhind, Pune, Maharashtra 411016, India, Tel.: +91-2025634021, Mob.: +91-9764447330, Fax: +9125650931, E-mail: tushar131189@gmail.com

Vikash Kumar
Nuclear Agriculture and Biotechnology Division (NA&BTD), Bhabha Atomic Research Centre, Mumbai, Maharashtra 400085, India, Tel.: +91-22-25592331, E-mail: vikash007barc@gmail.com

Vinay Kumar
Department of Biotechnology, Modern College of Arts, Science and Commerce, Savitribai Phule Pune University, Ganeshkhind, Pune, Maharashtra 411016, India, Mob.: +91-9767839708, E-mail: vinaymalik123@gmail.com

Manisha Kumari
Department of Biochemistry, Collage of Basic Sciences and Humanities, Punjab Agriculture University, Ludhiana, Punjab 141004, India, E-mail: manishabindra@gmail.com

Ashwini Malla
Plant Genetic Engineering Laboratory, Department of Biotechnology, Bharathiar University, Coimbatore, Tamil Nadu 641046, India, Mob.: +919963978146, E-mail: am08.loyola@gmail.com

Sarika Mathure
Cytogenetics Laboratory, Department of Botany, Savitribai Phule Pune University, Pune, Maharashtra 411007, India, Mob.: +917774000575, E-mail: sarika.mathure@gmail.com

Rukmini Mishra
Centre of Biotechnology, Siksha O Anusandhan University, Bhubaneswar, Odisha 751030, India,
Mob.: +91-9556458058, Email: rukmini.mishra@gmail.com

Chakravarthi Mohan
Molecular Biology Laboratory (LBM), Department of Genetics and Evolution (DGE),
Federal University of Sao Carlos (UFSCar), São Paulo 13565905, Brazil, Tel.: +551633518378,
Mob.: +5516996196465, E-mail: chakra3558@gmail.com

Jatindra Nath Mohanty
Centre of Biotechnology, Siksha O Anusandhan University, Bhubaneswar, Odisha 751030, India,
Mob.:+91-7751831025, E-mail: jatindranathmohanty@gmail.com

Suvendu Mondal
Nuclear Agriculture and Biotechnology Division (NA&BTD), Bhabha Atomic Research Centre,
Mumbai, Maharashtra 400085, India, Tel.: +91-22-25590779, E-mail: suvenduhere@yahoo.co.in

Pandiyan Muthuramalingam
Department of Biotechnology, Alagappa University, Karaikudi, Tamil Nadu 630003, India,
Tel: +91-4565-225215, Mob.: +91-9597771342, Fax: +91-4565-225202,
E-mail: pandianmuthuramalingam@gmail.com

Altafhusain B. Nadaf
Cytogenetics Laboratory, Department of Botany, Savitribai Phule Pune University, Pune, Maharashtra
411007, India, Tel.: +912025601439, Mob.: +917588269987, 9822878306, Fax: +912025601439,
E-mail: abnadaf@unipune.ernet.in

Satyabrata Nanda
Centre of Biotechnology, Siksha O Anusandhan University, Bhubaneswar, Odisha 751030, India,
Mob.: +91-7751831025, E-mail: sbn.satyananda@gmail.com

Ganesh Nikalje
Department of Botany, Savitribai Phule Pune University, Pune, Maharashtra 411007, India,
Mob.: +00-91-9969462817, E-mail: ganeshnikalje7@gmail.com

Subramanian Radhesh Krishnan
Department of Biotechnology, Alagappa University, Karaikudi, Tamil Nadu 630003, India,
Tel.: +91-4565-225215, Mob.: +91-9566422094, Fax: +914565225202, E-mail: radheshkrishnan.s@
gmail.com

Manikandan Ramesh
Department of Biotechnology, Alagappa University, Karaikudi, Tamil Nadu 630003, India,
Mob.: +91-9442318200, Fax: +914565225202, E-mail: mrbiotech.alu@gmail.com

Parmeshwar Kumar Sahu
Department of Genetics and Plant Breeding, Indira Gandhi Krishi Vishwavidyalaya, Raipur,
Chhattisgarh 492012, India, Mob.: +00-91-8103795885, E-mail: parmeshwarsahu1210@gmail.com

Ramalingam Sathishkumar
Plant Genetic Engineering Laboratory, Department of Biotechnology, Bharathiar University,
Coimbatore, Tamil Nadu 641046, India, Mob.: +91-9360151669, Fax: +91-4222422387,
E-mail: rsathish@buc.edu.in

Deepak Sharma
Department of Genetics and Plant Breeding, Indira Gandhi Krishi Vishwavidyalaya, Raipur,
Chhattisgarh 492012, India, Mob.: +91-9826647509, E-mail: deepakigkv@gmail.com

Deepak Shelke
Department of Botany, Amruteshwar Arts, Commerce and Science College, Vinzar, Velha, Pune, Maharashtra 412213, India, Mob.: +91-7620110669, E-mail: dpk.shelke1@gmail.com

Vishal Singh
Department of Agricultural Engineering, M. S. Swaminathan School of Agriculture, Centurion University of Technology and Management, Odisha, India, Mobile: +91-8093872582, +91-8348521736, E-mail: vishalsinghiitkgp87@gmail.com, vishalsinghiitkgp@cutm.ac.in

Prem Prakash Srivastav
Agricultural and Food Engineering Department, Indian Institute of Technology, Kharagpur, West Bengal 721302, India, Mob.: +91-9434043426, Tel.: +91-3222-283134, Fax: +91-3222-282224, E-mail: pps@agfe.iitkgp.ernet.in

Inchakalody P. Varghese
Plant Genetic Engineering Laboratory, Department of Biotechnology, Bharathiar University, Coimbatore, Tamil Nadu 641046, India, E-mail: vinchakalody@gmail.com

Deepak Kumar Verma
Agricultural and Food Engineering Department, Indian Institute of Technology, Kharagpur, West Bengal 721302, India, Tel.: +91-3222281673, Mob.: +91-7407170259, 7407170260, Fax: +91-3222282224, E-mail: deepak.verma@agfe.iitkgp.ernet.in, rajadkv@rediffmail.com

Gautam Vishwakarma
Nuclear Agriculture and Biotechnology Division (NA&BTD), Bhabha Atomic Research Centre, Mumbai, Maharashtra 400085, India, Tel.: +91-22-25593632, E-mail: gtmvish@barc.gov.in

Kantilal V. Wakte
Cytogenetics Laboratory, Department of Botany, Savitribai Phule Pune University, Pune, Maharashtra 411007, India, Mob.: +918888888385, E-mail: kanti999ster@gmail.com

ABBREVIATIONS

$^1O_2$	singlet oxygen
2,4-D	2,4-dichlorophenoxyaceticacid
2-AP	2-acetyl-1-pyrroline
2-DE	two-dimensional electrophoresis
2D-FDIGE	two-dimensional-fluorescense difference gel electrophoresis
2-DGE	two-dimensional gel electrophoresis
2D-PAGE	two-dimensional polyacrylamide gel electrophoresis
5'-IAF	5'-iodoacetamidofluorescein
AA	ascorbic acid
aa	amino acids
ABA	absisic acid
AbS	abiotic stress
AC	amylose content
ACE	area of chalky endosperm
ADT 43	Aduthurai 43
AFLPs	amplified fragment length polymorhisims
AHAS	acetohydroxyacid synthase
AI	aliphatic index
ALS	acetolactate synthase
AM	association mapping
amiRNAs	artificial miRNAs
ANOVA	analysis of variance
ApGSMT	*aphanothece halophytica* glycine sarcosine methyltransferase
ApSDMT	*aphanothece halophytica* sarcosine dimethylglycine methyltransferase
APX	ascorbate peroxidase
ARF	auxin response factor
ASA	ascorbic acid
ASA-GSH	ascorbate-glutathione
ASC	ascorbate
At	*Arabidopsis thaliana*
ATAF	Arabidopsis transcription activation factor

AtbZIP60	*Arabidopsis thaliana* transcription factor regulates stress signalling
AtDREB1A	*Arabidopsis thaliana* dehydration response element binding transcription factors
AtEm6	*Arabidopsis thaliana* late embryogenesis abundant gene
Athsp101	*Arabidopsis thaliana* heat shock protein 101
AtSAP	*Arabidopsis thaliana* stress associated protein
badh2/fgr	betaine aldehyde dehydrogenase gene
BAP	6-benzyl aminopurine
BB	bacterial blight
BCH	β-carotene hydroxylase
BCTV	beet curly top virus
Bd	*Brachypodium distachyon*
BEL	bentazon sensitive lethal
BLAST	basic local alignment search tool
BPH	brown plant hopper
BrUGE1	*Brassica rapa* UDP-glucose 4-epimerase 1
BS	biotic stress
Bt	*Bacillus thuringiensis*
CAM	calmodulin
CaMV	cauliflower mosaic virus
CaMV35S	cauliflower mosaic virus 35 S
CAPs	cleaved amplified polymorphism
CAR/DVB/PDMS	carboxen/divinyl-benzene/poly-dimethyl-siloxane
Cas9	CRISPR associated protein 9
CAT	catalase
CAT1	catalase 1
CBF	C-repeating binding factor
CBF3	core-binding factor
CBL	calcineurin B-like proteins
CBP	calcium-binding protein
CCU	crop's consumptive use
CDKA	cyclin-dependent kinase A
CDKB	cyclin-dependent kinase B
cDNA	complementary DNA
CDPK	calcium-dependent protein kinases
CG	candidate gene
CINV	cytosolic/plastidic/mitochondrial invertases
CO_2	carbon dioxide

COX	choline oxidase
CRISPR	clustered regularly interspaced short palindromic repeats
crRNA	CRISPR RNA
CRT	C repeat
CsZfp	*Camellia sinensis* zinc-finger proteins
Ct	chloroplast
Cu/Zn-Sod	copper/zinc superoxide dismutase
CUC	cup-shaped cotyledon
CWINV	cell wall invertase
DAO	diamine oxidase
DArT	diversity array technique
DD	differential display
DE	differentially expressed
DH	double haploids
DHA	dehydroascrobic acid
DHA	docosahexaenoic acid
DHAR	dehydroascorbate reductase
DHAr	DHA reductase
DIGE	2D-fluorescence difference gel electrophoresis
DIL	drought-induced lipid transfer protein
DMRT	Duncan's multiple range test
DNA	deoxyribo nucleic acid
DNPH	2,4-dinitrophenylhydrazine
DREB	dehydration responsive element binding
DREF1	drought-responsive ethylene response factor
DSB	DNA double strand break
dsDNA	double stranded DNA
DSR2	DUF966-stress-repressive gene 2
EAP	external antisense primer
EBE	effector binding element
EBI	European Bioinformatics Institute
EC	electrical conductivity
EC	extracellular
Eluc	click beetle luciferase
EMMA	efficient mixed model analysis
EMS	ethyl methanesulphonate
ENA	European Nucleotide Archive
EPSPS	5-enolpyruvylshikimate-3-phosphate synthase
ERFs	ethylene responsive factors
ESLpred	eukaryotic subcellular localization prediction

ESP	external sense primer
EST	expressed sequence tags
ET	ethylene
ExPASy	Expert Protein Analysis System
exRNA	extracellular RNA
FAO	Food and Agriculture Organization
Fd	ferrodoxin
FID	flame ionization detector
Fru	fructose
FT-IR	Fourier-transform infrared
FUE	fertilizer use efficiency
GA	gibberlic acid
GAPIT	Genome Association and Prediction Integrated Tool
GBS	genotyping by sequencing
GC approach	genomic control approach
GC	gas chromatography
GC	gel consistency
GC	genomic control
GC-FID	gas chromatography-flame ionization detector
GC-MS	gas chromatography-mass spectrometry
GFP	green fluorescent protein
Glc	glucose
GLM	Generalized Linear Model
GLOGS	genome-wide LOGistic mixed model/score test
Gly I	glyoxalase I
Gn	grain number
GO	gene ontology
GOLD	Genomes Online Database
GPX	glutathione peroxidase
GR	glutathione reductase
GRAVY	grand average of hydropathicity
gRNA	guide RNA
GS	genomic selection
Gs	grain size
GSH	glutathione
GSSG	glutathione disulfide
GST	glutathione S-transferase
GT	gelatization temperature
GUS	β-glucuronidase
GWAS	genome-wide association study

GWASpi	genome-wide association study pipeline
H₂O	water
H₂O₂	hydrogen peroxide
HA	haplotype analysis
HDR	homology directed repair
Hg	Mercury
HKs	histidine kinases
HMMSCAN	Hidden Markov Model SCAN
hptII	hygromycin resistance gene
HR	homologous recombination
HR-MAS-NMR	high-resolution magic-angle spinning nuclear magnetic resonance
HSF	heat shock factors
HSP	heat shock protein
Hsp101	heat shock protein 101
HS-SPME	head space-solid phase microextraction
Hv	*Hordeum vulgare*
IAA	indole-3-acetic acid
ICAT	isotope-coded affinity tags
ICE1	*Arabidopsis thaliana* transcription factor
IFAP	internal fragrant antisense primer
II	instability index
InDel	insertion and deletions
INSP	internal non-fragrant sense primer
IRRI	International Rice Research Institute
ISSRs	inter-simple sequence repeats
iTRAQ	isobarric tags for relative and absolute quantification
JA	jasmonic acid
KIN	kinetin
LC-EC-MS	liquid chromatography-electrochemistry-mass spectrometry
LC-MS	liquid chromatography-mass spectrometry
LC-NMR	liquid chromatography-nuclear magnetic resonance
LD	linkage disequilibrium
LD mapping	linkage disequilibrium mapping
LE	linkage equilibrium
LEA	late embryogenesis abundant
LOXs	lipoxygenases
Lp	*Leersia perrieri*
LS	Linsmaeir and Skoog

Ma	*Musa acuminata*
MAB	marker-associated breeding
MABC	marker-assisted backcross breeding
MAGIC	multiparent advanced generation intercross
MAGP	marker-assisted gene pyramiding
MALDI-TOF-TOF	matrix-assisted laser desorption/ionization-time of flight-time of flight
MAPKs	mitogen-activated protein kinases
MAPS	marker-assisted and phenotype selections
MAS	molecular marker-assisted selection
MAS-QTL	marker-assisted selection–quantitative trait loci
MCS	multiple cloning sites
MDA	malondialdehyde
MDA	monodehydro-ascorbate
MDAR	MDA reductase
MeMV	Merremia mosaic virus
MFS	major facilitator super
MIPS	myo-inositol phosphate synthase
miRNA	micro RNA
MLM	mixed linear model
MLMM	multilocus mixed model
MLO	mildew resistant locus
MnSOD	manganese superoxide dismutase
MPSS	massively parallel signature sequencing
mRNA	messenger RNA
MS	Murashige and Skoog
MTL	mean tidal level
mtlD	mannitol-1-phospho dehydrogenase
MtSAP	*Medicago truncatula* stress-associated protein
MudPIT	multidimensional protein identification methods
NAA	1-naphthaleneacetic acid
Naat	nicotinamide aminotransferase
NaCl	sodium chloride
NADPH	nicotinamide adenine dinucleotide phosphate
NAM	no apical meristem
NaN$_3$	sodium azide
NCBI	National Centre for Biotechnological Informations
NCED2	9-cis-epoxycarotenoid dioxygenase 2
ncRNAs	non-coding RNAs
NGS	next generation sequencing

NHEJ	non-homologous end-joining
NHX1	Na $(^+)$/H $(^+)$ exchanger
NILs	near isogenic lines
NiR	ferredoxin nitrite reductase
NMR	nuclear magnetic resonance
NMRS	nuclear magnetic resonance spectroscopy
NO	nitric oxide
NPBT	new plant breeding techniques
NPK1	nicotiana protein kinase 1
NPR	net photosynthetic rate
NUC	nuclease lobe
O_2^-	superoxide radical
O_2^{2-}	peroxide ions
OECD	Organization for Economic Co-operation and Development
OH	hydroxyl radical
Os	*Oryza sativa*
OsALS	rice acetolactate synthase
OsBADH2	*Oryza sativa* betaine aldehyde dehydrogenase 2
OsEPSPS	5-enolpyruvylshikimate-3-phosphate synthase
OsGL1-6	*Oryza sativa* fatty aldehyde decarbonylase 6
OsGW2	*Oryza sativa* grain width 2
OsHsp101	*Oryza sativa* heat shock protein 101
Osi	*Oryza sativa-indica*
OsMYBI4	*Oryza sativa* myeloblastosis transcription factors
OsNAC6	*Oryza sativa* NAC6 gene
OsOAT	*Oryza sativa* ornithine δ-amino-transferase
OsPDS	*Oryza sativa* phytoenedesaturase
OsPDS	rice sucrose transporter gene
OsSAP	*Oryza sativa* stress-associated proteins
OsSWEET	*Oryza sativa* sucrose-efflux transporter
OsSWEET13	R*ice waxy* gene
OsTGW6	*Oryza sativa* thousand-grain weight 6
P5CS	pyrroline 5-carboxylate synthetase
P5CS	delta 1-pyrroline-5-carboxylate synthetase
P5CSF129A	Δ1-pyrroline-5-carboxylate synthetase F129A
PAM	protospacer adjacent motif
PAO	polyamine oxidase
Pas	Polyamines
PCA	principle component analysis

pChlCOD	*Arthrobacter globiformis* choline oxidase
PCR	polymerase chain reaction
PDB	Protein Data Bank
PEG	polyethylene glycol
PEP	phosphoenolpyruvate
PEPC	phosphoenol-pyruvate carboxylase
PFAM	ProteinFAMily
PgNHX1	*Pennisetum glaucum* vacuolar Na $^+$/H $^+$ antiporter
pI	isoelectric point
PID	PAM-interacting domain
PL	panicle length
PLACE	plant cis-acting regulatory DNA elements
PlantCARE	plant cis-acting regulatory element
POX	peroxidase
Ppb	parts per billion
PPDB	Plant Proteomics Database
PPDK	pyruvate orthophosphate dikinase
PpEXP1	*Poa pratensis* expansins gene
PPI	protein–protein interaction
PRIN	Predicted Rice Interactome Network
Protox	protoporphyrinogen oxidase
PsCBL	*Pisum sativum* calcineurin B-like protein
PsCIPK	*P. sativum* calcineurin B-like interacting protein kinase
PSI	photosystem I
PSSRNA	plant short small RNA
PTGS	post-transcriptional gene silencing
PUFA	polyunsaturated fatty acid
PUT	putrescine quantitative trait loci
QTLs	quantitative trait loci
RAMPs	repeat-associated mysterious proteins
RAPD	random amplification of polymorphic DNA
RDCP1	RING domain-containing protein
REC	recognition lobe
RFLP	restriction fragment length polymorphism
RiceFOX DB	Rice Full-length cDNA Overexpressor Database
RILs	recombinant inbred lines
RIPP-DB	RIKEN Plant Phosphoproteome Database
RLCK253	receptor-like cytoplasmic kinase 253
RNA	ribonucleic acid
RNAi	RNA interference

RNS	reactive nitrogen species
ROC5	rice outermost cell-specific gene5
ROS	reactive oxygen species
rRNAs	ribosomal RNAs
RT	retention time
RuBisCo	ribulose-1,5-bisphosphate carboxylase/oxygenase
RuBP	ribulose-1,5-bisphosphate
RVD	repeat variable di-residues
SA	structure association
SA	salicylic acid
SAGE	serial analysis of gene expression
Saltol	salinity tolerance
SAMDC	S-adenosylmethionine decarboxylase
SAPs	stress-associated proteins
SAS	Statistical Analysis System
SBS	sequencing by synthesis
SC drop-out	synthetic complete drop-out
SCARs	sequence characterized amplified regions
Sd1	semidwarf gene
SDIR	salt- and drought-induced ring finger
SEM	scanning electron microscopy
SflAP	*Solanum lycopersicum* L. antiapoptotic
ShSAP	*Saccharum officinarum* stress associated protein
SIGnAL	Salk Institute Genomic Analysis Laboratory
siRNAs	small interfering RNAs
SlAGO7	*Solanumlycopersicon* ARGONAUTE7
SlSAP	*Solanum lycopersicum* stress associated protein
SM	secondary metabolite
SmAPX	*Solanum melongena* ascorbate peroxidase
SMB	secondary metabolite biosynthesis
snoRNAs	small nucleolar RNAs
SNPs	single nucleotide polymorphisms
snRNAs	small nuclear RNAs
SOD	superoxide dismutase
SOS	salt overly sensitive
SOS1	salt overly sensitive 1
SOS2	salt overly sensitive 2
SPD	spermidine
SPM	spermine
SPP	stromal processing peptidase

SPP	sucrose phosphatase
SPS	sucrose phosphate synthase
SPSS	Statistical Package for the Social Sciences
SRA	Sequence Read Archive
SSA	Single strand annealing
SsGDH	*Sclerotinia sclerotiorum* glutamate dehydrogenase
SSH	suppression subtractive hybridization
SSLP	simple sequence length polymorphisms
SSNs	sequence specific nucleases
SSR	simple sequence repeats
SSS	reactive sulfur species
*St*ALS1	*Solanum tuberosum* acetolactate synthase1
STK	serine/threonine-protein kinase
STPs	signal transduction pathways
STRING	search tool for the retrieval of interacting genes/proteins
STS	sequence tagged sites
Sub1	submergence 1
Suc	sucrose
Super SAGE	super serial analysis of gene expression
SUSY	sucrose synthase
T_0	primary transgenic plant
Ta	*Triticum aestivum*
TALE	transcription activator-like effector
TALEN	transcription activator-like effector nuclease
tAPx	thylakoid-bound APx
TASSEL	trait analysis by association, evolution and linkage
TDT	transmission disequilibrium test
TF	transcription factor
TGS	transcriptional gene silencing
TGW	thousand grain weight
TILLING	targeting induced local lesions in genomes
TIR	transport inhibitor response
TNF	tumor necrosis factor
TPP	trehalose-6-phosphate phosphatase
TPS	trehalose-6-phosphate synthase
tracrRNA	trans activating CRISPR RNA
tRNAs	transfer RNAs
TS	target sequence
TTC	triphenyl tetrazolium chloride
TYLCV	tomato yellow leaf curl virus

tzs	*trans*-zeatin secretion gene
UDP-Glc	UDP-glucose
Uns	unstable
URFTCMS	ultra-high resolution Fourier-transform-ion cyclotron MS
VHb	citreoscilla hemoglobin gene
VINV	vacuolar invertase
WHC	water holding capacity
WUE	water use efficiency
Wx	waxy gene
Xa21	Xanthomonas oryzae
ZFNs	zinc-finger nucleases
ZFPs	zinc-finger proteins
Zm	*Zea mays*

PREFACE

Rice is a staple food for more than two billion people worldwide, and in Asia it is the most preferred cereal crop. It provides 27% of dietary energy and 20% of dietary protein in the developing world. Besides being human food, it is also used as animal feed and provides the major source of income for rural people. Due to the Green Revolution during the 1950s and 1960s, a quantum leap in rice yield took place over the past three decades. Although increased food production did not eliminate poverty and hunger, the yield increase did help to avert famine and prevent a greater disruption of the food supply in Asia.

Today's global population of 7.4 billion is expected to reach over eight billion by 2020. In order to meet the increasing demand for food, especially rice, there is necessity to increase the production of rice by 25–50%. Despite the improved varieties and the usage of technologies, rice yield has remained stagnant for past three decades. Therefore, implication of biotechnology may enhance rice yield further and enable rice to grow in drought and saline conditions, and lead to more nutritious rice for reducing malnutrition.

This present book, *Rice Science: Biotechnological and Molecular Advancements,* addresses the recent biotechnological and molecular tools and techniques adopted for increasing rice production. The book has been divided into four parts. Part I: Role of Stress: Recent Trends and Advances in Rice Research consists of four chapters that cover the recent trends that are followed to overcome abiotic stress. It includes chapters discussing the use of the most recent technique, such as A20/AN1 zinc-finger proteins in improving rice productivity under abiotic stress, potent avenues open for conferring salt stress tolerance, and technological developments for combating abiotic stress tolerance. Part II: Biochemical Trends and Advances in Rice Research, consists of two chapters that describe the biochemical trends and advances in rice research. It covers assessment of aroma content in traditional aromatic rice varieties and biochemical evaluation of irrigated flooded transplanted and aerobic rice. Part III: Biotechnological Tools and Trends for Production and Rice Quality Improvement, consists of two chapters that take into account the biotechnological tools and trends for production and rice quality improvement. One chapter discusses the use of *Agrobacterium*-mediated genetic transformation, and second, the use of genetic markers and marker-assisted selection (MAS) as a tool for improvement of rice quality and production. Part IV: Molecular Advances and Trends for Quality Improvement in Rice, covers molecular advances and trends followed for quality improvement in rice. It includes three chapters, one discussing the most recent technique followed for genome editing, that is CRISPER/Cas and its applications in rice improvement; second, describing molecular advanced tools and trends in which genetic engineering and biotechnology are contributing to their role for quality improvement of rice crop; and third, chapter focuses on a high-resolution mapping technique used for mapping complex traits.

The contributions of the eminent scientists and researchers enabled us to present their work in the form of this book. We hope that this book will serve as a guide for researchers and students working in the area of rice biotechnology and molecular biology.

—Editors

PART I
Role of Stress: Recent Trends and Advances in Rice Research

PART I

CHAPTER 1

EMERGING TRENDS OF A20/ AN1 ZINC-FINGER PROTEINS IN IMPROVING RICE PRODUCTIVITY UNDER ABIOTIC STRESS

SUBRAMANIAN RADHESH KRISHNAN[1],
PANDIYAN MUTHURAMALINGAM[1,2], CHAKRAVARTHI MOHAN[3],
and MANIKANDAN RAMESH[1,4,*]

[1]*Department of Biotechnology, Alagappa University, Karaikudi, Tamil Nadu 630003, India, Tel.: +91 4565 225215, Mob.: +91 9566422094, Fax: +914565225202, E-mail: radheshkrishnan.s@gmail.com*

[2]*pandianmuthuramalingam@gmail.com, Mob.: +919597771342*

[3]*Molecular Biology Laboratory (LBM), Department of Genetics and Evolution (DGE), Federal University of Sao Carlos (UFSCar), São Paulo 13565905, Brazil, Tel.: +551633518378, Mob.: +5516996196465, E-mail: chakra3558@gmail.com*

[4]*Mob.: +91 9442318200, Fax: +914565225202*

Corresponding author. E-mail: mrbiotech.alu@gmail.com

1.1 INTRODUCTION

Plants are sessile and hence are confronted by the surrounding environment; this is a major constraint to their growth and development. Crop yield is severely affected by the climate change and rise in the temperature through global warming which drastically reduces the water withholding capacity of the soil and thereby crop production (Lesk et al., 2016). These stimuli are evoked by numerous molecular and biochemical crosstalk that activate the candidate component to resist stress. These actions are packed with stimulation of receptor molecules by stress followed by a series of molecular crosstalk that promote downstream process and activate the genes for its survival. Over the decades, plants have evolved to tackle these stresses by more sophisticated mechanisms than the animals (Qin et al., 2011; Rejeb et al., 2014). In crop plants, such as rice *Oryza sativa* L., production and quality are adversely hampered by biotic and abiotic stresses. There are various key proteins that are generated by the plant immune system against these stresses and are largely being regulated by their respective transcription factors (Kavar et al., 2008). Certain stress regulators act as transcription binding factors such as DRE/CRT; DREBs/ CBFs; HSF/heat shock proteins (HSPs) and MAPKs that are highly conserved domains in plants and hence are the key regions for researchers to explore abiotic stress resistance (Mizoi et al., 2012; Nakashima et al., 2012; Hussain Wani et al., 2016). Such a class of conserved domains are A20/AN1 stress-associated proteins (SAPs) that exists as zinc-finger domain proteins (ZFPs) (Mukhopadhyay et al., 2004; Kothari et al., 2016). ZFPs are rapidly evolving key research interests owing to their multiple genes, conservative nature, and wide host range compatibility; *O. sativa* (Mukhopadhyay et al., 2004), *Nicotiana tabacum* (Kanneganti and Gupta, 2008), *Arabidopsis thaliana* (Dixit and Dhankher, 2011a) and *Camellia sinensis* L (Paul and Kumar, 2015). The presence of these proteins has also been reported in other living systems such as bacteria and animals (Huang et al., 2004; Hishiya et al., 2006; Zhang et al., 2015). The first report on A20 zinc-finger domain and its multiple motifs (cys2/cys2) from human endothelial cells was reported by Dixit et al. (1990). These were characterized for their role to combat the effects induced by multiple stresses (Vij and Tyagi, 2007). Up till now 18 members of A20/AN1 SAPs from different rice species have been reported (Dansana et al., 2014). In this chapter, *OsiSAP8* was used as the candidate to explain the importance of ZFPs and their role in abiotic stress tolerance by comparing it with the available omics-based resources.

1.2 DEDUCING THE KNOWLEDGE ON SAPS BY IN SILICO ANALYSIS: AN OMICS-BASED APPROACH

With the available whole genome sequences and the advancement in the omics research, understanding the bioinformatics and its related tools has become a prerequisite to gain insights into the molecular crosstalk. Newly developed ultrahigh throughput omics approaches in genomics and interactomics pave way to various emerging systematic/analytical applications. Moreover, in rice, the advancement of omics resources has allowed us to understand the molecular biological properties of individual species. Combinatorial knowledge from omics-based sources is used in identifying the insights of molecular systems and biological properties that accelerate gene mining and its functional characters. A conceptual flow chart is depicted for the overall understanding of omics-based prediction of the stress regulative pathways (Fig. 1.1). Numerous softwares are available to study the protein structure, function, ontology, and the relative pathway.

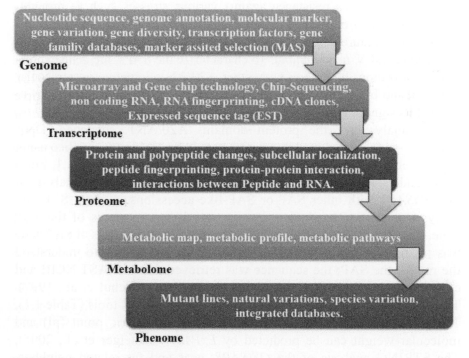

FIGURE 1.1 (See color insert.) A conceptual model called "omic space" with layers ranging from the "genome" to "phenome."

Source: Adapted from Toyoda and Wada (2004).

Many researchers are interested in SAPs or A20/AN1 ZFPs from heterologous systems such as *A. thaliana, Nicotiana benthamiana, O. sativa* and *Solanum lycopersicum* (Huang et al., 2004, 2008; Vij and Tyagi, 2006; Kanneganti and Gupta, 2008; Dixit and Dhankher, 2011a; Liu et al., 2011; Giri et al., 2013; Paul and Kumar, 2015). Hitherto, these model systems have been well explored but can also be extrapolated with that of rice. They were able to analyze only specific functions by expressing these genes in crop plants whose genome has been sequenced. However, the key function or the pathway by which these SAPs function is still unpredictable (Giri et al., 2013). The knowledge on the functional A20/AN1 domains of plants is available, yet the method by which it resists abiotic stress remains unclear. In this chapter, we provide an overall relation of SAPs based on the available reports and elucidate the omics data with respect to *OsiSAP8*. The possible predictions using available bioinformatic tools are given in Table 1.1. We contemplate *OsiSAP8* gene which was characterized by Kanneganti and Gupta (2008) for its resistance against various stresses such as drought, salinity, cold, and effects of various metals. The gene was reported to be multigenic in nature and to combat stress when it was first characterized in *A. thaliana* and *N. benthamiana*. To characterize the SAPs, the gene expression, localization, homology with other A20/AN1 domains, its molecular function, and biological metabolisms have to be given foremost priority to unravel its signaling mechanism.

The analysis of the protein domains A20/AN1, sequence, DNA sequence, and the predicted interactome could be performed in silico using several bioinformatics tools and databases mentioned in Table 1.1. For a better understanding of SAP genes, we performed an in silico analysis of the *OSiSAP8* with other SAP or SAP-like accessions. A basic STRING analysis of *OsiSAP8* revealed the conserved domain status of the SAP gene. Based on the available resources, a preliminary analysis of SAP gene was predicted using various plant informatics tools so as to understand the role of the SAP; the sequence was retrieved from BLAST-NCBI and compared with *A. thaliana* and other crop types (Altschul et al., 1997). This was further analyzed using various available omics tools (Table 1.1). The protein structure and properties such as isoelectric point (pI) and molecular weight can be predicted by *ExPASy* (Gasteiger et al., 2005). The STRING analysis of the *OSiSAP8*, near and far related neighborhood proteins/gene interaction was deduced (Fig. 1.2A). Furthermore, the closely related protein/genes that shared homology with *OsiSAP8* were predicted (Fig. 1.2B) (Szklarczyk et al., 2015) (Table 1.1). The most closely related proteins were of Osl_19520, Osl_33775, Os_19519, Osl_24236 and rab5A. Here the protein–protein interactions are predicted based on the previous datasets about genomic context, high-throughput experiments, and co-expression.

TABLE 1.1 Computational Analysis Tools Used for Various Omics Approaches.

Database Name	Purpose	URL	References
BLAST – rice genome database	Local alignment prediction	http://www.ncbi.nlm.nih.gov/Blast/Genome/PlantBlast.shtml	(Altschul et al., 1997)
ExPASy – protparam	Protein properties prediction	http://www.expasy.org/ http://web.expasy.org/protparam/	(Gasteiger et al., 2005)
Interproscan	Putative domain identification	http://www.ebi.ac.uk/interproscan/	(Quevillon et al., 2005)
Gramene	Comparative resource for grass genomic information like BLAST, SSR, gene diversity, chromosomal location	http://www.gramene.org/	(Tello-Ruiz et al., 2016)
EBI	Amino acid sequence identity determination	http://www.ebi.ac.uk/blast2	–
ClustalW	Multiple sequence alignment	http://www.ebi.ac.uk/clustalW	–
PlantCARE and PLACE	Cis-acting regulatory elements in the promoter region prediction	http://bioinformatics.psb.ugent.be/webtools/plantcare/html/ http://www.dna. affrc.go.jp/PLACE	(Higo et al., 1999; Lescot et al., 2002)
Primer 3	Primer prediction	http://bioinfo.ut.ee/primer3–0.4.0/	(Untergasser et al., 2012)
SWISS model	Homology modelling and structural information of the protein	https://swissmodel.expasy.org/	(Arnold et al., 2006; Biasini et al., <link rid="bib4">2014</link>)
Phyre2	Homology modelling and structural information of the protein	http://www.sbg.bio.ic.ac.uk/phyre2/	(Sakhanokho and Kelley, 2009)
PDB		http://www.rcsb.org/pdb/home/home.do	(Berman et al., 2002)
STRING	Protein–protein interaction and gene–protein interaction	http://string-db.org/newstring_cgi/show_network_section.pl	(Szklarczyk et al., 2015)
PRIN		http://bis.zju.edu.cn/prin/	(Gu et al., 2011)

TABLE 1.1 *(Continued)*

Database Name	Purpose	URL	References
Phytozome v10.2	Comparative genomics, retrieve the corresponding genomic, transcript, coding sequence and chromosomal position prediction	https://phytozome.jgi.doe.gov/pz/portal. html	(Goodstein et al., 2012)
Circos	Comparative mapping and genome visualization	http://circos.ca/	(Krzywinski et al., 2009)
Uniprot KB	Proteomic information	http://www.uniprot.org/	
HMMSCAN	PFAM domain prediction	http://www.ebi.ac.uk/Tools/hmmer/ search/hmmscan	(Finn et al., 2011)
SCANProsite		http://prosite. expasy.org/scanprosite/	(de Castro et al., 2006)
RiceFOX DB	Hormone profile	http://ricefox.psc.riken.jp/	(Maruyama et al., 2011)
Yale Plant genomics	Epigenome analysis in rice and maize	http://plantgenomics.biology.yale.edu/	(Wang et al., 2011)
SIGnAL	Epigenome analysis in *Arabidopsis*	http://signal.salk.edu/cgi-bin/methylome http://neomorph.salk.edu/epigenome/ epigenome.html	(Lister et al., 2008)
PPDB	Plant proteomics	http://ppdb.tc.cornell.edu/	(Sun et al., 2009)
RIPP-DB	Plant phosphoproteome database	http://phosphoproteome.psc.database. riken.jp	(Nakagami et al., 2010)
Gene investigator	Processed transcriptomic data	https://www.genevestigator.com/gv/	–
SRA	NGS–plant genomics	http://www.ncbi.nlm.nih.gov/sra	–
ENA		http://www.ebi.ac.uk/ena/home	–

TABLE 1.1 *(Continued)*

Database Name	Purpose	URL	References
DDBJ-SRA		http://trace.ddbj.nig.ac.jp/dra/index_e.shtml	—
PSSRNAit	Non coding RNA	http://plantgrn.noble.org/pssRNAit/Analysis.gy	—
Golm metabolome database	Metabolome profile	http://gmd.mpimp-golm.mpg.de/	—
EPIC website	Chromatin epigenome	https://www.plant-epigenome.org/links.	—
Cello2GO	High-resolution subcellular localization prediction and gene ontology	http://cello.life.nctu.edu.tw/cello2go/	(Yu et al., 2014)
Wolf Psort 2	High-resolution subcellular localization prediction	http://wolfpsort.org/	(Horton et al., 2007)
Bacello	Low-resolution localization prediction	http://gpcr.biocomp.unibo.it/bacello/	(Pierleoni et al., 2006)
ESLPred2		http://www.imtech.res.in/raghava/eslpred2/	
SubLoc		http://www.bioinfo.tsinghua.edu.cn/SubLoc/	(Hua and Sun, 2001)

Source: Adapted from Toyoda and Wada (2004)

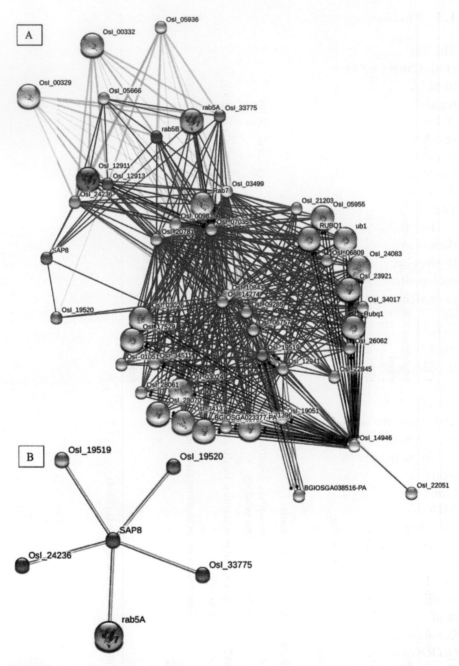

FIGURE 1.2　(See color insert.) Multigenic nature of *OsiSAP8* predicted by string analysis.

1.3   PRESENCE OF SAPS IN OTHER CROPS/PLANTS

The first report on rice system *OsiSAP1* was reported by Mukhopadhyay et al. (2004) a ZFP with an A20/AN1 domain that showed high expression in pre-pollination stage and root cells. The ZFPs exist as a multiple gene regulated system that are conserved domains with A20/AN1 (Giri et al., 2013; Sharma et al., 2015) and act in a stipulated way to combat abiotic stress in plants. Several crops have been reported to resist stress especially at small-scale production or vegetative state (Gaudin et al., 2013) and many lead events are under field trials but still there is a vast gap in understanding the stress tolerance and the mode of the gene/protein action during stress. The SAPs and similar proteins exist in a wide host range.

The role of SAPs, which has around 12 variants, is well studied on *A. thaliana* (Ströher et al., 2009; Dixit and Dhankher, 2011b) all contributing to stress regulation by E3 ubiquitination and accumulation of key metals during stress. Ströher et al. (2009) also reported the redox-dependent action to regulate stress by *AtSAP12*. The *Aeluropus littoralis*-derived *AlSAP* under stress affected the expression of drought transcription factors such as DREB, osmoregulant proline by alteration in *P5CS* gene, superoxide dismutase and MYBCORE factors that are key regulators during abiotic stress in plants. Role of rice SAPs have been reported to provide a good stress endurance in both tobacco and *Arabidopsis* (Giri et al., 2013). Regulation of the transcription factors was observed in cotton when expressed by *GaZnF* which gives a lead that there might be a regulatory action of SAP proteins in the promoter region of abiotic stress genes (Zahur et al., 2012).

Thus, it is nearly unpredictable if these SAP genes work directly or indirectly to resist/avoid abiotic or biotic stress. The SAPs, *CsZfp* from *Camellia sinensis L.* were found to be resistant against H_2O_2, NaCl and ABA-induced stress (Huang et al., 2008; Paul and Kumar, 2015). *CsZfp* was observed to have pI of 6.50 with 18.44 kDa protein and a grand average of hydropathicity (GRAVY) value of -0.334. Charrier et al. (2012) were first to report the role of *MtSAP1*, a SAP variant in osmotic regulation, which produced nitric oxide to combat salt and osmotic stress at the seedling stage. Various other reports about SAP variants have been done in many crops such as *S. lycopersicum* (Solanke et al., 2009), Musaspp cv. *Karibale Monthan* (Sreedharan et al., 2012), *Zea mays* (Xuan et al., 2011), *OsDOG* and *ZFP177 O. sativa* (Huang et al., 2008; Liu et al., 2011) and *Gossypium arboreum* (Zahur et al., 2012). Some key findings of SAPs and their orthologs are mentioned in Table 1.2.

TABLE 1.2 A20/AN1 Zinc-Finger Proteins in Different Crop/Plant Species that Belong to SAP Gene Family.

Gene	Species	Functions/Region	Abiotic stress	References
MtSAP1	Medicago truncatula	Accumulation of nitric oxide	Osmotic and salt stress	(Charrier et al., 2012)
A20/AN1 Zinc-finger domain containing protein (CsZfp)	Camellia sinensis L. (Evergreen tree tea)	Maximum expression in root and fruit, upregulation of H_2O_2 and NaCl	Abiotic stresses (polyethylene glycol, H_2O_2, NaCl and ABA)	(Paul and Kumar, 2015)
AlSAP (OsSAP9-ortholog in rice)	Aeluropus littoralis	Spatio-temporal regulation	Dehydration, salt, heat, ABA, SA, wounding and tissue-specific expression	(Donaldo et al., 2015)
SlSAP gene family (SlSAP1–SlSAP13)	Solanum lycopersicum	Almost eight genes were intronless and induced by multiple stresses	Desiccation, cold, salt, abscisic acid, wounding, heat, submergence and oxidative stresses	(Solanke et al., 2009)
ShSAP1	Saccharum officinarum	Sugarcane maturation	Salt, drought, ABA, GA	(Li et al., 2011)
AtSAP (AtSAP1–AtSAP12)	Arabidopsis thaliana	Developing ubiquitin ligase activity. Accumulation of metals in roots and shoots	Water deficit, salt, Zn, Mg, Ni, high temperature and redox dependent, E3 ligase activity	(Ströher et al., 2009; Dixit and Dhankher, 2011b)
MusaSAP1	Musaspp cv. Karibale Monthan	Lesser membrane damage as indicated by malondialdehyde content. Probable improvement in stress resistance by the gene.	Salt, water-deficit and oxidative stress	(Sreedharan et al., 2012)
OsiSAP (OsiSAP1–SAP18)	Oryza sativa	Intronless genes, multiple stress inducible	ABA, cold, desiccation, drought, heavy metals, salt, submergence and wounding	(Mukhopadhyay et al., 2004; Vij and Tyagi, 2006; Giri et al., 2013)

TABLE 1.2 *(Continued)*

Gene	Species	Functions/Region	Abiotic stress	References
ZmAN13	*Zea mays*	*ZmAN13* fused with GFP showed localization at cell cytoplasm and nucleus	Induced by cold and ABA Downregulated by NaCl	(Xuan et al., 2011)
OsDOG	*O. sativa*	Negatively regulates GA homeostasis and maintains the plant cell elongation in rice	Induced by gibberellin	(Liu et al., 2011)
ZFP177	*O. sativa* japonica	Localization at cytoplasm of leaf and root cells	Induced by cold, drought, heat and H_2O_2. Downregulated by salt stress	(Huang et al., 2008)
GaZnF (has a high sequence similarity with ZFPs)	*Gossypium arboreum*	A high gene expression was observed on stress treated root tissues	Induced by drought, salt and heavy metal stresses	(Zahur et al., 2012)

1.4   HOMOLOGY AND ORTHOLOGS

1.4.1   HOMOLOGY OF SAPS WITHIN SPECIES/ORGANISMS

Plant A20/AN1 ZFP shares a similar homology with tumor necrosis factor (TNF) from human umbilical vein gene (Dixit et al., 1990). Cell proliferation, inflammation, innate, and adaptive immunity were associated with A20 proteins (Lee, 2000). Here the structural information of *OsiSAP8* and its homology modeling was done using SWISS model (Arnold et al., 2006; Biasini et al., 2014) from Protein Data Bank (PDB) (Kelley et al., 2015). The gene expression analyses with SAPs are reported to be associated with multiple roles to combat abiotic stress (Sharma et al., 2015). Plants produce many stress proteins by their defense system to resist abiotic or biotic stress. Unlike abiotic stresses, it is easier to tackle biotic stresses as they are regulated by plethora of genes (Kanneganti and Gupta, 2008). It was pointed out by Mukhopadhyay et al. (2004) that *OsSAP* proteins may carry out their function through protein–protein interaction. A relation with human TNF was found to be negatively regulated by ZNF216, a SAP variant. Here the *OsiSAP8* was analyzed for homology and could match up to 96 and 89% with A20 and AN1 domains respectively of maize ZFP. The homology of *OSiSAP8* was predicted with respect to *Citrus, Camellia, Medicago, Prunus,* and *Phaseolus*. The maize was very closely related with 89 and 96% homology for A20/AN1 domains of *OsiSAP8*. Other than this, *Medicago, Prunus,* and *Phaseolus* had around 70–80% homology with *OsiSAP8* (Table 1.3). Similarly, the orthologs of SAPs were predicted using GRAMENE (Tello-Ruiz et al., 2016). A homology of 94% was noted with *Leersia perrieri* positioned at 6th chromosome. Further in silico analysis of SAP8 by Phytozome (Goodstein et al., 2012) depicted the presence at chromosomes 3 and 6 (data not shown). Here we noticed the presence of similar protein sequences at chromosomes 3 and 6 of closely related species (Table 1.4).

TABLE 1.3   *OsiSAP8* and its Homology with Other Plant Species.

Plant species	NCBI ID	A20 Domain of OsiSAP8 (%)	AN1 Domain of OsiSAP8 (%)
Maize	AAS00453.1	96	89
Citrus	ABL67658.1	87	81
Camellia	ABI31653.1	72	88
Medicago	ABN08135.1	72	77
Prunus	AAD38146.1	78	80
Phaseolus	AA33773	73	73

TABLE 1.4 *OsiSAP8* Orthologs and Their Similarity.

Gene hit	Source	Gramene ID	Chr	Start	End	Similarity with *OsiSAP8* (%)
SAP8	*Osi*	ASM465v1	6	25595032	25595490	100
SAP2	*At*	AT1G51200.3	1	18986014	18986208	77.3
BRADI3G07060	*Bd*	BRADI3G07060.2	3	5230795	5231307	83
MLOC_20.848	*Hv*	MLOC_20,848.1	6	283693841	283,694,101	82.8
LPERR06G17490	*Lp*	LPERR06G17490.1	6	16162780	16163292	94.7
SAP8	*Ma*	GSMUA_Achr3G11830_001	3	8767736	8767939	86.8
Traes_6DS_066F7755D	*Ta*	Traes_6DS_066F7755D.1	6	64971639	64971639	82.8
GRMZM2G058866	*Zm*	GRMZM2G058866.1	4	233973914	233974426	82.5

Osi: *Oryza sativa indica*; At: *Arabidopsis thaliana*; Bd: *Brachypodium distachyon*; Hv: *Hordeum vulgare var. distichum*; Lp: *Leersia perrieri*; Ma: *Musa acuminata subsp. malaccensis*; Ta: *Triticum aestivum*; Zm: *Zea mays*. Chr: *chromosome*.

1.4.2   ORTHOLOGS OF SAPS WITHIN SPECIES/ORGANISMS

Orthologs are genes present in different species that evolved from common ancestor by speciation. Generally, orthologs resemble same molecular function with the course of evolution. Identification of orthologs is easy by available whole genome sequence and bioinformatics tools. Here some reference genes of SAPs or A20/AN1 domain-like proteins were tested against *OsiSAP8* which gave a good similarity index with *L. perrieri, Musa acuminata* subsp. *malaccensis* (Table 1.4).

1.4.3   SAPS IN COMBATING ABIOTIC STRESS

SAPs are proteins with A20/AN1 zinc-finger domains that are responsive to multiple environmental stresses ranging from cold, salt, drought, submergence, wounding, and several heavy metals. This complex nature is a unique and superior feature which distinguishes them from the other stress-associated genes. In addition, their overexpression confers enhanced abiotic stress tolerance in numerous crop plants (Mukhopadhyay et al., 2004; Huang et al., 2008; Kanneganti and Gupta, 2008). Furthermore, an SAP1 gene from sugarcane has been reported to have role in plant development (Li et al., 2011). In this section, we have compared the different subcellular localization of SAPs and their orthologs to have a clear cut idea on their mode of action by protein sorting, cell signaling, and protein–protein interaction at the time of stress defense mechanism. The subcellular localization and the protein properties were observed to have an analytical role on SAPs existence (Fig. 1.3).

Various subcellular localizations were predicted using various bioinformatics tools (Hua and Sun, 2001; Pierleoni et al., 2006; Horton et al., 2007; Yu et al., 2014). The prediction was based on amino acid sequence, functional domains, protein sorting, and signaling (Table 1.5). *OsiSAP8* showed nucleus, mitochondria and cytoplasm as the point of localization; the close ally *AtSAP2* was noticed to be present over extracellular membrane apart to the later three. Few novel accessions were predicted which showed key localization places like A0A0D9WS37 (secretory); A8QZ71, W5H1C8, and M0V9H3 were localized at same places as that of *OsiSAP* ZFPs. The protein properties of SAPs such as length of amino acids sequence, molecular weight, instability index, aliphatic index, GRAVY, stability and pI (Table 1.6) were observed in the reference genes. Almost all the proteins showed similar values and

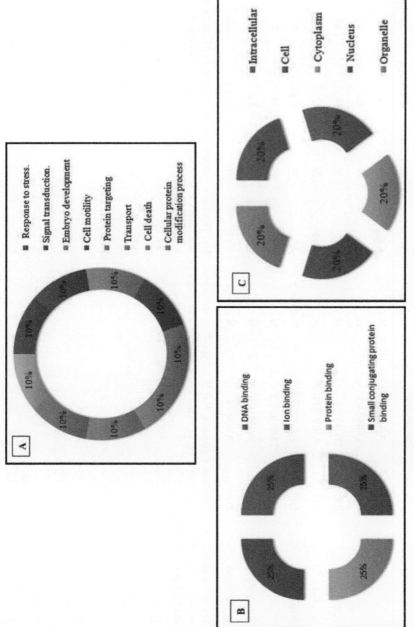

FIGURE 1.3 (See color insert.) Gene ontology of *OsiSAP8* gene. (A) Biological process of reference SAP genes, (B) molecular function of reference SAP genes, and (C) cellular component present in reference SAP genes.

TABLE 1.5 Subcellular Localization.

Gene	Source	Gene ac.no (protein sequence)		Commercially available online tools used			
			Bacello	Wolf Psort 2	ESLPred2	Cello2go	Subloc
SAP8	*Osi*	AAQ84334.1	N	Ct	N	N	Mt
SAP2	*At*	Q8H0X0	N	Ct	N	Ec	Mt
BRADI3G07060	*Bd*	I1HYB8	N	Ct	N	N	Mt
MLOC_20,848	*Hv*	M0V9H3	N	Cyto	N	Ct	Mt
LPERR06G17490	*Lp*	A0A0D9WS37	Secretory	Mt	N	Cyto	N
SAP8	*Ma*	M0SDV0	N	Ct	N	N	N
Traes_6DS_066F7755D	*Ta*	W5H1C8	N	Ct	N	Ec	Mt
GRMZM2G058866	*Zm*	A8QZ71	N	Cyto	N	Cyto	Mt

Osi: Oryza sativa-indica; At: Arabidopsis thaliana; Bd: Brachypodium distachyon; Hv: Hordeum vulgare var. distichum; Lp: Leersia perrieri; Ma: Musa acuminata subsp. malaccensis; Ta: Triticum aestivum; Zm: Zea mays. N: Nucleus; Ct: Chloroplast; Ec: Extracellular membrane; Mt: Mitochondria; Cyto: cytoplasm.

TABLE 1.6 Protein Properties of SAP Genes.

Gene id	Source	Aa	M.wt	II	Ai	GRAVY	S/Uns	pI
AAQ84334.1	*Osi*	171	18401.1	27.12	66.26	−0.386	S	8.63
Q8H0X0	*At*	173	18428.9	42.56	59.25	−0.265	Un	7.99
I1HYB8	*Bd*	171	18513.1	32.29	59.88	−0.405	S	8.47
M0V9H3	*Hv*	184	19646.6	33.17	69.51	−0.193	S	8.76
A0A0D9WS37	*Lp*	202	21884.2	30.00	69.11	−0.362	S	8.94
M0SDV0	*Ma*	102	10689.0	9.90	63.24	−0.467	S	9.15
W5H1C8	*Ta*	193	20470.5	32.78	65.80	−0.162	S	8.62
A8QZ71	*Zm*	171	18307.9	31.15	64.62	−0.323	S	8.28

aa: Amino acids sequence length; M.wt: molecular weight; II: instability index; AI: aliphatic index; GRAVY: grand average of hydropathy; S/Uns: stable/unstable; pI: isoelectric point.

there was not much difference among the values. The functional mode and percentage of metabolic profile of reference SAP gene was analyzed. The key factors such as signal transduction, cell apoptosis, protein sorting, protein binding, and DNA and ion bindings were hypothesized.

In plants during stress, their roles extend in diverse functions such as ubiquitination, redox sensing, hormone metabolism, and gene regulation (Ströher et al., 2009; Liu et al., 2011; Zahur et al., 2012; Charrier et al., 2012; Giri et al., 2013). Owing to these significant functions, there has been a keen interest in manipulation of these gene families to impart stress tolerance in crop plants in general and rice in particular, which is evident through the numerous publications in the recent years. Transcriptomic studies have illustrated that *OsiSAP1* overexpression affected several endogenous genes that were part of stress response, signal transduction, transcriptional activation, membrane transport, plant growth, development, and metabolism (Dansana et al., 2014). Though the precise mode of action of the ZFPs is not clear, it is presumed that they may be involved in the downregulation of pathway associated with stress-linked injury such as cell death by ubiquitinylation of key proteins thereby targeting their degradative process (Vij and Tyagi, 2006).

Lately, *OsiSAP1* overexpression was known to indirectly affect the expression of other genes by acting as ubiquitin ligase thereby imparting stress tolerance (Kothari et al., 2016). Among the eight SAP genes, M0SDV0 were identified smallest protein with 102 amino acid (aa) sequence, although the biggest one was A0A0D9WS37 (202 aa). The molecular weights of the proteins according to protein size ranged from 10.6 kDa (M0SDV0) to 21.8 kDa (A0A0D9WS37). pI of proteins were M0SDV0 (9.15) and A0A0D9WS37 (8.94). There was no major variation between the observed proteins of the SAP reference genes when compared with that of *OsiSAP8*.

1.5   FUTURE PERSPECTIVES

Considerable progress has been made to enhance abiotic stress tolerance in rice. It has been suggested by geneticists that abiotic stress tolerance is a trait controlled by multiple loci involving a plethora of genes. However, there is still a huge gap between basic research and developing abiotic stress-tolerant crop plants. First, there should be proper assessment of the environmental stress tolerance according to the agricultural production. There is a considerable difference in the transgenic crops in pot and field trials related to abiotic stress. Moreover, though several authors have reported transgenic rice with enhanced abiotic stress resistance, no transgenic rice with improved tolerance is currently approved (Gaudin et al., 2013). Events that are tolerant in pot trials fail in field experiments which illustrate the poor assessment of environmental factors. Most rice transgenic studies are undertaken and assessed under controlled environments with very few reports having detailed analyses on field experiments imparting abiotic stress. Second major factor is the yield potential. Stress-tolerant plants are not always high yielding which is the primary necessity for breeders (Qin et al., 2011). To combine high yielding potential coupled with high abiotic stress tolerance while preventing any pleiotropic effects, tunable promoters that have restricted and precise gene regulation in a spatio-temporal or tissue specific manner are the appropriate choice. A successful transformation strategy relies more on the appropriate use of the promoter which is the key regulator determining the strength and specificity of gene expression (Chakravarthi et al., 2015). Several studies on overexpressing genes under the control of constitutive or stress inducible promoters to develop stress tolerant rice have been discussed. However, use of constitutive promoters has also led to alterations in phenotypes which may be overcome by using stress inducible or tissue-specific promoters (Liu et al., 1998; Kasuga et al., 2004; Takasaki et al., 2010; Khan et al., 2016).

Despite several stress inducible promoters being used to produce abiotic stress-tolerant rice (Liu et al., 1998), there exists the necessity for novel and unique promoters that have precise control over transgene expression. With the rice genome sequence availability it has become far easier for the researchers and the only challenge that lies ahead is the functional annotation of genes and validation of promoters to accelerate crop improvement through biotechnological tools. Functional genomics is feasible through the employment of multiple tools including global transcript profiling coupled with using mutants and transgenics to characterize innumerable genes in a shorter time-frame.

1.6 SUMMARY AND CONCLUSION

OSiSAPs are genes that encode A20/AN1 ZFPs to resist multiple abiotic stresses (Mukhopadhyay et al., 2004). There are about 18 members of SAPs that have been characterized to combat abiotic stresses (Vij and Tyagi, 2006). The SAPs are reported to be induced by multiple abiotic stresses. Overexpressing the gene in rice conferred resistance to salinity, drought, dehydration and cold (Mukhopadhyay et al., 2004). These proteins have been reportedly found to have similar functions in regulating the immune system in humans. They act as ubiquitin ligases that stand as a common link between plants and animals. Presence of cysteine residues at the A20/AN1 domain at variable numbers in different rice SAPs speculates their role in the redox sensing role. Although there are many reports on SAPs of rice and *A. thaliana,* the mechanism and the signaling of how they fight the abiotic stress is yet to be deciphered. Considering this, a comparative transcriptomic profile of a wide array of SAPs and bioinformatic applications on the other hand play a pivotal role in omics-based studies which may significantly help us in understanding the molecular crosstalk of SAP genes and also aid in identifying the novel stress responsive genes. This gives us a holistic approach regarding the behavior and/or effect of the same gene(s) (whose role in tolerance to a particular stress is known) to multiple stresses (which is the actual situation in the agricultural field). Integrating the outcome of such multiple web-based omics databases and applications are expected to serve as a platform to identify the biological properties of the grass species especially in rice, molecular signals promoting comparative genomics, and functional analysis of the specific genes and uncharacterized genes. Thus, here we explored the knowledge on rice SAPs in order to decipher the key dynamics in the behavior of this protein during stress so as to utilize its potential to improve the crop yield.

Although there are many reports that are being presented with respect to stress biology and its application on the field, the future and commercialization of these depends on the safety and the reliability of the transgenic crops (Khan et al., 2016). In this chapter, we highlighted the importance of the SAP genes present in the rice crop to have clear cut knowledge on its mode of action and self-defense mechanism to tackle the abiotic stresses. The most predominantly used genes for abiotic stress are DREB transcription factors, HSPs, Na^+/H^+ antiporters of vacuolar membrane and many more. However, the relative future of this system is nearly unpredictable because of erupting multiple/combined abiotic stresses by the nature due to the human intervention. SAP genes could be a suitable substitute owing to its multigenic nature. They have established themselves with a novel A20/AN1 ZFPs that impart tolerance to multiple abiotic stresses in plants (Mukhopadhyay et al., 2004). The sense of ubiquitination and redox potential has been deciphered. The protein is naive of tyrosine residue at the key binding site during ubiquitination and on the other hand it has a strong redox potential with a key interaction of RLCK253 (Sharma et al., 2015). This opens a new scenario on pathogenic resistance of the SAP gene. Moreover, the homology of these with animal A20/AN1 ZFPs improves the understanding of SAPs as enhancer of innate immunity. Thus, understanding the gene regulatory networks coupled with key signaling processes would enable us to unravel the intricate mechanisms of abiotic stress tolerance in crops thereby employing this information to develop novel varieties with better stress tolerance benefitting mankind.

KEYWORDS

- Bioinformatics
- cell proliferation
- comparative mapping and genomics
- constitutive promoter
- gene regulation
- hormone metabolism
- in silico predictions
- Medicago truncatula
- peptide finger printing
- RNA finger printing
- subcellular localization
- transcriptional activation and factors
- transcriptome
- ultrahigh-throughput omics
- zinc-finger domain and protein

REFERENCES

Altschul, S. F.; Madden, T. L.; Schäffer, A. A.; Zhang, J.; Zhang, Z.; Miller, W.; Lipman, D. J. Gapped BLAST and PSI-BLAST: A New Generation of Protein Database Search Programs. *Nucleic Acids Res.* **1997**, *25*, 3389–3402.

Arnold, K.; Bordol, L.; Kopp, J.; Schwede, T.; The SWISS-MODEL Workspace: A Web-Based Environment for Protein Structure Homology Modelling. *Bioinformatics* **2006**, *22*, 195–201.

Berman, H. M.; Battistuz, T.; Bhat, T. N.; Bluhm, W. F.; Philip, E.; Burkhardt, K.; Feng, Z.; Gilliland, G. L.; Iype, L.; Jain, S.; Fagan, P.; Marvin, J.; Padilla, D.; Ravichandran, V.; Thanki, N.; Weissig, H.; Westbrook, J. D. The Protein Data Bank. *Acta Crystallogr. D Biol. Crystallogr.* **2002**, *58*, 899–907.

Biasini, M.; Bienert, S.; Waterhouse, A.; Arnold, K.; Studer, G.; Schmidt, T.; Kiefer, F.; Cassarino, T. G.; Bertoni, M.; Bordoli, L.; Schwede, T. SWISS-MODEL: Modelling

Protein Tertiary and Quaternary Structure Using Evolutionary Information. *Nucleic Acids Res.* **2014**, *42,* 1–7.

de Castro, E.; Sigrist, C. J. A.; Gattiker, A.; Bulliard, V.; Langendijk-Genevaux, P. S.; Gasteiger, E.; Bairoch, A.; Hulo, N. Scanprosite: Detection of PROSITE Signature Matches and ProRule-Associated Functional and Structural Residues in Proteins. *Nucleic Acids Res.* **2006**, *34,* W362–W365.

Chakravarthi, M.; Philip, A.; Subramonian, N. Truncated Ubiquitin 5′ Regulatory Region from Erianthus Arundinaceus Drives Enhanced Transgene Expression in Heterologous Systems. *Mol. Biotechnol.* **2015**, *57,* 820–835.

Charrier, A.; Planchet, E.; Cerveau, D.; Gimeno-Gilles, C.; Verdu, I.; Limami, A. M.; Lelièvre, E. Overexpression of a Medicago Truncatula Stress-Associated Protein Gene (MtSAP1) Leads to Nitric Oxide Accumulation and Confers Osmotic and Salt Stress Tolerance in Transgenic Tobacco. *Planta* **2012**, *236,* 567–577.

Dansana, P. K.; Kothari, K. S.; Vij, S.; Tyagi, A. K. Osisap1 Overexpression Improves Water-Deficit Stress Tolerance in Transgenic Rice by Affecting Expression of Endogenous Stress-Related Genes. *Plant Cell Rep.* **2014**, *33,* 1425–1440.

Dixit, A. R.; Dhankher, O. P. A Novel Stress-Associated Protein "AtSAP10" from Arabidopsis Thaliana Confers Tolerance to Nickel, Manganese, Zinc, and High Temperature Stress. *PloS One* **2011a**, *6,* e20921.

Dixit, A. R.; Dhankher, O. P. A Novel Stress-Associated Protein "AtSAP10 " from Arabidopsis Thaliana Confers Tolerance to Nickel, Manganese, Zinc, and High Temperature Stress. 6, 2011b.

Dixit, V. M.; Green, S.; Sarma, V.; Holzman, L. B.; Wolf, F. W.; O'Rourke, K.; Ward, P. A.; Prochownik, E. V.; Marks, R. M. Tumor Necrosis Factor-Alpha Induction of Novel Gene Products in Human Endothelial Cells Including a Macrophage-Specific Chemotaxin. *J. Biol. Chem.* **1990**, *265,* 2973–2978.

Donaldo, R. B.; Walid, M.; Mieulet, D. The Promoter of the AlSAP Gene from the Halophyte Grass *Aeluropus Littoralis* Directs a Stress-Inducible Expression Pattern in Transgenic Rice Plants. *Plant Cell Rep.* **2015**, 34(10), 1791–1806.

Finn, R. D.; Clements, J.; Eddy, S. R. HMMER Web Server: Interactive Sequence Similarity Searching. *Nucleic Acids Res.* **2011**, *39,* 29–37.

Gasteiger, E.; Hoogland, C.; Gattiker, A.; Duvaud, S.; Wilkins, M. R.; Appel, R. D.; Bairoch, A. Protein Identification and Analysis Tools on the ExPASy Server. In *The Proteomics Protocols Handbook*; Walker, J. M., Ed.; Humana Press, NY, USA, 2005; pp 571–607.

Gaudin, A. C. M.; Henry, A.; Sparks, A. H.; Slamet-Loedin, I. H. Taking Transgenic Rice Drought Screening to the Field. *J. Exp. Bot.* **2013**, *64,* 109–117.

Giri, J.; Dansana, P. K.; Kothari, K. S.; Sharma, G.; Vij, S.; Tyagi, A. K. SAPs as Novel Regulators of Abiotic Stress Response in Plants. *BioEssays†⁻: News and Reviews in Molecular, Cellular and Developmental Biology* **2013**, *35,* 639–648.

Goodstein, D. M.; Shu, S.; Howson, R.; Neupane, R.; Hayes, R. D.; Fazo, J.; Mitros, T.; Dirks, W.; Hellsten, U.; Putnam, N.; Rokhsar, D. S. Phytozome: a Comparative Platform for Green Plant Genomics. *Nucleic Acids Res.* **2012**, *40,* 1178–1186.

Gu, H.; Zhu, P.; Jiao, Y.; Meng, Y.; Chen, M. PRIN: a Predicted Rice Interactome Network. *BMC Bioinf.* **2011**, *12,* 13.

Higo, K.; Ugawa, Y.; Iwamoto, M.; Korenaga, T. Plant Cis-Acting Regulatory DNA Elements (PLACE) Database: 1999. *Nucleic Acids Res.* **1999**, *27,* 297–300.

Hishiya, A.; Iemura, S.; Natsume, T.; Takayama, S.; Ikeda, K.; Watanabe, K. A Novel Ubiquitin-Binding Protein ZNF216 Functioning in Muscle Atrophy. *EMBO J.* **2006,** *25,* 554–564.

Horton, P.; Park, K. J.; Obayashi, T.; Fujita, N.; Harada, H.; Adams-Collier, C. J.; Nakai, K. WoLF PSORT: Protein Localization Predictor. *Nucleic Acids Res.* **2007,** *35,* 585–587.

Hua, S.; Sun, Z. Support Vector Machine Approach for Protein Subcellular Localization Prediction. *Bioinformatics* **2001,** *17,* 721–728.

Huang, J.; Teng, L.; Li, L.; Liu, T.; Li, L.; Chen, D.; Xu, L-G.; Zhai, Z.; Shu, H.-B. ZNF216 is an A20-Like and I kappa-B Kinase Gamma-Interacting Inhibitor of NF kappa-B Activation. *J. Biol. Chem.* **2004,** *279,* 16847–16853.

Huang, J.; Wang, M. M.; Jiang, Y.; Bao, Y. M.; Huang, X.; Sun, H.; Xu, D. Q.; Lan, H. X.; Zhang, H. S. Expression Analysis of Rice A20/AN1-Type Zinc Finger Genes and Characterization of ZFP177 that Contributes to Temperature Stress Tolerance. *Gene* **2008,** *420,* 135–144.

Hussain Wani, S.; Ritturaj Kushwaha, H.; Wang, H.; Wang, H.; Shao, H.; Tang, X. Recent Advances in Utilizing Transcription Factors to Improve Plant Abiotic Stress Tolerance by Transgenic Technology. Frontiers in Plant Science Article Front. *Plant Sci.* **2016,** *7,* 1–13.

Kanneganti, V.; Gupta, A. K. Overexpression of OsiSAP8, a Member of Stress Associated Protein (SAP) Gene Family of Rice Confers Tolerance to Salt, Drought and Cold Stress in Transgenic Tobacco and Rice. *Plant Mol. Biol.* **2008,** *66,* 445–462.

Kasuga, M.; Miura, S.; Shinozaki, K.; Yamaguchi-Shinozaki, K. A Combination of the Arabidopsis DREB1a Gene and Stress-Inducible rd29A Promoter Improved Drought- and Low-Temperature Stress Tolerance in Tobacco by Gene Transfer. *Plant Cell Physiol.* **2004,** *45,* 346–350.

Kavar, T.; Maras, M.; Kidrič, M.; Šuštar-Vozlič, J.; Meglič, V. Identification of Genes Involved in the Response of Leaves of Phaseolus Vulgaris to Drought Stress. *Mol. Breeding* **2008,** *21,* 159–172.

Kelley, L. A.; Mezulis, S.; Yates, C. M.; Wass, M. N.; Sternberg, M. J. E. The Phyre2 Web Portal for Protein Modeling, Prediction and Analysis. *Nat. Protoc.* **2015,** *10,* 845–858.

Khan, M. S.; Khan, M. A.; Ahmad, D. Assessing Utilization and Environmental Risks of Important Genes in Plant Abiotic Stress Tolerance. *Front Plant Sci.* **2016,** *7,* 1–13.

Kothari, K. S.; Dansana, P. K.; Giri, J.; Tyagi, A. K. Rice Stress Associated Protein 1 (OsSAP1) Interacts with Aminotransferase (OsAMTR1) and Pathogenesis-Related 1a Protein (OsSCP) and Regulates Abiotic Stress Responses. *Front Plant Sci.* **2016,** *7,* 1–16.

Krzywinski, M.; Schein, J.; Birol, I.; Connors, J.; Gascoyne, R.; Horsman, D.; Jones, S. J.; Marra, M. A. Circos: an Information Esthetic for Comparative Genomics. *Genome Res.* **2009,** *19*(9), 1639–1645.

Lee, E. G. Failure to Regulate TNF-Induced NF-kappa B and Cell Death Responses in A20-Deficient Mice. *Science* **2000,** *289,* 2350–2354.

Lescot, M.; Déhais, P.; Thijs, G.; Marchal, K.; Moreau, Y.; Van de Peer, Y.; Rouzé, P.; Rombauts, S. PlantCARE, a Database of Plant Cis-acting Regulatory Elements and a Portal to Tools for in Silico Analysis of Promoter Sequences. *Nucleic Acids Res.* **2002,** *30,* 325–327.

Lesk, C.; Rowhani, P.; Ramankutty, N. Influence of Extreme Weather Disasters on Global Crop Production. *Nature* **2016,** *529,* 84–87.

Li, X.; Cai, W.; Zhang, S.; Xu, L.; Chen, P.; Wang, J. Cloning and Expression Pattern of a Zinc Finger Protein Gene ShSAP1 in *Saccharum officinarum* (*Shengwu Gongcheng Xuebao). Chin. J. Biotechnol.* **2011,** *27,* 868–875.

Lister, R.; O'Malley, R. C.; Tonti-Filippini, J.; Gregory, B. D.; Berry, C. C.; Millar, A. H.; Ecker, J. R. Highly Integrated Single-Base Resolution Maps of the Epigenome in Arabidopsis. *Cell* **2008**, *133*, 523–536.

Liu, Q.; Kasuga, M.; Sakuma, Y.; Abe, H.; Miura, S.; Yamaguchi-Shinozaki, K.; Shinozaki, K. Two Transcription Factors, DREB1 and DREB2, with an EREBP/AP2 DNA Binding Domain Separate Two Cellular Signal Transduction Pathways in Drought- and Low-Temperature-Responsive Gene Expression, Respectively, in Arabidopsis. *Plant Cell* **1998**, *10*, 1391–1406.

Liu, Y.; Xu, Y.; Xiao, J.; Ma, Q.; Li, D.; Xue, Z.; Chong, K. OsDOG, a Gibberellin-Induced A20/AN1 Zinc-Finger Protein, Negatively Regulates Gibberellin-Mediated Cell Elongation in Rice. *J. Plant Physiol.* **2011**, *168*, 1098–1105.

Maruyama, K. Y.; Todaka, D. A.; Mizoi, J. U.; Yoshida, T. A.; Kidokoro, S. A.; Matsukura, S. A.; Takasaki, H. I.; Sakurai, T. E.; Yamamoto, Y. O. Y.; Yoshiwara, K. Y.. Identification of Cis-acting Promoter Elements in Cold- and Dehydration-Induced Transcriptional Pathways in Arabidopsis, Rice, and Soybean. *DNA Res.* **2011**, *19*(1), 1–13.

Mizoi, J.; Shinozaki, K.; Yamaguchi-Shinozaki, K. AP2/ERF Family Transcription Factors in Plant Abiotic Stress Responses. *Biochim. Biophys. Acta Gene Regul. Mech.* **2012**, *1819*, 86–96.

Mukhopadhyay, A.; Vij, S.; Tyagi, A. K. Overexpression of a Zinc-Finger Protein Gene from Rice Confers Tolerance to Cold, Dehydration, and Salt Stress in Transgenic Tobacco. *Proc. Natl. Acad. Sci. U. S. A.* **2004**, *101*, 6309–6314.

Nakagami, H.; Sugiyama, N.; Mochida, K.; Daudi, A.; Yoshida, Y.; Toyoda, T.; Tomita, M.; Ishihama, Y.; Shirasu, K. Large-Scale Comparative Phosphoproteomics Identifies Conserved Phosphorylation Sites in Plants. *Plant Physiol.* **2010**, *153*, 1161–1174.

Nakashima, K.; Takasaki, H.; Mizoi, J.; Shinozaki, K.; Yamaguchi-Shinozaki, K. NAC Transcription Factors in Plant Abiotic Stress Responses. *Biochim. Biophys. Acta Gene Regul. Mech.* **2012**, *1819*, 97–103.

Paul, A.; Kumar, S. An A20/AN1-Zinc-Finger Domain Containing Protein Gene in Tea is Differentially Expressed During Winter Dormancy and in Response to Abiotic Stress and Plant Growth Regulators. *Plant Gene* **2015**, *1*, 1–7.

Pierleoni, A.; Martelli, P. L.; Fariselli, P.; Casadio, R. BaCelLo: a Balanced Subcellular Localization Predictor. *Bioinformatics* **2006**, *22*(14), 408–416.

Qin, F.; Shinozaki, K.; Yamaguchi-Shinozaki, K. Achievements and Challenges in Understanding Plant Abiotic Stress Responses and Tolerance. *Plant Cell Physiol.* **2011**, *52*, 1569–1582.

Quevillon, E.; Silventoinen, V.; Pillai, S.; Harte, N.; Mulder, N.; Apweiler, R.; Lopez, R. Interproscan: Protein Domains Identifier. *Nucleic Acids Res.* **2005**, *33*, 116–120.

Rejeb, I.; Pastor, V.; Mauch-Mani, B. Plant Responses to Simultaneous Biotic and Abiotic Stress: Molecular Mechanisms. *Plants* **2014**, *3*, 458–475.

Sakhanokho, H. F.; Kelley, R. Y. Influence of Salicylic Acid on in Vitro Propagation and Salt Tolerance in *Hibiscus acetosella* and *Hibiscus moscheutos* (cv "Luna Red "). *Afr. J. Biotechnol.* **2009**, *8*, 1474–1481.

Sharma, G.; Giri, J.; Tyagi, A. K. Rice OsiSAP7 Negatively Regulates ABA Stress Signalling and Imparts Sensitivity to Water-Deficit Stress in Arabidopsis. *Plant Sci.* **2015**, *237*, 80–92.

Solanke, A. U.; Sharma, M. K.; Tyagi, A. K.; Sharma, A. K. Characterization and Phylogenetic Analysis of Environmental Stress-Responsive Sap Gene Family Encoding A20/AN1 Zinc Finger Proteins in Tomato. *Mol. Genet. Genomics* **2009**, *282*, 153–164.

Sreedharan, S.; Shekhawat, U. K. S.; Ganapathi, T. R. MusaSAP1, a A20/AN1 Zinc Finger Gene from Banana Functions as a Positive Regulator in Different Stress Responses. *Plant Mol. Biol.* **2012**, *80*, 503–517.

Ströher, E.; Wang, X. J.; Roloff, N.; Klein, P.; Husemann, A.; Dietz, K. J. Redox-Dependent Regulation of the Stress-Induced Zinc-Finger Protein SAP12 in Arabidopsis Thaliana. *Mol. Plant* **2009**, *2*, 357–367.

Sun, Q.; Zybailov, B.; Majeran, W.; Friso, G.; Olinares, P. D. B.; van Wijk, K. J. PPDB, the Plant Proteomics Database at Cornell. *Nucleic Acids Res.* **2009**, *37*, 969–974.

Szklarczyk, D.; Franceschini, A.; Wyder, S.; Forslund, K.; Heller, D.; Huerta-Cepas, J.; Simonovic, M.; Roth, A.; Santos, A.; Tsafou, K. P.; Kuhn, M.; Bork, P.; Jensen, L. J.; Von Mering, C. STRING v10: Protein-Protein Interaction Networks, Integrated over the Tree of Life. *Nucleic Acids Res.* **2015**, *43*, D447–D452.

Takasaki, H.; Maruyama, K.; Kidokoro, S.; Ito, Y.; Fujita, Y.; Shinozaki, K.; Yamaguchi-Shinozaki, K.; Nakashima, K. The Abiotic Stress-Responsive NAC-Type Transcription Factor OsNAC5 Regulates Stress-Inducible Genes and Stress Tolerance in Rice. *Mol. Genet. Genomics* **2010**, *284*, 173–183.

Tello-Ruiz, M. K.; Stein, J.; Wei, S.; Preece, J.; Olson, A.; Naithani, S.; Amarasinghe, V.; Dharmawardhana, P.; Jiao, Y.; Mulvaney, J.; Kumari, S.; Chougule, K.; Elser, J.; Wang, B.; Thomason, J.; Bolser, D. M.; Kerhornou, A.; Walts, B.; Fonseca, N. A.; Huerta, L.; Keays, M.; Tang, Y. A.; Parkinson, H.; Fabregat, A.; McKay, S.; Weiser, J.; D'Eustachio, P.; Stein, L.; Petryszak, R.; Kersey, P. J.; Jaiswal, P.; Ware, D. Gramene 2016: Comparative Plant Genomics and Pathway Resources. *Nucleic acids Res.* **2016**, *4*(44), D1133–D1140.

Toyoda, T.; Wada, A. Omic Space: Coordinate-Based Integration and Analysis of Genomic Phenomic Interactions. *Bioinformatics* **2004**, *20*, 1759–1765.

Untergasser, A.; Cutcutache, I.; Koressaar, T.; Ye, J.; Faircloth, B. C.; Remm, M.; Rozen, S. G. Primer3-New Capabilities and Interfaces. *Nucleic Acids Res.* **2012**, *40*, 1–12.

Vij, S.; Tyagi, A. K. Genome-Wide Analysis of the Stress Associated Protein (SAP) Gene Family Containing A20/AN1 Zinc-Finger(S) in Rice and Their Phylogenetic Relationship with Arabidopsis. *Mol. Genet. Genomics* **2006**, *276*, 565–575.

Vij, S.; Tyagi, A. K. Emerging Trends in the Functional Genomics of the Abiotic Stress Response in Crop Plants. *Plant Biotechnol. J.* **2007**, *5*, 361–380.

Wang, J.; Kong, L.; Zhao, S.; Zhang, H.; Tang, L.; Li, Z.; Gu, X.; Luo, J.; Gao, G. Rice-Map: a New-Generation Rice Genome Browser. *BMC Genomics* **2011**, *12*, 165.

Xuan, N.; Jin, Y.; Zhang, H.; Xie, Y.; Liu, Y.; Wang, G. A Putative Maize Zinc-Finger Protein Gene, ZmAN13, Participates in Abiotic Stress Response. *Plant Cell Tissue Organ Cult (PCTOC)* **2011**, *107*, 101–112.

Yu, C.-S.; Cheng, C.-W.; Su, W.-C.; Chang, K.-C.; Huang, S.-W.; Hwang, J.-K.; Lu, C.-H. CELLO2GO: a Web Server for Protein SubCELlular LOcalization Prediction with Functional Gene Ontology Annotation. *PloS One* **2014**, *9*, e99368.

Zahur, M.; Maqbool, A.; Irfan, M.; Jamal, A.; Shahid, N.; Aftab, B.; Husnain, T. Identification and Characterization of a Novel Gene Encoding Myb-Box Binding Zinc Finger Protein in *Gossypium arboreum*. *Biol. Plant* **2012**, *56*, 641–647

Zhang, J.; Li, Y.; Jia, H.; Li, J.; Huang, J.; Lu, M.; Hu, J.-J. The Heat Shock Factor Gene Family in *Salix Suchowensis*: a Genome-Wide Survey and Expression Profiling During Development and Abiotic Stresses. *Front Plant Sci.* **2015**, *6*, 1–14.

CHAPTER 2

POTENT AVENUES FOR CONFERRING SALINITY TOLERANCE IN RICE

TUSHAR KHARE[1] and VINAY KUMAR[1,2,*]

[1]*Department of Biotechnology, Modern College of Arts, Science and Commerce, Savitribai Phule Pune University, Ganeshkhind, Pune, Maharashtra 411016, India, Tel.: +91–2025634021, Mob.: +91 9764447330, Fax: +9125650931, E-mail: tushar131189@gmail.com*

[2]*Mob.: +91–9767839708*

[*]*Corresponding author. E-mail: vinaymalik123@gmail.com*

2.1 INTRODUCTION

Rice is one of the major crops with noteworthy share in food grain production globally. However, most of the cultivated rice varieties are glycophytic. With the prompt increase in primary rice consumers and the abating soil and water quality in the rice cultivation fields, there is a need to comprehend the adaptive response of this important crop toward hyper saline environments. With the eventual objective to raise rice plant with better suitability toward soil salinity, rigorous efforts are on worldwide commissioning physiological, biochemical, and molecular tools to accomplish this task. Despite several studies, our knowledge on the salt stress response mechanisms associated with rice is imperfect due to high complexity of responses, including ionic and an osmotic component, triggering many morphological, physiological and metabolic deviations coupled with oxidative damages. Though the generated information regarding salinity responses in rice is far from complete, but there are substantial evidences covering the salinity-induced responses on cells, tissues, organs, and whole plant levels. Hence, along with potent biotechnological advancement, particularly in post-genomic era, this information can be employed efficiently for improvement of rice for better salinity tolerance and yield. Biotechnological tools such as manipulating direct/indirect action genes and non-coding RNAs present sound platforms for developing salt-tolerant rice cultivars. Besides, the omics approached especially transcriptomics, proteomics and metabolomics technologies are emerging avenues to develop new rice varieties to combat salinity burden with increment in yield. This chapter presents a description of the recent applications of biotechnological tools in improvement of salt tolerance in rice varieties followed by reviewing attempts by crop biotechnologists and breeders which will help to enlighten the different avenues for better crop production in saline lands leading to fulfillment of global food requirements.

2.2 SALINITY STRESS: A SERIOUS THREAT TO RICE PRODUCTION

Soil salinity is becoming a global threat to plant growth and crop yields (Rengasamy, 2010). Approximately 900 million ha of land is imperiled to salinity including sodic and saline soils (OECD/FAO, 2012), which counts nearly 7–8% of the total land area of the world. Rice is an important crop worldwide, which is mainly cultivated in the Asia and acts as a staple food for more than half the world's population (FAO, 2013a). The total rice produced worldwide is around 675 million t, contributing approximately 27% to the total crop production (FAO, 2013b). Although rice serves as one of the most abundant crop species throughout the world, it is one of the most salt-sensitive crop among the grain crops (Khare et al., 2015; Kumar and Khare, 2016).

Rice shows variable degrees of salt tolerance at different developmental stages, with germination phase being relatively salt tolerant and reproductive stages as most salt sensitive ones, directly disturbing the yield (Singh et al., 2008; Kumar et al., 2009). Under the field conditions, one-third yield reduction can be observed when cultivated at electrical conductivity (EC) of 2 dS/m (low salinity level) and nearly halved at 4 dS/m (moderate salinity level) (Grattan et al., 2002), and at 3 dS/m level, 12% yield reduction has been reported per dS/cm (Mohammadi et al., 2013). Under such conditions, plants may adapt to mild persistent or short-term moderate salinity levels by re-scheduling complex and interconnected various physiological and developmental processes (Munns and Tester, 2008). Normal physiology of the plant and hence the yield is maintained by a higher rate of new leaf and inflorescence formation, higher panicle initiation with enhanced grain-filling period (Schmidt et al., 2014; Munns, 2005). However, when the cultivation environment provides the salinity levels above the threshold of the growing rice variety, the abovementioned adaptive mechanisms get compromised and cause lower productivity (Munns, 2005). The effects are more severe during the reproductive phase and grain-filling phase as the plants have very limited alternate mechanisms to alleviate the accumulation of salt-sensitive tissues at this terminal phase of the lifecycle (Munns, 2005; Mittler and Blumwald, 2010; Kumar and Khare, 2016). This collectively results into higher floret sterility, higher number of partially filled or unfilled grains, and ultimately the reduced total yield (Beckles and Thitisaksakul, 2014; Roessner and Beckles, 2012).

The response of the cultivated rice plants to such saline environment depends on various factors therefore the outcome may not always be as expected (Thitisaksakul et al., 2015). Moreover, as the whole rice grain is consumed with minimal processing compared to the other crops, the cultivation environment intensely dictates the grain post-harvest qualities (Sun et al., 2011; Beckles and Thitisaksakul, 2014). Thus, the salinity is one of the prime causes for rice yield reduction. It is a global issue of concern as it is gradually threatening the serenity of food requirements from ever-increasing population.

2.3 PHYSIOLOGICAL, BIOCHEMICAL, AND MOLECULAR RESPONSES OF RICE TO SALINITY STRESS

Plant response to soil salinity is natural, random and complex event. To combat the salinity-induced growth reductions and yield losses, such responses have been studied thoroughly on physiological, biochemical and molecular levels not only in rice but in many important crops also. As rice is considered to be a salt-sensitive crop, salinity has been proved to be an inductive element for biochemical and physiological alterations in the plants causing growth inhibition and yield reduction (Fukuda et al., 2007; Rao et al., 2008). Several physiological parameters have been studied to evaluate the differential physiological responses of salt-sensitive, moderately tolerant and tolerant rice varieties to differential salinity levels at different developmental stages of the plants (Kumar et al., 2007, 2008; Danai-Tambhale et al., 2011; Kumar and Khare, 2016). These parameters include germination rate, plant height (root length and shoot length), total biomass produces (fresh weight and dry weights), leaf injury (in terms of electrolytic leakage), stomatal conductance, and distribution of ions (principally Na^+, Cl^-, and K^+)

with their ratios (Kumar et al., 2008; Bhowmik et al., 2009; Haq et al., 2009; Ali et al., 2014a).

The initiation of the osmotic disproportion indicates the adverse effects of soil salinity, which is characterized by a drop in osmotic potential followed by ionic imbalance and ion toxicity. As all the effects are considered to initiate at cellular level, two cellular organelles namely chloroplast and mitochondria have proved to be the most vulnerable to the detrimental effects of salinity (Rahman et al., 2000). Therefore, total chlorophyll content, chlorophyll fluorescence (Fv/Fm) and membrane permeability are the potential parameters showing the inhibitory effects of salinity on the photosynthetic machinery (Baker, 2008). Ultrastructure observations also revealed the disruption of chloroplastids and swollen thylakoids, which are also responsible for reduced photosynthetic efficiency under saline conditions (Rahman et al., 2000). It also affects the total leaf area and shows some profound anatomical changes in the leaf (Wankhade et al., 2010, 2013). Under severe saline conditions these effects can be observed in vascular tissues as well apart from the regular leaf mesophylls. The harmful effect of salt accumulation is mainly fought by rice via salt exclusion (Garcia et al., 1997), selective ion uptake and by maintaining Na^+/K^+ ratio (Rahman et al., 2000). A strong association between yield, sodium concentration, and Na^+/K^+ ratio in rice is reported by many breeders. Along with the mentioned parameters, some more agronomic morpho-physiological studies are considered such as tiller number, leaf area, panicle length and number, relative water content, relative growth rate, which are useful to differentiate the response between different rice cultivars based on their salt tolerance capabilities (Zeng and Shannon, 2000; Kumar and Khare, 2016). Till date, many rice varieties have been screened under many of the mentioned physiological parameters, which have enabled to build up a strong database about numerous physiological strategies developed by rice in response to salt stress.

The salinity in field usually results in three-fold more damage due to conditions such as osmotic stress, ionic imbalance and oxidative stress which are result of excess salt exposure (Khare et al., 2015). This three-fold detrimental process results in excessive generation of reactive oxygen species (ROS) due to low stomatal conductance under low water potential, low rate of electron transport in cellular organelles and over accumulation of sodium and chloride (Türkan and Demiral, 2009). These ROS are prime cause of oxidative damages to lipid membrane and other necessary macromolecules including pigments, proteins and nucleic acids; which eventually incites cell death (Khare et al., 2015). These ROS also act as signaling molecules

(Qureshi et al., 2013; Yildiztugay et al., 2014; Ismail et al., 2014). To combat the complex stressed situation, plants react with set of different mechanisms such as selective accumulation, exclusion and compartmentalization of ions at whole plant level as well as cellular level, synthesis of compatible osmolytes and initiation of enzymatic and non-enzymatic antioxidant machinery for effective osmotic and ionic balancing and ROS scavenging (Munns and Tester, 2008; Kumar et al., 2009; Qureshi et al., 2013; Yildiztugay et al., 2014). The osmotic adjustment is accomplished by amassing high concentrations of inorganic ions or low-molecular-weight organic solutes (Kumar et al., 2010). It includes low-molecular-weight sugars, sugar alcohols (polyols) and organic acids, nitrogenous compounds (amino/imino acids, amides, proteins and quaternary ammonium compounds) (Kumar et al., 2016). Proline accumulation in rice has been reported in numerous cases (Kumar et al., 2009, Kumar and Khare, 2015). Hydrophilic low-molecular-weight antioxidants, such as ascorbic acid (AsA), reduced glutathione (GSH), flavonoids, phenolics, and lipophilic low-molecular-weight antioxidants, such as tocopherols and carotenoids, possess ability to quench and detoxify the ROS. Both these categories are considered as non-enzymatic antioxidants and are reported in rice (Khan and Panda, 2008; Khare et al., 2015). Soluble sugars and starch also act as osmoticum in many plants. Improved sugar content in shoots as well as increment in root starch content is observed in rice cultivated under salinity. This strategy can be correlated to maximum storage of reserve food to sustain more efficiently under salt stress (Amirjani, 2011). Moreover, on similar basis, the storage of nitrogen to combat salinity is facilitated by altered protein profiles of rice (Kumar et al., 2009). This may include overexpression, downregulation or de novo synthesis of some new proteins. On the other hand, enzymatic antioxidants such as superoxide dismutase (SOD), catalase (CAT), peroxidase (POX) (directly reacts with ROS to maintain their low levels) and glutathione reductase (GR), ascorbate peroxidase (APX) (acts to regenerate oxidized antioxidants) are also produced by rice to maintain redox state of the plant.

Apart from the physiological and biochemical parameters, molecular factors in rice are also considered significant in salinity-related responses, as ultimately these studies lead to the development of potent transgenics by exploring salt-responsive genes. The course of salinity imparts series of complex responses which differ in inter-varietal manner. Hence, genetic analysis of many rice cultivars have been performed to screen the salt tolerance level using molecular markers such as restriction fragment length polymorphism (RFLP), random amplification of polymorphic DNA (RAPD), simple sequence length polymorphisms (SSLP), simple sequence repeats

(SSR) and so forth (Rikke and Johnson, 1998; Kanawapee et al., 2011, Ali et al., 2014b) as well as positional cloning and insertional mutagenesis (Ron and Weller, 2007; Salvi and Tuberosa, 2005). The data generated via these experiments along with newly emerged and progressively developing bioinformatics tools can be employed to identify the potential salt-responsive genes as well as whole expression profiles of different rice varieties under salt stress.

The controlled genetic mechanism acts as pillar in induced signaling pathway against salinity. Several regulatory genes including histidine kinases (HKs), calcium-dependent protein kinases (CDPKs), mitogen-activated protein kinase (MAPKs) and several genes in salt overly sensitive (SOS) pathway (Zhu 2002; Xiong and Yang, 2003; Haq et al., 2008). In recent years, quantitative trait loci (QTL) analysis of the rice cultivars has helped to map several QTLs on several rice chromosomes. It includes QTLs in response to sodium and potassium uptake (Lin et al., 2004), and shoot and root growth (Sabouri and Sabouri, 2008; Mohammadi et al., 2013; Koyama et al., 2014). Identification of such QTLs opened new frontier in understanding the salinity-induced molecular responses as well as in development of salt-tolerant rice varieties. The small length microRNAs (miRNAs) are also emerging as new tool in salt-tolerant crop development as several differentially expressed miRNAs are identified in rice which can be employed to develop salt-tolerant rice cultivars.

2.4 POTENT BIOTECHNOLOGICAL TOOLS FOR CONFERRING SALINITY TOLERANCE IN RICE

2.4.1 DIRECT ACTION GENES

Salinity stress is responsible for various changes from cellular to whole plant level, which includes differential expression of genes that are directly involved in the process to boost the adaptive mechanisms of the plants for sustenance under the saline conditions. The production of ROS and osmotic imbalance are the eventual consequences of the salinity-induced cellular damages. The activation of the antioxidant machinery is a counter-action by the plants, which lifts the endogenous levels of enzymatic and non-enzymatic antioxidants (Miller et al., 2010; Gill and Tuteja, 2010). Such direct action genes are prime targets of researchers. These can be effectively employed to improve salt tolerance in rice by using different biotechnological tools. The direct action genes which are ordinarily overexpressed in transgenic rice include genes coding for different anti-oxidant enzymes including SOD, CAT, POX, and GR or genes coding for production of osmo-protectants/compatible solutes (proline, polyamines, sugar alcohols etc.).

Zhao and Zhang (2006) transferred Glutathione S-transferase (GST) and CAT genes from *Suaeda salsa* to rice (cv. Zhonghua No. 11) via *Agrobacterium*-mediated transformation under the control of CaMV-35S promoter. The transformed lines showed reduced oxidative damages under applied salt stress. The transgenic lines showed improved GST and CAT activity along with higher SOD activity. SOD is the first line of plant enzymatic defense against stressful conditions. Cytosolic copper/zinc superoxide dismutase (*Cu/Zn-Sod*) from mangrove species *Avicennia marina* was cloned into pCAMBIA1300 (binary vector) and overexpressed in *indica* rice (var. Pusa Basmati-1) (Prashanth et al., 2008). The transformed plants performed strongly under 150 mM salinity stress compared to untransformed counterparts which wilted under the same stress environment. The selenium non-containing enzyme glutathione peroxidase (GPX) is also an important enzyme in ROS detoxification system. The GPX gene from *Nelumbo nucifera* L. (*NcGPX*) was successfully introduced in rice cultivar Yuetai B (Diao et al., 2014). This integration in the rice genome significantly improved the salt tolerance level of the transformed plants compared to the wild type. Dehydroascorbate reductase (DHAR) is a key enzyme for maintenance of redox status of ascorbate in cell. The transgenic line (cv. Ilmi) overexpressing cDNA encoding DHAR (*OsDHAR1*) was developed by Kim et al. (2014) that was regulated by maize ubiquitin promoter. The transformed plants showed enhanced DHAR activity and AsA/DHA ratio along with increased levels of enzymes involved in ascorbate-glutathione cycle. They also showed better plant growth, ion leakage and quantum yield collectively giving the better performance under salt stress.

To improve salt tolerance, bacterial genes were also used as a source. On these lines Moriwaki et al. (2008) introduced *E. coli* catalase gene (*katE*) into BR5 rice cultivar via *Agrobacterium*-mediated transformation. The transformed T_1 and T_2 lines showed higher CAT activities and showed high tolerance to salinity compared to non-transgenic plants. Similar approach was used to improve salt tolerance in rice cv. Nipponbare by Nagamiya et al. (2007).

Under hyper-saline environment, soil osmotic pressure exceeds that of the plant cells, which eventually reduces water uptake by plants. To counteract this situation, plants accumulate compatible osmolytes, such as amino acids, polyols, sugars and sugar alcohols, amines and so forth, as a key strategy. Hence, the genes encoding such molecules have also been employed on many occasions for salinity tolerance improvement. Choline oxidase (COX) is an enzyme which catalyzes the formation of glycine betaine, an important osmolyte, from choline. Glycine betaine has been reported to provide increased tolerance against variety of stress environment. The *codA* gene from *Arthrobacter globiformis* is successfully inserted into Pusa Basmati 1, *indica* rice (Mohanty et al., 2002). The resultant transgenics showed higher production of glycine betaine in R_0 and R_1 plants with increased survival and performance under salt stress (150 mM). Similarly, Sakamoto and Murata (1998) also used the *codA* gene from same organism to transform Nipponbare rice. The transgenic plants showed better recovery after the exposure to the salinity than the non-transformed plants. Su et al. (2006) demonstrated the integration of COX gene from *Arthrobacter pascens* into *japonica* rice (TNG67) using stress inducible stress promoters. The saline growth conditions improved the glycine betaine accumulation in the transgenic lines with greater level of salt tolerance. Rapid accumulation of the proline is also one of the first processes for plants under stresses. As an osmoprotectant, proline acts as osmotic balancer to protect intracellular structures, proteins and enzymes (Kumar et al., 2010). Therefore, genes encoding the enzymes involved in biosynthesis of proline are usually targeted to produce transgenics. *P5CS* gene from *Vigna aconitifolia*, from proline biosynthesis pathway was successfully introduces in *indica* rice cultivar ADT 43 via *Agrobacterium*-mediated transformation. The transgenic plants were able to survive at 200 mM of NaCl with improved proline production whereas, normal plants died within 10 days under the same conditions. Similar method was implemented by Kumar et al. (2010), who incorporated *P5CSF129A* gene in KJT-3 rice under the control of *CaMV 35S* promoter.

Polyamines, which mainly include spermine, spermidine and putrescine are also low-molecular-weight compatible osmolytes which are used largely to combat stress environment by plants. S-adenosylmethionine decarboxylase (SAMDC) is a key enzyme from polyamine biosynthetic pathway. The cDNA encoding for SAMDC from *Tritordeum* was introduced in *japonica* rice cultivar TNG 67 (Roy and Wu, 2002). Transgenic plants showed three-to four-fold increased levels of spermine and spermidine with high seedling growth and salt tolerance level. Non-reducing disaccharide trehalose is also a compatible osmolyte which can be observed in many prokaryotes and eukaryotes under the stress environments. The regulated expression of *E. coli* trehalose biosynthetic genes namely *otsA* and *otsB* (as fusion gene) were achieved in transgenic rice (cv. PB-1). Higher accumulation of trehalose was observed under salt as well as many other abiotic stress environments (Garg et al., 2002). On the similar line, trehalose-producing transgenic rice, expressing gene bifunctional fusion (TPSP) of the trehalose-6-phosphate synthase (TPS) gene and trehalose-6-phosphate phosphatase (TPP) gene of *E. coli*, was produced by Jang et al. (2003). Transgenic plants exhibited better performance under abiotic stresses including salinity. The enhanced abiotic stress tolerance can be also achieved by increased accumulation of sugar alcohols such as mannitol. The *E. coli* mannitol-1-phospho dehydrogenase (*mtlD*), involved in mannitol synthesis in plants was successfully introduced in rice to produce several putative transgenic rice plants (cv. PB-1). Transgenic plants exhibited improved drought and salinity stress tolerance which was attributed to mannitol accumulation (Pujni et al., 2007). Though various direct action genes have been prime target for transgenic technologies, their exogenous application has also been employed in recent times to decrease the salinity-induced cellular damage and eventual yield enhancement (Akram et al., 2015; Salama and Mansour, 2015; Bhusan et al., 2016; Savvides et al., 2016). A summarized list of these genes and their overexpression for imparting salinity tolerance in resultant transgenic rice varieties has been given in Table 2.1.

2.4.2 NON-CODING RNAS

Non-coding RNAs (ncRNAs) is a term employed to the RNA molecules which does not code for any protein but still possesses vital functions (Shriram et al., 2016). They include highly abundant and functionally important RNAs including transfer RNAs (tRNAs) and ribosomal RNAs (rRNAs) beside, small nucleolar RNAs (snoRNAs), miRNAs, small interfering RNAs (siRNAs), small nuclear RNAs (snRNAs), extra cellular RNAs (exRNAs) and the long ncRNAs, besides some yet to be characterized classes of them. Thousands of ncRNAs have been identified using transcriptomic and bioinformatics tools and the number is rising rapidly. These ncRNAs genes produce functional RNA molecules instead of proteins. Recent evidences have suggested the fact that majority of the complex genome of the higher level organisms is transcribed into ncRNAs, many of which are alternatively spliced/administered into smaller products. In higher eukaryotes, these stem-loop structures-derived tiny non-coding molecules are considered as the ubiquitous repressors of gene expression as they regulate the expression of target gene(s) and degrade and/or inhibit protein production (Akdogan

TABLE 2.1 Summarized List of Overexpressed Direct Action Genes for Improved Salt Tolerance of Resultant Transgenic Rice Varieties.

Gene	Source organism	Transformed rice variety	Improved character	Reference
GST and CAT	Suaeda salsa	Zhonghua No. 11	Salinity and oxidative stress tolerance	Zhao and Zhang (2006)
Cu/Zn-SodI	Avicennia marina	Pusa Basmati-1	Salinity, drought and oxidative stress tolerance	Prashanth et al. (2008)
NcGPX	Nelumbo nucifera	Yuetai B	Salinity tolerance	Diao et al. (2014)
OsDHAR1	Oryza sativa	Ilmi	Salinity tolerance	Kim et al. (2014)
katE	Escherichia Coli	BR5	Salinity tolerance, increased catalase activity	Moriwaki et al. (2008)
katE	E. coli	Nipponbare	Salinity tolerance, increased catalase activity	Nagamiya et al. (2007)
codA	Arthrobacter globiformis	Pusa Basmati-1	Salinity tolerance, increased glycine betaine production	Mohanty et al. (2002)
codA	A. globiformis	Nipponbare	Improved recovery performance after salinity exposure	Sakamoto and Murata (1998)
COX	Arthrobacter pascens	TNG67	Salinity tolerance, increased glycine betaine production	Su et al. (2006)
P5CS	Vigna aconitifolia	ADT 43	Salinity tolerance, increased proline production	Karthikeyan et al. (2011)
P5CSF129A	Vigna aconitifolia	KJT-3	Salinity tolerance, increased proline production	Kumar et al. (2010)
SAMDC	Tritordeum	TNG67	Increased seedling growth, spermine and sperdine production and salt tolerance.	Roy and Wu (2002)
otsA and otsB	E. coli	PB-1	Higher accumulation of the trehalose, improved salinity tolerance	Garg et al. (2002)
TPS and TPP	E. coli	Nakdong	Increased tolerance to salt, drought and cold	Jang et al. (2003)
mtlD	E. coli	PB-1	Increased mannitol accumulation, salinity and drought stress	Pujni et al. (2007)

et al., 2015). These specialized RNA molecules are concealed internal controllers of various gene expression levels (Eddy, 2001). The ncRNAs modulates many plant cellular responses to the surrounding environment by direct regulation of their target genes involved in crop growth and productivity. Hence due to their agricultural importance these small molecules have become spotlights in crop improvement programs in post-genomic era (Shin and Shin, 2016).

Plants trigger protective mechanisms to counteract the harsh and adverse environmental conditions. These mechanisms involve a network of genetic regulations including altered expression of large proportion of genes by transcriptional and/or translational regulations (Ku et al., 2015). Plants upregulate the protective genes while downregulating the negative regulators (Shriram et al., 2016). Amongst various plant ncRNAs, particularly miRNAs and siRNAs, are regarded as critical post-transcriptional regulators of stress responses (Bologna and Voinnet, 2014). Several ncRNAs, particularly miRNAs, have been identified as salt stress responsive in various plant species (Ding et al., 2011; Barrera-Figueroa et al., 2012; Liu and Zhang, 2012).

As plants in field condition face combinations of more than one abiotic stress conditions, such as salinity, heat, drought and oxidative stress, various small RNA species help the plants to adjust with such environment by regulating the endogenous level of regulatory ncRNAs in response. Many miRNA families have been studied in rice as well as in many other economically important crops with respect to their roles in stress responses and tolerance. For instance, miR393 is a conserved miRNA family present in monocot and dicot plants which shows differential expression levels to various environmental stresses.

The research in past decade has provided substantial evidence that miRNAs have abilities to be potential targets for genetic manipulations to confer abiotic stress tolerance to plants. Various methods are being worked out for miRNA manipulations such as overexpression or repression of stress-responsive miRNAs and/or their target protein-coding messenger RNAs, miRNA-resistant target genes, besides artificial miRNAs (amiRNAs) (Shriram et al., 2016).

Recently, Cui et al. (2015) overexpressed *miR156* (the first documented plant miRNA), and the resultant rice transgenics showed enhanced cold tolerance. Whereas overexpression of *miR172* in *Arabidopsis* resulted in improved tolerance to water deficit and salt stress (Li et al., 2016). Similarly, enhanced cold tolerance in engineered rice was attributed to the overexpression of *miRNA319* employing the RNAi strategy (Yang et al., 2013). Gao et al. (2011) transformed *miR393* into rice under control of CaMV 35S promoter. Under the course of salinity (150 mM) for 15 days, transgenic rice plants showed improved sensitivity to salt treatment compared to the non-transformed lines. Similarly, Xia et al. (2012) raised the rice lines transformed with *osmiR393*. The transformed plants showed increased number of tillers and early flowering and also reduced tolerance to salt and hyposensitivity to auxins. Though rice is the crop from which highest number of miRNAs are identified and analyzed for different kinds of abiotic stresses; yet the lack of information on regulatory mechanisms and technical limitations to miRNA information have led to the lesser number of successful reports about miRNA engineering-mediated salinity tolerance in rice. Therefore, this sound platform seems yet to be explored fully for producing salt-tolerant crops including rice.

2.4.3   OMICS APPROACHES

Recent advances in the field of genomics, proteomics, transcriptomics, and metabolomics have led to new understandings of intricate phenomena underlying salinity stress responses and tolerance mechanisms in plants (Sinha et al., 2015). These large-scale experiments, collectively known as omics approaches, refer to the study of large sets of biological molecules for detection of genes (genomics), mRNAs (trnascriptomics), proteins (proteomics) and metabolites (metabolomics) (Soda et al., 2015). These potential approaches are being looked as sound platforms for developing salt-tolerant crops (Das et al., 2015).

Owing to the accessibility of the whole genome sequence of rice (Matsumoto et al., 2007), functional genomics of rice salinity tolerance has seen rapid developments in recent years (Das et al., 2015). Numerous molecular markers have been successfully used to identify salt-tolerant rice cultivars (Dhar et al., 2011); however, the advent of next-generation sequencing (NGS) has made sequencing-based genotyping much more efficient, accurate and fast. Salinity stress responses in rice are controlled by several QTLs (Thomson et al., 2010; Wang et al., 2013). Various key salinity tolerance QTLs with significant effects have been identified from rice cultivars such as *qSKC1*, *qSNC7*, and *Saltol* (Lin et al., 2004; Thomson et al., 2010). However, there are only a few successful occasions where salt-tolerant rice cultivars have been developed using these identified QTLs-based molecular breeding approaches (Wang et al., 2016). Although

QTL mapping has resulted in several QTL linked with salinity tolerance in rice (Prasad et al., 2000; Koyama et al., 2014; Gregorio et al., 2002; Lin et al., 2004), the development of advanced genomics tools has resulted in better and deeper understanding of salt tolerance linked gene networks (Soda et al., 2015; Das et al., 2015). Various genomics tools including whole genome sequence, expressed sequence tags and microarrays have proved highly significant in our understanding of salinity responses and tolerance mechanisms in rice.

Transcriptomics, the study of complete set of RNA transcripts (the transcriptome) that are produced by the genome (also known as gene expression profiling) has accelerated our understandings of genetic mechanisms that control the plant responses to environmental cues (Zhou et al., 2016). Usually, the comparison of transcriptomes of stressed and normal plants felicitates the identification of salinity stress responsive genes.

The transcriptome sequencing of *Porteresia coarctata,* a wild salt-tolerant rice variety, generated large number of high-quality reads, followed by functional annotation revealed the presence of genes involved in diverse cellular processes as well as encoding transcription factors (Garg et al., 2014). Further, more than 15,000 genes were identified to be involved in salinity and/or submergence responses as revealed by differential gene expression analysis (Garg et al., 2014). Similarly, Venu et al. (2013) used massively parallel signature sequencing (MPSS) and sequencing by synthesis (SBS) approaches to investigate leaf and root transcriptomes of *japonica* rice cultivar Nipponbare under various abiotic stresses including cold, drought and salinity. Various genes up/downregulated under specific or common stress factors were identified as revealed by clustering analysis. Authors identified highly conserved *cis*-regulatory elements in the promoter of upregulated genes (Venu et al., 2013).

In a recent study, Zhou et al. (2016) carried out transcriptome profiling of leaves and roots of a wild rice variety growing under salinity as well as normal conditions at seedling stage. The obtained profiles were analyzed using Illumina HiSeq 2000 platform, and the results revealed that more than 6800 transcripts from leaves and around 5000 transcripts from roots expressed differentially under salinity stress (Zhou et al., 2016). The authors reported co-localization of numerous salinity-induced genes onto the fine-mapped salt tolerance related QTLs that could be used as candidates for gene cloning and elucidation of salt tolerance mechanisms in rice. All these and several other transcriptomic studies in rice with regard to salinity stress reveal that such data generated has the potential to serve as a resource for unraveling the causal mechanisms of salinity stress responses and devising the strategies to engineer salt tolerance in rice.

As discussed above, transcriptomics has opened potent avenues for its use in understanding salinity stress responses and conferring salinity tolerance. However, changes in gene expression at transcript level do not necessarily reflect at protein level (Aghaei and Komatsu, 2013). Therefore, study of proteins at large scale, that is proteomics, has emerged as an efficient omics tool in recent years. The study of proteins represents the shortest as well as a direct way to define the role of expressed genes (Das et al., 2015). Proteome is also an essential link between transcriptome and metabolome (Lata, 2015). Studying the proteome holds significance since proteins are directly involved in plant stress responses. This has resulted in a popularity gain for plant proteomics which mostly relies on two-dimensional gel electrophoresis (2-DGE) protein separation (Barkla et al., 2013). Recent advancements in 2-DGE technique and mass spectrometry analysis, including MS/MALDI-TOF/TOF enable deeper and more accurate analysis of the proteome and post-translationally modified proteins (Mostek et al., 2015). Investigators have identified the stress-linked proteins using proteomics approaches, particularly with the use of 2-DGE along with the application of 5'-iodoacetamidofluorescein (5'-IAF) or 2D-fluorescence difference gel electrophoresis (DIGE) (Cuddihy et al., 2009; Fu et al., 2008). Studying the protein alterations in response to salinity stress is crucial and proteomic analysis holds significance in providing novel aspects of underlying protection mechanisms in plants against the salinity stress as revealed by recent studies (Nam et al., 2012; Liu et al., 2014; Zhang, 2014). The proteome analyses are looked upon for highlighting the genes and/or pathways involved in imparting salt tolerance in rice (Zhang, 2014).

Owing to the fact that the metabolites are final products of cellular reactions reflecting the reaction of biological systems to environmental fluxes (Royuela et al., 2000), fingerprinting salt stress-regulated metabolites has, therefore, gained unprecedented attention in recent years (Mann et al., 2016). The database thus generated via accounting the up/downregulated cellular metabolites as a response to environmental stress including soil salinity may provide a sound platform for devising novel strategies to impart salinity stress tolerance in major crops. Various plant metabolites such as sugars (trehalose, sucrose, fructan), polyols including sugar alcohols (mannitol, sorbitol, pinitol, glycerol, galactinol), amino acids (proline, glutamate, glutamine, alanine) and their derivatives (ectoine and hydroxyectoine), quaternary amines (glycine betaine, polyamines, and dimethyl sulfonioproprionate) are involved in salt stress responses in plants including rice (Kumar et al., 2016). Higher plants synthesize a vast array of primary as well as secondary metabolites and among them the latter hold more significance as secondary metabolites are involved in ecological interactions between plants and their respective environments (Mann et al., 2016). There is a recent advocacy that plants accumulate a range of osmolytes to maintain cellular osmoticum, which seems to be a universal response (Anwar et al., 2016). Metabolomics is being seen as a potent tool for studying plant responses to environmental perturbations (Bundy et al., 2009).

There has been creditable technological advent in the field of metabolomics, where various tools and techniques are being used, such as gas chromatography-mass spectrometry (GC-MS), liquid chromatography-mass spectrometry (LC-MS), liquid chromatography-electrochemistry mass spectrometry (LC-EC-MS), nuclear magnetic resonance spectroscopy (NMRS), LC-NMR, along with the modern and high-throughput methods such as Fourier-transform infrared (FT-IR) and ultra-high-resolution FT-ion cyclotron mass spectrometry (URFTCMS); however, NMR and GC/LC-MS are being used most often.

Various investigations focused on rice metabolomics have been carried out, for instance to determine the types and quality of metabolites that endorse the seed germination (Shu et al., 2008), metabolic profiling at different developmental stages (Tarpley et al., 2005) as well as for observing the natural variations in metabolite profiles of different rice varieties (Kusano et al., 2007). Similarly, to gain the knowledge about the salinity-induced alterations in the metabolite accumulation or processing, Fumagalli et al. (2009) examined the total root and shoot metabolite profiles of two rice cultivars (Nipponbare and Arborio) via liquid state high-resolution magic-angle spinning NMR (HR-MAS-NMR) coupled with multivalent statistical analysis. Authors advocated that the altered metabolism of some vital metabolites such as glutamate, aspartate, proline, valine, lactate, malate and so forth, play vital roles in salt tolerance. Besides, there are contemporary trends to integrate all the omics approaches, owing to the overlapping and interdependent nature of each of these three approaches, with an aim to identify key genes/pathways responsive to salinity stress which would eventually help in developing salinity stress tolerant crops including rice. Figure 2.1 depicts the overview of potent avenues for conferring salinity tolerance in rice.

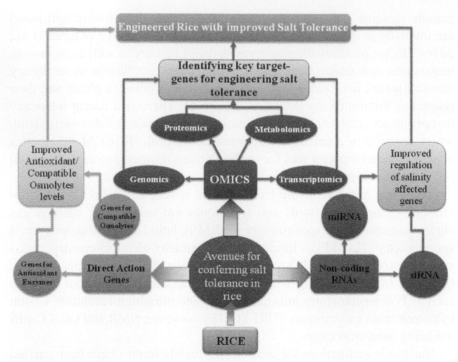

FIGURE 2.1   (**See color insert.**) Potent avenues for conferring salinity tolerance in rice.

2.5   CONCLUSION AND FUTURE PERSPECTIVES

Salinity stress is one of the severe threats to global rice production in terms of both yield and quality of the produce. Despite enormous amount of research, our understanding of the salt stress responses and resistance mechanisms in rice is imperfect due to high complexity of responses. Nevertheless, the biotechnological tools to engineer salt tolerance in rice have got momentum in recent years with the availability of its complete and high-resolution genome sequence. Accordingly, new and potent tools particularly transcriptomics, proteomics and metabolomics have emerged as sound platforms for understanding the salinity stress responses in rice as well as using these tools for producing salt-tolerant rice. In addition, the ncRNAs have also come into view as critical post-transcriptional regulators of gene expression that plants assign to respond to salinity stress. The ncRNAs seem to hold the key as potential targets for genetic manipulations to engineer salt tolerance in rice, and there are positive leads in the attempts made to overexpress the selected ncRNAs (particularly miRNAs) for ncRNA-mediated enhanced salinity stress tolerance of transgenic plants. Therefore, the use of omics tools to supplement genetic engineering and breeding techniques presents potent avenues for deciphering the salt stress response mechanisms and to impart salinity tolerance in targeted rice plants.

KEYWORDS

- Artificial miRNAs
- bioinformatics tools
- chlorophyll content and fluorescence
- direct action genes
- enzymatic antioxidants
- GC-MS
- genetic engineering
- HR-MAS-NMR
- LC-EC-MS
- LC-NMR
- messenger RNAs
- miRNA
- Na+/K+ ratio
- next generation sequencing
- NMRS
- omics tool
- oxidative damages
- plant proteomics
- post-harvest qualities
- QTL mapping
- RFLP; saline soils
- salinity responses and tolerance
- salt stress
- signaling molecules
- siRNA
- transcription factors
- whole-genome sequence

REFERENCES

Aghaei, K.; Komatsu, S. Crop and Medicinal Plants Proteomics in Response to Salt Stress. *Front. Plant Sci.* **2013,** *4,* 8.

Akdogan, G.; Tufekci, E. D.; Uranbey, S.; Unver, T. miRNA-Based Drought Regulation in Wheat. *Funct. Integr. Genomics* **2015,** *16,* 221–233.

Akram, N. A.; Noreen, S.; Noreen, T.; Ashraf, M. Exogenous Application of Trehalose Alters Growth, Physiology and Nutrient Composition in Radish (*Raphanus sativus* L.) Plants Under Water-Deficit Conditions. *Braz. J. Bot.* **2015,** *38,* 431–439.

Ali, M. N.; Ghosh, B.; Gantait, S.; Chakraborty, S. Selection of Rice Genotypes for Salinity Tolerance through Morpho-Biochemical Assessment. *Rice Sci.* **2014a,** *21,* 288–298

Ali, M. N.; Yeasmin, L.; Gantait, S.; Goswami, R.; Chakraborty, S. Screening of Rice Land Races for Salinity Tolerance at Seedling Stage through Morphological and Molecular Markers. *Physiol. Mol. Biol. Plants* **2014b,** *20,* 411–423

Amirjani, M. R. Effect of Salinity Stress on Growth, Sugar Content, Pigments and Enzyme Activity of Rice. *Int. J. Bot.* **2011,** *7,* 73–81.

Anwar, K.; Lakra, N.; Singla-Pareek, S. L.; Pareek, A. Investigating Abiotic Stress Response Machinery in Plants, the Metabolomic Approach. In *Innovative Saline Agriculture;* Dagar, J. C., Sharma, P. C., Sharma, D. K., Singh, A. K., Eds.; Springer: India, 2016; pp 303–319.

Baker, N. R. Chlorophyll Fluorescence: a Probe of Photosynthesis in Vivo. *Annu. Rev. Plant Biol.* **2008,** *59,* 89–113.

Barkla, B. J.; Castellanos-Cervantes, T.; Diaz de León, J. L.; Matros, A.; Mock, H. P.; Perez-Alfocea, F.; et al. Elucidation of Salt Stress Defense and Tolerance Mechanisms of Crop Plants Using Proteomics-Current Achievements and Perspectives. *Proteomics* **2013,** *13,* 1885–1900.

Barrera-Figueroa, B. E.; Gao, L.; Wu, Z.; Zhou, X.; Zhu, J.; Jin, H.; et al. High Throughput Sequencing Reveals Novel and Abiotic Stress-Regulated microRNAs in the Inflorescences of Rice. *BMC Plant Biol.* **2012,** *12,* 132.

Beckles, D. M.; Thitisaksakul, M. How Environmental Stress Affects Starch Composition and Functionality in Cereal Endosperm. *Starch-Starke* **2014,** *66,* 58–71.

Bhowmik, S. K.; Titov, S.; Islam, M. M.; Siddika, A.; Sultana, S.; et al. Phenotypic and Genotypic Screening of Rice Genotypes at Seedling Stage for Salt Tolerance. *Afr. J. Biotechnol.* **2009,** *8,* 6490–6494.

Bhusan, D.; Das, D. K.; Hossain, M.; Murata, Y.; Hoque, M. A. Improvement of Salt Tolerance in Rice (*Oryza sativa* L.) by Increasing Antioxidant Defense Systems Using Exogenous Application of Proline. *Aust. J. Crop Sci.* **2016,** *10,* 50.

Bologna, N. G.; Voinnet, O. The Diversity, Biogenesis, and Activities of Endogenous Silencing Small RNAs in Arabidopsis. *Annu. Rev. Plant Biol.* **2014,** *65,* 473–503.

Bundy, P. J. G.; Davey, M. P.; Viant, M. R. Environmental metabolomics: a critical review and future perspectives. *Metabolomics,* **2009,** *5,* 3–21.

Cuddihy, S. L.; Baty, J. W.; Brown, K. K.; Winterbourn, C. C.; Hampton, M. B. Proteomic Detection of Oxidized and Reduced Thiol Proteins in Cultured Cells. *Methods Mol. Biol.* **2009,** *519,* 363–375.

Cui, N.; Sun, X.; Sun, M.; Jia, B.; Duanmu, H.; Lv, D.; Duan, X.; Zhu, Y. Overexpression of *OsmiR156k* Leads to Reduced Tolerance to Cold Stress in Rice (*Oryza sativa*). *Mol. Breed.* **2015,** *35,* 214.

Danai-Tambhale, S.; Kumar, V.; Shriram, V. Differential Response of Two Scented Indica Rice (Oryza sativa) Cultivars Under Salt Stress. J. Stress Physiol. Biochem. **2011**, 7, 387–397.

Das, P.; Nutan, K. K.; Singla-Pareek, S. L.; Pareek, A. Understanding Salinity Responses and Adopting 'Omics-Based' Approaches to Generate Salinity Tolerant Cultivars of Rice. Front. Plant Sci. **2015**, 6, 712.

Dhar, R.; Sägesser, R.; Weikert, C.; Yuan, J.; Wagner, A. Adaptation of Saccharomyces cerevisiae to Saline Stress through Laboratory Evolution. J. Evol. Biol. **2011**, 7, 1135–1153.

Diao, Y.; Xu, H.; Li, G.; Yu, A.; Yu, X.; Hu, W.; Zheng, X.; Li, S.; Wang, Y.; Hu, Z. Cloning a Glutathione Peroxidase Gene from Nelumbo nucifera and Enhanced Salt Tolerance by Overexpressing in Rice. Mol. Biol. Rep. **2014**, 41, 4919–4927.

Ding, Y.; Chen, Z.; Zhu, C. Microarray-Based Analysis of Cadmium-Responsive microRNAs in Rice (Oryza sativa). J. Exp. Bot. **2011**, 62, 3563–3573.

Eddy, S. R. Non–Coding RNA Genes and the Modern RNA World. Nat. Rev. Genet. **2001**, 2, 919–929.

FAO (Food and Agriculture Organization of the United Nations). FAO Statistical Yearbook – World Food and Agriculture, 2013a, p 132. www.fao.org/docrep/018/i3107e/i3107e03.pdf (accessed Nov 20, 2013)

FAO (Food and Agriculture Organization of the United Nations). Global Cereals Forecast to Increase by 7 Percent in 2013. 2013b. Accessed on 20 July 2014 http://www.fao.org/ asiapacific/rap/home/news/detail/en/?news uid=180032.

Fu, C.; Hu, J.; Liu, T.; Ago, T.; Sadoshima, J.; Li, H. Quantitative Analysis of Redox-Sensitive Proteome with DIGE and ICAT. J. Proteome Res. **2008**, 7, 3789–3802.

Fukuda, T.; Saltu, A.; Wasaki, J.; Shinana, T.; Osaki, M. Metabolic Alterations Proposed by Proteome in Rice Roots Grown Under a Low P and High Al Concentrations Under Low pH. Plant Sci. **2007**, 172, 1157–1165.

Fumagalli, E.; Baldoni, E.; Abbruscato, P.; Piffanelli, P.; Genga, A.; Lamanna, R.; et al. NMR Techniques Coupled with Multivariate Statistical Analysis: Tools to Analyze Oryza sativa Metabolic Content Under Stress Conditions. J. Agron. Crop Sci. **2009**, 195, 77–88.

Gao, P.; Bai, X.; Yang, L.; Lv, D.; Pan, X.; Li, Y.; Cai, H.; Ji, W.; Chen, Q.; Zhu, Y. Osa-MIR393: A Salinity- and Alkaline Stress-Related microRNA Gene. Mol. Biol. Rep. **2011**, 38, 237–242.

Garcia, A.; Rizzo, C. A.; Ud-Din, J.; Bartos, S. L.; Senadhira, D.; et al. Sodium and Potassium Transport to the Xylem are Inherited Independently in Ice, and the Mechanism of Sodium: Potassium Selectivity Differs between Rice and Wheat. Plant Cell Environ. **1997**, 20, 1167–1174.

Garg, A. K.; Kim, J. K.; Owens, T. G.; Ranwala, A. P.; Choi, Y. D.; Kochian, L. V.; Wu, R. J. Trehalose Accumulation in Rice Plants Confers High Tolerance Levels to Different Abiotic Stresses. Proc. Natl. Acad. Sci. U. S. A. **2002**, 99, 15898–15903.

Garg, R.; Verma, M.; Agrawal, S.; Shankar, R.; Majee, M.; Jain, M. Deep Transcriptome Sequencing of Wild Halophyte Rice, Porteresia coarctata, Provides Novel Insights Into the Salinity and Submergence Tolerance Factors. DNA Res. **2014**, 21, 69–84.

Gill, S. S.; Tuteja, N. Reactive Oxygen Species and Antioxidant Machinery in Abiotic Stress Tolerance in Crop Plants. Plant Physiol. Biochem. **2010**, 48, 909–930.

Grattan, S. R.; Zeng, L.; Shannon, M. C.; Roberts, S. R. Rice is More Sensitive to Salinity Than Previously Thought. Calif. Agric. **2002**, 56, 189–198.

Gregorio, G. B.; Senadhira, D.; Mendoza, R. D.; Manigbas, N. L.; Roxas, J. P.; Guerta, C. Q. Progress in Breeding for Salinity Tolerance and Associated Abiotic Stresses in Rice. *Field Crops Res.* **2002,** *76,* 91–101.

Haq, T. U.; Akhtar, J.; Gorham, J.; Steele, K. A.; Khalid, M. Genetic Mapping of QTLs, Controlling Shoot Fresh and Dry Weight Under Salt Stress in Rice (*Oryza sativa* L) Cross Between Co39 × MOROBEREKAN. *Pak. J. Bot.* **2008,** *40,* 2369–2381.

Haq, T. U.; Akhtar, J.; Nawaz, S.; Ahmad. R. Morpho-Physiological Response of Rice (*Oryza sativa* L) Varieties to Salinity Stress. *Pak. J. Bot.* **2009,** *41,* 2943–2956.

Ismail, A.; Takeda, S.; Nick, P. R. Life and Death Under Salt Stress: Same Players, Different Timing? *J. Exp. Bot.* **2014,** *65*(12), 2963–2979.

Jang, I. C.; Oh, S. J.; Seo, J. S.; Choi, W. B.; Song, S. I.; Kim, C. H.; Kim, Y. S.; Seo, H. S.; Choi, Y. D.; Nahm, B. H.; Kim, J. K. Expression of a Bifunctional Fusion of the *Escherichia Coli* Genes for Trehalose-6-Phosphate Synthase and Trehalose-6-Phosphate Phosphatase in Transgenic Rice Plants Increases Trehalose Accumulation and Abiotic Stress Tolerance Without Stunting Growth. *Plant Physiol.* **2003,** *131,* 516–524.

Kanawapee, N.; Sanitchon, J.; Srihaban, P.; Theerakulpisut, P. Genetic Diversity Analysis of Rice Cultivars (*Oryza sativa* L.) Differing in Salinity Tolerance Based on RAPD and SSR Markers. *Electron. J. Biotechnol.* **2011,** *14,* 1–17.

Karthikeyan, A.; Pandian, S. K.; Ramesh, M. Transgenic *Indica* Rice Cv. ADT 43 Expressing a Δ^1-Pyrroline-5-Carboxylate Synthetase (*P5CS*) Gene from *Vigna aconitifolia* Demonstrates Salt Tolerance. *Plant Cell Tissue Organ. Cult.* **2011,** *107,* 383–395.

Khan, M. H.; Panda, S. K. Alterations in Root Lipid Peroxidation and Antioxidative Responses in Two Rice Cultivars Under Nacl-Salinity Stress. *Acta Physiol. Plant* **2008,** *30,* 81–89.

Khare, T.; Kumar, V.; Kishor, P. K. Na+ and Cl⁻ Ions Show Additive Effects Under NaCl Stress on Induction of Oxidative Stress and the Responsive Antioxidative Defense in Rice. *Protoplasma* **2015,** *252,* 1149–1165.

Kim, Y. S.; Kim, I. S.; Shin, S. Y.; Park, T. H.; Park, H. M.; Kim, Y. H.; Lee, G. S.; Kang, H. G.; Lee, S. H.; Yoon, H. S. Overexpression of Dehydroascorbate Reductase Confers Enhanced Tolerance Yo Salt Stress in Rice Plants (*Oryza sativa* L. Japonica). *J. Agron. Crop Sci.* **2014,** *200,* 444–456.

Koyama, M. L.; Levesley, A.; Koebner, R. M. D.; Flowers, T. J.; Yeo, A. R. Quantitative Trait Loci for Component Physiological Traits Determining Salt Tolerance in Rice. *Plant Physiol.* **2014,** *125,* 406–422.

Ku, Y. S.; Wong, J. W. H.; Mui, Z.; Liu, X.; Hui, J. H. L.; Chan, T. F.; Lam, H. M. Small RNAs in Plant Responses to Abiotic Stresses: Regulatory Roles and Study Methods. *Int. J. Mol. Sci.* **2015,** *16,* 24532–24554.

Kumar, V.; Khare, T. Differential Growth and Yield Responses of Salt-Tolerant and Susceptible Rice Cultivars to Individual (Na⁺ and Cl⁻) and Additive Stress Effects of NaCl. *Acta. Physiol. Plant* **2016,** *38,* 170.

Kumar, V.; Shriram, V.; Jawali, N.; Shitole, M. G. Differential Response of Indica Rice Genotypes to NaCl Stress in Relation to Physiological and Biochemical Parameters. *Arch. Agron. Soil Sci.* **2007,** *53,* 581–592.

Kumar, V.; Shriram, V.; Nikam, T. D.; Jawali, N.; Shitole, M. G. Sodium Chloride-Induced Changes in Mineral Nutrients and Proline Accumulation in Indica Rice Cultivars Differing in Salt Tolerance. *J. Plant Nutr.* **2008,** *31,* 1999–2017.

Kumar, V.; Shriram, V.; Nikam, T. D.; Jawali, N.; Shitole, M. G. Antioxidant Enzyme Activities and Protein Profiling Under Salt Stress in Indica Rice Genotypes Differing in Salt Tolerance. *Arch. Agron. Soil Sci.* **2009,** *55,* 379–394.

Kumar, V.; Shriram, V.; Kavi Kishor, P. B.; Jawali, N.; Shitole, M. G. Enhanced Proline Accumulation and Salt Stress Tolerance of Transgenic Indica Rice by Over Expressing P5CSF129A Gene. *Plant Biotechnol. Rep.* **2010,** *4,* 37–48.

Kumar, V.; Khare, T. Individual and Additive Effects of Na+ and Cl- Ions on Rice Under Salinity Stress. *Arch. Agron. Soil Sci.* **2015,** *61,* 381–395.

Kumar, V.; Shriram, V.; Hussain, M. A.; Kavi Kishor, P. B. Engineering Proline Metabolism for Enhanced Plant Salt Stress Tolerance. In *Managing Salinity Tolerance in Plants: Molecular and Genomic Perspectives;* Wani, S. H., Hussain, M. A., Eds.; CRC Press: Boca Ratton, 2016; pp 353–372.

Kusano, M.; Fukushima, A.; Kobayashi, M.; Hayashi, N.; Jonsson, P.; Moritz, T.; et al. Application of a Metabolomics Method Combining One-Dimensional and Two-Dimensional Gas Chromatography-Time-of-Flight/Mass Spectrometry to Metabolic Phenotyping of Natural Variants in Rice. *J Chromatogr. B Anal. Biomed. Life Sci.* **2007,** *855,* 71–79.

Lata, C. H. Advances in Omics for Enhancing Abiotic Stress Tolerance in Millets. *Proc. Indian Natl. Sci. Acad.* **2015,** *81,* 397–417.

Li, W.; Wang, T.; Zhang, Y.; Li, Y. Overexpression of Soybean *mir172c* Confers Tolerance to Water Deficit and Salt Stress, But Increases Aba Sensitivity in Transgenic *Arabidopsis thaliana. J. Exp. Bot.* **2016,** *67,* 175–194.

Lin, H. X.; Zhu, M. Z.; Yano, M. J.; Gao, P.; Liang, Z. W.; Su, W. A.; Hu, X. H.; Ren, Z. H.; Chao, D. Y. QTLs for Na+ and K+ Uptake of the Shoots and Roots Controlling Rice Salt Tolerance. *Theor. Appl. Genet.* **2004,** *108,* 253–260.

Liu, Q.; Zhang, H. Molecular Identification and Analysis of Arsenite Stress-Responsive miRNAs in Rice. *J. Agric. Food Chem.* **2012,** *60,* 6524–6536.

Liu, C. W.; Chang, T. S.; Hsu, Y. K.; Wang, A. Z.; et al. Comparative Proteomic Analysis of Early Salt Stress-Responsive Proteins in Roots and Leaves of Rice. *Proteomics* **2014,** *14,* 1759–1775.

Mann, A.; Bishi, S. K.; Mahatma, M. K.; Kumar, A. Metabolomics and Salt Stress Tolerance in Plants. In *Managing Salinity Tolerance in Plants: Molecular and Genomic Perspectives;* Wani, S. H., Hossain, M. A., Eds.; CRC Press: Boca Ratton, 2016; pp 251.

Matsumoto, T.; Wing, R. A.; Han, B.: Sasaki, T. Rice Genome Sequence: The Foundation for Understanding the Genetic Systems. In: *Rice Functional Genomics Challenges, Progress and Prospects,* Upadhyaya, N. M. (Ed.), Springer-Verlag, NY, USA, 2007, pp 5–20.

Miller, G.; Suzuki, N.; Ciftci-Yilmaz, S.; Mittler, R. Reactive Oxygen Species Homeostasis and Signalling During Drought and Salinity Stresses. *Plant Cell Environ.* **2010,** *33,* 453–467.

Mittler, R.; Blumwald, E. Genetic Engineering for Modern Agriculture: Challenges and Perspectives. *Annu. Rev. Plant Biol.* **2010,** *61,* 443–462.

Mohammadi, R.; Mendioro, M. S.; Diaz, G. Q.; Gregorio, G. B.; Singh, R. K. Mapping Quantitative Trait Loci Associated with Yield and Yield Components Under Reproductive Stage Salinity Stress in Rice (*Oryza sativa* L.). *J. Genet.* **2013,** *92,* 433–443.

Mohanty, A.; Kathuria, H.; Ferjani, A.; Sakamoto, A.; Mohanty, P.; Murata, N.; Tyagi, A. Transgenics of an Elite *Indica* Rice Variety Pusa Basmati 1 Harbouring the codA Gene are Highly Tolerant to Salt Stress. *Theor. Appl. Genet.* **2002,** *106,* 51–57.

Moriwaki, T.; Yamamoto, Y.; Aida, T.; Funahashi, T.; Shishido, T.; Asada, M.; Prodhan, S. H.; Komamine, A.; Motohashi, T. Overexpression of the *Escherichia Coli* Catalase Gene, *katE,* Enhances Tolerance to Salinity Stress in the Transgenic Indica Rice Cultivar, BR5. *Plant Biotechnol. Rep.* **2008,** *2,* 41–46.

Mostek, A.; Börner, A.; Badowiec, A.; Weidner, S. Alterations in Root Proteome of Salt-Sensitive and Tolerant Barley Lines Under Salt Stress Conditions. *J. Plant Physiol.* **2015**, *174*, 166–176.

Munns, R. Genes and Salt Tolerance: Bringing Them Together. *New Phytol.* **2005**, *167*, 645–663.

Munns, R.; Tester, M. Mechanisms of Salinity Tolerance. *Annu. Rev. Plant Biol.* **2008**, *59*, 651–681.

Nagamiya, K.; Motohashi, T.; Nakao, K.; Prodhan, S. H.; Hattori, E.; Hirose, S.; Ozawa, K.; Ohkawa, Y.; Takabe, T.; Takabe, T. Enhancement of Salt Tolerance in Transgenic Rice Expressing an *Escherichia Coli* Catalase Gene, *katE. Plant Biotechnol. Rep.* **2007**, *1*, 49–55.

Nam, M. H.; Huh, S. M.; Kim, K. M.; Park W. J.; et al. Comparative Proteomic Analysis of Early Salt Stress-Responsive Proteins in Roots of SnRK2 Transgenic Rice. *Proteome Sci.* **2012**, *10*, 25.

OECD/FAO (Organization for Economic Co-operation and Development / Food and Agriculture Organization of the United Nations). OECD-FAO Agricultural Outlook 2012-2021. 2012. https://reliefweb.int/sites/reliefweb.int/files/resources/OECD_FAO_Ag-outlook2012_ENG.pdf (accessed July 20, 2014).

Prasad, S. R.; Bagali, P. G.; Hittalmani, S.; Shashidhar, H. E. Molecular Mapping of Quantitative Trait Loci Associated with Seedling Tolerance to Salt Stress in Rice (*Oryza sativa* L.). *Curr. Sci.* **2000**, *78*, 162–164.

Prashanth, S. R.; Sadhasivam, V.; Parida, A. Over Expression of Cytosolic Copper/Zinc Superoxide Dismutase from a Mangrove Plant *Avicennia marina* in Indica Rice Var Pusa Basmati-1 Confers Abiotic Stress Tolerance. *Transgenic Res.* **2008**, *17*, 281–291.

Pujni, D.; Chaudhary, A.; Rajam, M. V. Increased Tolerance to Salinity and Drought in Transgenic Indica Rice by Mannitol Accumulation. *J. Plant Biochem. Biotechnol.* **2007**, *16*, 1–7.

Qureshi, M. I.; Abdin, M. Z.; Ahmad, J.; Iqbal, M. Effect of Long-Term Salinity on Cellular Antioxidants, Compatible Solute and Fatty Acid Profile of Sweet Annie (*Artemisia annua* L.). *Phytochemistry* **2013**, *95*, 215–223.

Rahman, M. S.; Matsumuro, T.; Miyake, H.; Takeoka, Y. Salinity Induced Ultrastructural Alterations in Leaf Cells of Rice (*Oryza sativa* L). *Plant Prod. Sci.* **2000**, *3*, 422–429.

Rao, P. S.; Mishra, B.; Gupta, S. R.; Rathore, A. Reproductive Stage Tolerance to Salinity and Alkalinity Stresses in Rice Genotypes. *Plant Breed* **2008**, *127*, 256–261.

Rengasamy, P. Soil Processes Affecting Crop Production in Salt-Affected Soils. *Funct. Plant Biol.* **2010**, *37*, 613–620.

Rikke, B. A.; Johnson, T. E. Towards the Cloning of Genes Underlying Murine QTLs. *Mamm. Genome* **1998**, *9*, 963–968.

Roessner, U.; Beckles, D. M. Metabolomics for Salinity Research. In *Plant Salt Tolerance: Methods and Protocols (Method in Molecular Biology);* Shabala, S., Cuin, T. A., Eds.; Humana Press: New York, 2012.

Ron, M.; Weller, J. I. From QTL to QTN Identification in Livestock Winning by Points Rather Than Knock-Out. *Anim. Genet.* **2007**, *38*, 429–439.

Roy, M.; Wu, R. Overexpression of S-Adenosylmethionine Decarboxylase Gene in Rice Increases Polyamine Level and Enhances Sodium Chloride-Stress Tolerance. *Plant Sci.* **2002**, *163*, 987–992.

Royuela, M.; Gonzalez, A.; Gonzalez, E. M.; Arrese-Igor, C.; Aparicio-Tejo, P. M.; Gonzalez-Murua, C. Physiological Consequences of Continuous, Sublethal Imazethapyr Supply to Pea Plants. *J. Plant Physiol.* **2000**, *157*, 345–354.

Sabouri, H.; Sabouri, A. New Evidence of QTLs Attributed to Salinity Tolerance in Rice. *Afr. J. Biotechnol.* **2008,** *7,* 4376–4383.

Sakamoto, A.; Murata, A. N. Metabolic Engineering of Rice Leading to Biosynthesis of Glycinebetaine and Tolerance to Salt and Cold. *Plant Mol. Biol.* **1998,** *38,* 1011–1019.

Salama, K. H.; Mansour, M. M. Choline Priming-Induced Plasma Membrane Lipid Alterations Contributed to Improved Wheat Salt Tolerance. *Acta. Physiol. Plant.* **2015,** *37,* 1–7.

Salvi, S.; Tuberosa, R. To Clone or Not to Clone Plant QTLs: Present and Future Challenges. *Trends Plant Sci.* **2005,** *10,* 297–304.

Savvides, A.; Ali, S.; Tester, M.; Fotopoulos, V. Chemical Priming of Plants Against Multiple Abiotic Stresses: Mission Possible? *Trends Plant Sci.* **2016,** *21,* 329–340.

Schmidt, R.; Caldana, C.; Mueller-Roeber, B.; Schippers, J. H. M. The Contribution of SERF1 to Root-To-Shoot Signaling During Salinity Stress in Rice. *Plant Signaling Behav.* **2014,** *9,* e27540.

Shin, S. Y.; Shin, C. Regulatory Non-Coding RNAs in Plants: Potential Gene Resources for the Improvement of Agricultural Traits. *Plant Biotechnol Rep.* **2016,** *10,* 35–47.

Shriram, V.; Kumar, V.; Devarumath, R.; Khare, T. S.; Wani, S. H. Micrornas as Potential Targets for Abiotic Stress Tolerance in Plants. *Front. Plant Sci.* **2016,** *7,* 817.

Shu, X. L.; Frank, T.; Shu, Q. Y.; Engel, K. H. Metabolite Profiling of Germinating Rice Seeds. *J. Agric. Food Chem.* **2008,** *56,* 11612–11620.

Singh, A. K.; Ansari, M. W.; Pareek, A.; Singla-Pareek, S. L. Raising Salinity Tolerant Rice: Recent Progress and Future Perspectives. *Physiol. Mol. Biol. Plants* **2008,** *14,* 137–154.

Sinha, S.; Kukreja, B.; Arora, P.; Sharma, M.; Pandey, G. K.; Agarwal, M.; Chinnusamy, V. The Omics of Cold Stress Responses Plants. In: *Elucidation of abiotic stress signaling in plants;* Pandey, G. K., Ed.; Springer: New York, 2015; pp 143–194.

Soda, N.; Wallace, S.; Karan, R. Omics Study for Abiotic Stress Responses in Plants. *Adv. Plants Agric. Res.* **2015,** *2*(1), 00037.

Su, J.; Hirji, R.; Zhang, L.; He, C.; Selvaraj, G.; Wu, R. Evaluation of the Stress-Inducible Production of Choline Oxidase in Transgenic Rice as a Strategy for Producing the Stress-Protectant Glycine Betaine. *J. Exp. Bot.* **2006,** *57,* 1129–1135.

Sun, M. M.; Abdula, S. E.; Lee, H. J.; Cho, Y. C.; Han, L. Z.; Koh, H. J.; Cho, Y. G. Molecular Aspect of Good Eating Quality Formation in Japonica Rice. *PLoS One* **2011,** *6,* e18385.

Tarpley, L.; Duran, A.; Kebrom, T.; Sumner, L. Biomarker Metabolites Capturing the Metabolite Variance Present in a Rice Plant Developmental Period. *BMC Plant Biol.* **2005,** *5,* 8.

Thitisaksakul, M.; Tananuwong, K.; Shoemaker, C. F.; Chun, A.; Tanadul, O.; Labavitch, J. M.; Beckles, D. M. Effects of Timing and Severity of Salinity Stress on Rice (*Oryza sativa* L.) Yield, Grain Composition, and Starch Functionality. *J. Agric. Food Chem.* **2015,** *63,* 2296–2304.

Thomson, M. J.; de Ocampo, M.; Egdane, J.; et al. Characterizing the Saltol Quantitative Trait Locus for Salinity Tolerance in Rice. *Rice* **2010,** *3,* 148–160.

Türkan, I.; Demiral, T. Recent Developments in Understanding Salinity Tolerance. *Environ. Exp. Bot.* **2009,** *67,* 2–9.

Venu, R. C.; Sreerekha, M. V.; Madhav, M. S.; Nobuta, K.; Mohan, K. M.; Chen, S.; Jia, Y.; Meyers, B. C.; Wang, G. L. Deep Transcriptome Sequencing Reveals the Expression of Key Functional and Regulatory Genes Involved in the Abiotic Stress Signaling Pathways in Rice. *J. Plant Biol.* **2013,** *56,* 216–231.

Wang, Z.; Chen, Z.; Cheng, J.; Lai, Y.; Wang, J.; Bao, Y.; Huang, J.; Zhang, H. QTL Analysis of Na$^+$ and K$^+$ Concentrations in Roots and Shoots Under Different Levels of NaCl Stress in Rice (*Oryza sativa* L.). *PLoS One* **2013**, *7*, e51202.

Wang, W. S.; Zhao, X. Q.; Li, M.; Huang, L. Y.; Xu, J. L.; Zhang, F.; Cui, Y. R.; Fu, B. Y.; Li, Z. K. Complex Molecular Mechanisms Underlying Seedling Salt Tolerance in Rice Revealed by Comparative Transcriptome and Metabolomic Profiling. *J. Exp. Bot.* **2016**, *61*, 405–419.

Wankhade, S. D.; Bahaji A; AndrésI, M.; Cornejo, M. J. Phenotypic Indicators of NaCl Tolerance Levels in Rice Seedlings: Variations in Development and Leaf Anatomy. *Acta. Physiol. Plant.* **2010**, *32*, 1161–1169.

Wankhade, S. D.; Cornejo, M. J.; Mateu-Andrés, I.; Sanz, A. Morphophysiological Variations in Response to NaCl Stress During Vegetative and Reproductive Development of Rice. *Acta. Physiol. Plant.* **2013**, *35*, 323–333.

Xia, K.; Wang, R.; Ou, X.; Fang, Z.; Tian, C.; Duan, J.; Wang, Y.; Zhang, M. *OsTIR1* and *OsAFB2* Downregulation Via *OsmiR393* Overexpression Leads to More Tillers, Early Flowering and Less Tolerance to Salt and Drought in Rice. *PLoS One* **2012**, *7*, e30039.

Xiong, L.; Yang, Y. Disease Resistance and Abiotic Stress Tolerance in Rice are Inversely Modulated by an Abscisic Acid-Inducible Mitogen Activated Protein Kinase. *Plant Cell* **2003**, *15*, 745–759.

Yang, C.; Li, D.; Mao, D.; Liu, X. U. E.; Ji, C.; Li, X.; Zhao, X.; Cheng, Z.; Chen, C.; Zhu, L Overexpression of microRNA319 Impacts Leaf Morphogenesis and Leads to Enhanced Cold Tolerance in Rice (*Oryza sativa* L.). *Plant Cell Environ.* **2013**, *36*, 2207–2218.

Yildiztugay, E.; Ozfidan-Konakci, C.; Kucukoduk, M. Modulation of Osmotic Adjustment and Antioxidant Status in Salt-Stressed Leaves of *Thermopsis turcica*. *Acta. Physiol. Plant.* **2014**, *36*, 125–138.

Zeng, L.; Shannon, M. C. Salinity Effects on Seedling Growth and Yield Components of Rice. *Crop Sci.* **2000**, *40*, 996–1003.

Zhang, J. Salinity Affects the Proteomics of Rice Roots and Leaves. *Proteomics* **2014**, *14*, 1711–1712.

Zhao, F.; Zhang, H. Salt and Paraquat Stress Tolerance Results from Co-Expression of the *Suaeda salsa* Glutathione S-Transferase and Catalase in Transgenic Rice. *Plant Cell Tiss. Organ Cult.* **2006**, *86*, 349–358.

Zhou, Y.; Yang, P.; Cui, F.; Zhang, F.; Luo, X.; Xie, J. Transcriptome Analysis of Salt Stress Responsiveness in the Seedlings of Dongxiang Wild Rice (*Oryza rufipogon* Griff.). *PLoS One* **2016**, *11*, e0146242.

Zhu, J. K. Salt and Drought Stress Signal Transduction in Plants. *Ann. Rev. Plant Biol.* **2002**, *53*, 247–273.

CHAPTER 3

SALT STRESS RESPONSES OF GLYCOPHYTIC RICE AND HALOPHYTIC RICE: PHYSIOLOGICAL, BIOCHEMICAL, AND MOLECULAR ASPECTS

DEEPAK SHELKE[1,*], GANESH NIKALJE[2], and PARMESHWAR KUMAR SAHU[3]

[1]Department of Botany, Amruteshwar Arts, Commerce and Science College, Vinzar, Velha, Pune, Maharashtra 412213, India, Mob.: +00-91-7620110669

[2]Department of Botany, Savitribai Phule Pune University, Pune, Maharashtra 411007, India, Mob.: +00-91-9969462817, E-mail: ganeshnikalje7@gmail.com

[3]Department of Genetics and Plant Breeding, Indira Gandhi KrishiVishwavidyalaya, Raipur, Chhattisgarh 492012, India, Mob.: +00-91-8103795885, E-mail: parmeshwarsahu1210@gmail.com

*Corresponding author. E-mail: dpk.shelke1@gmail.com

3.1 INTRODUCTION

Rice is one of the important staple food crops upon which almost half of the world population is subsisting (Verma et al., 2012, 2013, 2015). However, it is most susceptible to salt stress. To improve salt tolerance, basic study of salt tolerance mechanism in rice and other salt-tolerant plants especially halophytes, will be very helpful (Flowers and Colmer, 2008). In this context, halophytic plants have served as candidate plants as they are endowed with specific traits such as salt hair, salt glands, succulence and a set of specific genes which enable them to adapt to adverse conditions (Sengupta and Majumder, 2010). A comparative account of salt tolerance mechanism in the domesticated rice (*Oryza sativa* L.) and halophytic rice (*Porteresia coarctata*) will greatly help researchers. *Porteresia* spp. can withstand up to 30–40 dSm salinity without any adverse effects (Jagtap et al., 2006). Besides, it can withstand submergence under saline water for quite long periods. Since both these species belong to the same *Poaceae* family, genetic manipulation of salt tolerance genes and hybridization programmes are possible and can be carried out. Here, we emphasized on physiological, biochemical and molecular differences in salt tolerance mechanism of these species.

Increase in soil salinity leads to soil infertility and its degradation. Most of the agricultural land (almost 20%) is suffering from soil salinity due to poor irrigation water (FAO, 2009). Sodium is one of the major elements in both saline as well as sodic soil which adversely impacts agricultural growth and productivity (IRRI, 2011). As per FAO (2009) report, there is a need to increase world agricultural production by 70% by 2050 to feed increasing population. Upon exposure to salt, plant faces low water availability, affected lateral bud development and cell expansion (Munns and Tester, 2008). Salt inhibits each aspect of plant growth and development, membrane damage, ion homeostasis, enhanced lipid production and generation of toxic reactive oxygen species (ROS). Many molecular and genetic studies have given insights into protection mechanism used by plants to cope with detrimental effects of excessive salt (Zhu, 2002; Munns and Tester, 2008; Hauser and Horie, 2010).

Most of the agricultural crops are susceptible to salinity and therefore their improvement is very essential to increase their salt tolerance and yield. To improve stress tolerance in crops, it is necessary to have basic knowledge of salt tolerance in salt-tolerant plants. Halophytes are the true salt-tolerant plants which are native to saline soils. The categories of halophytes and glycophytes are based on salt-tolerance ability of plants. As per classification system both halophytes and glycophytes may differ in their salt tolerance but they are kept in same family, for example, *Arabidopsis* and *Thellungiella* both are crucifers. *Arabidospis* is salt sensitive (glycophyte) while *Thellungiella* is salt tolerant (halophyte) (Gong et al., 2005). Since, out main objective is to improve salt tolerance in crops we emphasized on rice and its wild relative salt-tolerant *P. coarctata*.

Rice ranks first in most salt-sensitive cereals (Munns and Tester, 2008). It is world's most important cereal and primary source of food which serves more than half of the population. It is widely cultivated all over the world but due to its sensitivity to salt stress, its production is limited in salt affected areas. To improve rice production, development of rice varieties suitable for saline soils is most needed. The International Rice Research Institute (IRRI) has screened almost 90,000 varieties but the donor varieties were traditional cultivars and tall varieties (Yeo et al., 1990) and degree of improved tolerance was not sufficient. Moreover, IRRI evaluation reveals that chances of getting to stresses is almost 50% more in wild than cultivated species. In this sense, *P. coarctata* is one of the wild relative of rice with high degree of salt tolerance and mine of salt tolerance-related genes.

3.2   HABIT AND HABITAT

The genus *Oryza* includes two cultivated (2n=24) and 22 wild species (2n=24, 48). Among them, *O. sativa* is cultivated rice and *P. coarctata* is a wild rice. Rice is a well-characterized crop. It requires a uniform temperature (about 25°C) throughout its life cycle means during sowing, growth and harvesting. The annual average rainfall should be more than 100–200 cm (Anonymous, 2000). At the time of sowing and early stage of growth, the rice fields should be flooded with water. It grows well in non-porous deep fertile clay or loamy soil with high water holding capacity (Anonymous, 2000). On the other hand, *Porterisia* is a wild species, native of estuarine and deltaic saline environments. It is a tetraploid monotypic genus (2n=48) (Sengupta and Majumder, 2010). It is ecologically important grass which forms a mat over coastline and prevents soil erosion. This vegetation is limited to soil which gets inundated twice a day with saline water of to 40 dSm^{-1} and acts a pioneer species in succession of mangroves forest (Jagtap et al., 2006). Every day *Porteresia* gets submerged by tidal water for 10–11 h. These areas have 1.10 m mean tidal level (MTL) from the base. The plant has very hard stem and thick leaves and grows up to 1 m height.

3.3   MORPHOLOGY AND ANATOMY UNDER SALT STRESS

Both rice and *Porteresia* grow well under submerged conditions. Under salt influence the plant height, total tillers, root and shoot length, dry weight decreases significantly in rice. The continuous stress results in necrosis and subsequent death of rice plants. But *Porteresia* does not show any adverse effect on growth. *P. coarctata* has distinct morphology which enables it to cope with salt stress (Sengupta and Majumder, 2009). It possess underground rhizome which gives rise to numerous leafy shoots from each node. Rhizoid-like structures arise from rhizome that help in anchorage and absorption. This extensive rhizomatous system helps to absorb water under high salinity. Unlike rice, it has waxy leaves and surfaces shows alternate ridges and furrows. The seeds of *Porteresia* dehisce prematurely and hence propagate by vegetative mean. In total presence of salt hairs, extensive rhizomatous systems, rhizoid-like root hairs and scanty panicles with dehiscent seeds are unique characters of *Porteresia* over *Oryza* which enables former to thrive well under salt stress at morphological level (Sengupta and Majumder, 2009).

On the basis of CO_2 compensation points and PEP carboxylase activity *Porteresia* is grouped into C_3 plant (Garcia, 1992). Anatomical study of this halophyte revealed that adaxial surfaces of leaves have unicellular salt hairs and numerous prominent ridges and furrows which runs longitudinally down the surface. These hairs are unicellular, thin, small with blunt tips and do not have any distinct basal and cap cells which are meant for salt excretion. Unlike other halophytes these salt hairs burst after some period (Liphschitz et al., 1974). X-ray analysis of salt hairs showed that there are two types of hairs; first one found at upper surface which are finger shaped and other one is at lower surface which are peg like (Flowers et al., 1990). The upper surface hairs do not rupture at high salt concentration and continues to secrete excessive salt while the lower surface salt hairs ruptures and fall down at high salinity and reforms. These hairs help to maintain intercellular sodium ion content and sodium:potassium ratio (Sengupta and Majumder, 2009). Flower et al. (1990) quantitatively showed that under salt stress (25% artificial saline water for 6 weeks) *Porteresia* can maintain as low as Na:K ratio 0.7:34.

The young roots of *Porteresia* have usual structures of epidermis, cortex and stele as like *Oryza* but after maturity they develop sclerotic pith (Rich-haria and Roy, 1965) Under strong influence of tidal waves plant develops pseudo-tap roots which give rise to fibrous roots at their tips and sobole (underground stem meant for vegetative propagation) (Latha et al., 2004).

3.4   PHYSIOLOGICAL RESPONSE AND SALT TOLERANCE

Being a halophyte *Porteresia* responds better to increasing salinity. The growth parameters such as root and shoot length ratio, root and shoot dry weight, total tillers, relative water content were adversely affected in rice under increasing salinity while in *Porteresia* leaf biomass increases upto 400 mM NaCl treatment (Khan et al., 1997). Generally at higher salinity crops lose their relative water content almost 60–70% but *Porteresia* retains 90% tissue water content in high salinity. This ability plays a key role in osmotic balance under salinity stress. The efficient salt exclusion and osmotic adjustment helps *Porteresia* to reduce cellular water potential than external water potential even under high salinity and water enters into cell. This kind of balance can be obtained by inducing salt-responsive genes (Flowers et al., 1990). Under salinity, photosynthetic machinery of rice varieties gets damaged and inactivates the photoreaction centres while *Poreresia* manages to remain photosynthetically competent. There is only minor blockage of some reaction centre but it does not affect plant health. The salinity stress declines overall performance index of salt-sensitive IR64 rice variety by 25% and in salt-tolerant Pokkali rice, there was meager effect while the *Porteresia* did not show any adverse effect in 200 mM NaCl treatment. After exposure to 400 mM NaCl the IR64 variety lost its vitality and overall performance index of Pokkali declined by 15% over control while *Porteresia* retained 88% efficiency (Sengupta and Majumder, 2009). The phenomenological flux models show salt tolerance mechanism in salt sensitive IR 64 rice variety, salt-tolerant Pokkali rice variety and the halophyte *P. coarctata*. In IR 64, with increasing salt concentration, the number of reaction centers decreases drastically. At 400 mM NaCl treatment, almost all reaction centers get blocked and plant dies. In Pokkali, up to 200 mM NaCl, highest trapping of utilizable energy per cross section was observed and it showed considerable retention of electron transport ability which may be probable reason for its salt tolerance over IR 64. But at 400 mM NaCl concentration, more than 50% reaction centers were inactivated and the electron transport and energy trapping lost. In contrast to IR64 and Pokkali there was no alteration in phenomenological fluxes or specific fluxes under low or high salinity stress in *Porteresia* (Sengupta and Majumder, 2009). This ability of *Porteresia* gives advantage over rice varieties to withstand against high salinity.

Rice plants transport excess sodium and chloride ions from root to shoot via apoplastic pathway (Yeo et al., 1987; Anil et al., 2005; Krishnamurthy et al., 2009). The accumulation of Na^+ ions in shoot is due to the passive leakage of Na^+ in to xylem. If rice plants successfully translocated 99% of sodium to sequester in to leaves, within 7 days almost 500 mM Na^+ will be accumulated in apoplast which will lead to cell dehydration and stomatal closure (Yeo and Flowers, 1986). The apoplastic Na^+ accumulation is negatively correlated with salt tolerance in all rice varieties (Krishnamurthy et al., 2011). The salt-tolerant Pokkali variety maintains low Na^+/K^+ ratio which makes it salt tolerant (Ray and Islam, 2008). *Porteresia* gets the benefit of having salt hairs over other rice varieties. These plants have the tremendous ability to maintain Na:K ratio as low as 0.7 when challenged with 25% artificially saline water having Na:K ratio of 34 for 6 weeks (Flower et al., 1990). Under high salt load, the excessive Na^+ ions are efficiently secreted through salt hairs as confirmed by x-ray microanalysis (Flower et al., 1990). Unlike other halophytes *Porteresia* accumulates more Na^+ in roots as compared to aerial parts (Garcia, 1992). The salt hairs are highly selectable for Na^+ ions.

3.5   BIOCHEMICAL RESPONSE AND SALT TOLERANCE

Salinity stress creates biochemical and metabolic changes in plants. The changing soil conditions adversely affect plant metabolism, cellular homeostasis and uncoupling of essential physiological processes (Munns and Tester, 2008). These metabolic changes lead to overproduction of ROS, reactive nitrogen species (RNS) or reactive sulfur species (RSS) (Suzuki and Mittler, 2006). Under salt stress the NADPH-dependent superoxide synthase produces superoxide ($^-O_2$) which is rapidly converted into H_2O_2 (hydrogen peroxide) by an enzyme superoxide dismutase. This H_2O_2 is further converted in to highly reactive hydroxyl ions in presence of metal ions or H_2O and O_2 by enzymes catalase and ascorbate peroxidase. The hydroxyl ions disrupt metabolic systems which lead to programmed cell death (Halliwell and Gutteridge, 1989). The different ROS molecules damage lipids, proteins and also DNA which are key components of cell metabolism (Casano et al., 1994). The ROS species are the result of oxidative stress which is responsible for major macromolecules of the cell. Plants have evolved different strategies to cope up with these toxic species. Plants are equipped with various antioxidant enzymes such as superoxide dismutase, catalase, ascorbate peroxidase, glutathione reductase, and so forth; non-enzymatic antioxidants such as ascorbate, glutathione; and osmolytes such as proline, glycine betaine, sugars like sucrose, trehalose, and so forth (Suprasanna et al., 2016). The sensitive varieties or rice like IR 64, IR 29 and salt-tolerant species Pokkali differ in their ROS accumulation under salinity stress. The Pokkali accumulate less ROS as compared to IR 64. This may be due to the fact that Pokkali possess efficient ROS scavenging system as compared to IR 64 (El-Shabrawi et al., 2010).

Glutathione and ascorbate are the non-enzymatic antioxidants and redox buffers which are key players of ascorbate glutathione cycle. They help plants to acclimatize to their environment (Noctor et al., 2002). The balance of GSH/GSSG and ASC/DHA decides the fate of plants acclimation to salinity or any oxidative stress. The reduced form of glutathione possesses redox buffering action and protects cell from oxidative damage (Ogava, 2005). The ascorbate directly gets oxidized by superoxide or acts as reducing agent of chromosyl radicle of oxidized tocopherol. This way ascorbate prevents lipid peroxidation by scavenging hydroxyl, peroxyl, and alkoxyl radicals (Halliwell and Gutteridge, 1989). The salt tolerant Pokkali variety constitutively maintains higher ratio of reduced forms of both glutathione and ascorbate than oxidized form under salt stress and also efficiently detoxifies methylglyoxal as compared to salt sensitive IR 64 variety. These results can be inferred as these redox buffers play potential role in salt tolerance mechanism of Pokkali variety (Vaidyanathan et al., 2003; Tausz et al., 2004).

The antioxidant enzymes play important role in ROS scavenging. The superoxide dismutase enzyme acts as a primary line of defence against oxidative stress. It converts superoxide into H_2O_2 which acts as a signaling molecule and activates downstream ROS scavenging machinery (Bhattacharjee and Mukherjee, 1997; Dionisio-Sese and Tobita, 1998). Catalase is one of the H_2O_2 detoxifying enzymes and mostly associated with peroxisomes where it removes H_2O_2 formed during photorespiration (El-Shabrawi et al., 2010). The ascorbate peroxidase utilizes ascorbate as a substrate to detoxify H_2O_2. It has several isoforms situated throughout the cell. In contrast glutathione, peroxide utilized reduced glutathione as a substrate to detoxify H_2O_2. The glutathione reductase protects cell from oxidative stress by maintaining reduced state of glutathione and generation of reduced ascorbate. This ensures smooth operation of ascorbate glutathione cycle. Along with these antioxidant enzymes, Different non-chloroplastic peroxidases and glutathione peroxidase plays an important role in ROS detoxification (Sreenivasulu et al., <link rid="bib51">2004</link>). These antioxidant enzymes show differential regulation under salt stress in contrasting varieties of rice. The salt tolerant Pokkali variety has highly efficient antioxidant enzyme machinery. Most of the enzymes were up regulated under salt stress. The catalase shows higher activity under both stressed and non-stressed conditions which marks the well preparedness of this variety for stress. The two isoforms of ascorbate peroxides show upregulation while IR 64 shows downregulation under salt stress. Same pattern was observed in activity of glutathione reductase (El-Shabrawi et al., 2010). In *Porteresia,* no such reference is available. Since, salt directly affects ion homeostasis and over produces ROS and *Porteresia* is a salt-tolerant species so it may have the effective antioxidant machinery for control of ROS in cell. Ghosh et al., (2001) conducted an experiment where he observed that purified fructose-1,6 bisphosphatase from *Porteresia* was highly salt tolerant to sodium chloride, potassium chloride, and ammonium chloride as compared to IR26 cultivar of rice. This enzyme catalyzes regeneration of RuBP in Calvin cycle which enables plant to continue photosynthesis under water stress when RuBisCo levels decrease.

Osmolytes are the low-molecular-weight organic compounds which protect plants from adverse stress conditions and maintain osmotic balance in plants cells. Generally amino acids such as proline, cysteine; sugars such as sucrose, trehalose, raffinose; and tertiary ammonium compounds are involved in salt to adaptation mechanism of plants (Slama et al., 2015; Suprasanna et al., 2016). The proline acts as a molecular chaperon and buffer which maintains the pH of the cytosolic redox status of the cell (Verbruggen and Hermans, 2008; Kido et al., 2013). It also acts as an antioxidant and scavenges singlet oxygen and reactive radicals (Smirnoff and Cumbes, 1989; Matysik et al., 2002). The salt-tolerant KDML105 and Sangyod varieties showed enhanced proline level as compared to sensitive Pathumthani 1 and Black Sticky varieties. The non-reducing disaccharide, trehalose is rare sugar which can protect proteins and cell membranes from denaturation under adverse environmental stress (Elbein et al., 2003). In rice transgenic plants, it is noted that the overproduction of trehalose showed tolerance to salt, drought and cold stress (Garg et al., 2002). It acts as a membrane stabilizer under salt stress and decreases ion leakage and lipid peroxidation rate in root and increases K^+/Na^+ ratio in leaves of maize (Zeid, 2009). The pinitol is the methylated derivative of myoinositol which is abundantly found in *Porteresia* but absent in rice (Sengupta et al., 2008). This is a compatible solute which provides osmotic adjustment by decreasing the osmotic potential and improving water retention capacity (Le Rudulier and Bouillard, 1983). Moreover, inositol and its derivatives are actively engaged in many cellular processes such as growth regulation, membrane biogenesis, signal transduction, ion channel physiology, and membrane dynamics (Loewus and Murthy, 2000). The different strategies used by *Porteresia* and rice under salt stress are depicted in Figure 3.1.

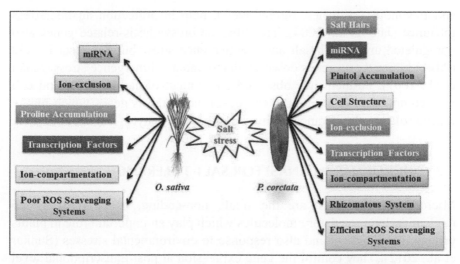

FIGURE 3.1 **(See color insert.)** Strategies involved in salt tolerance mechanism of halophyte *Porteresia* and glycophyte rice.

3.6 TRANSCRIPTOMICS IN PORTERESIA

The transcriptomic analysis of *P. coarctata* under submergence and salt treatment has been carried out (Garg et al., 2013). Since it is a wild relative of rice, the sequence similarity was expected but low similarity with rice proteins has been observed as compared to other wild relatives of rice. Therefore, a substantial number of genes specific to *Porteresia* have been obtained which can be further explored for improving rice salt tolerance. They have found 15,158 genes responsive to salinity and/or submergence stress. These genes were involved in photosynthesis, secondary metabolite biosynthesis, hormone biosynthesis, carbohydrate metabolism, cell wall-related enzymes, and transport similar to other plants like *Arabidopsis*, rice, and halophytes. Interestingly, in *Porteresia,* it was seen that certain transcription factor families showed expansion or contraction in relation to salt stress. The transcription factor (TF) families such as NAC, MYB and WRKY, bZIP, bHLH, HSF and AP2-EREBP were differentially regulated under salt and submergence stress (Fukao et al., 2006; Xu et al., 2006; Jung et al., 2010; Niroula et al., 2012). The former three TFs were important for salinity and later were crucial for submergence stress. Authors have also looked into metabolic pathways and found enrichment of gene involved in biosynthesis of hydroxycinnamic acid serotonin amides and phenylpropanoids which acts as antioxidants and flavonoids, stilbenes, lignins which help in protection against stress tolerance (Jang et al., 2004). The ethylene biosynthesis-related genes also upregulated under both salt and submergence stress but the abscisic acid (ABA) biosynthesis genes downregulated under submergence stress (Saika et al., 2007). Further, they observed enrichment of carbohydrate and cell wall components (Suberin and cellulose) biosynthesis genes which play a crucial role in salt tolerance of this wonderful halophyte (Koiwa, 2009).

3.7   MICRORNAS (MIRNAS) FOR SALT TOLERANCE

MicroRNAs (miRNAs) are the small, non-coding, endogenous, 17–26 nucleotide-long regulatory molecules which play an important role in plants growth, development and also response to environmental stresses (Sunkar et al., 2012). They control the gene expression at post-transcriptional level and play a pivotal role in stress response (Mondal et al., 2015). miRNAs mostly target transcription factors (Jones-Rhoades et al., 2006). There are only few reports of role of miRNA sin salt tolerance of halophytes and all of them are from dicots. Mondal et al., (2015) attempted to study role of small RNAs in salt tolerance of a monocotyledonous species *Porteresia*. They prepared two small RNA libraries from control and salt-treated leaves (24 h treatment of 450 mM NaCl). From these two libraries and search for publically available transcriptomic data of *Oryza coarctata*. This resulted into identification of 386 known and 95 novel miRNAs. Nine miRNAs such as as oco-miR166e-3p, oco-miR169g, oco-miR169o, oco-miR393a, oco-mi396c, ocomiR020-3p, oco-miR014-3p, oco-miR160b, and oco-miR395d were found salt responsive and GO and KEGG pathway analysis found that these miRNAs have 22 potential targets. Several *Porteresia* specific reads were found in libraries. Interestingly, the expression of known salt-responsive miRNAs was contradictory like oco-miR393 and oco-miR161 are well-known salt-responsive miRNA but in *Porteresia,* their expression was less than 100 transcripts while oco-miR159b, oco-miR166c-3p, oco-miR396e-5p, and oco-miR168a-5p miRNA showed high expression under stress. This reveals that miRNAs act in species specific manner irrespective of high sequence similarity. The predicted miRNA-target interaction showed that salinity induces calcium signaling pathway related miRNA, oco-miR1432-5p which targets calcium-binding protein (CBP) and oco-miR164d targets Calmodulin (CAM); MAPK signaling pathway miRNAs oco-miR528-3p targets serine/threonine-protein kinase 38 (STK) and oco-miR159a targets heat shock protein 72 (HSP72). Hormone-mediated signaling oco-miR167d targets auxin response factor (ARF) and oco-miR393b targets transport inhibitor response (TIR1) (Mondal et al., 2015). In response to this stress, oco-miR079-3p and oco-miR529b targets; major facilitator super family protein (MFS) and cation transporter HKT7 (HKT7) genes are involved in ion homeostasis (Aranda-Sicilia et al., 2012). The oco-miR087-5p targets peroxidase (POX) which is an antioxidant enzyme and oco-miR046-5p.

3.8 DEVELOPMENT OF TRANSGENIC FOR IMPROVED SALT TOLERANCE

For development of transgenic plants with improved salt tolerance, halophytes are the best choice. They are native of saline areas and possess special adaptive mechanism to salt stress. However, due to lack of genomic information, these plants cannot be utilized for development of transgenic plants. Transcriptomic analysis and miRNA sequencing of *Porteresia* has opened door of mine of salt tolerance related genes. The myo-inositol phosphate synthase (MIPS) is the fully characterized gene which improved salt tolerance in *Brassica* and rice (Das-Chatterjee et al., 2006). Further, PcINO1, McIMT1 and Serine-rich protein improved salt tolerance in tobacco and finger millet (Table 3.1). The differentially regulated genes and miRNAs under salt stress in *Porteresia* can be further used for transgenic development in crops.

TABLE 3.1 *Porteresia* Genes Used for Improving Salt Tolerance in Plants.

Gene	Transgenic plant	Improved trait	Reference
myo-Inositol phosphate synthase (MIPS)	Brassica, rice	Salt stress	Das-Chatterjee et al. (2006)
PcINO1 and McIMT1	Tobacco	Salt stress	Patra et al. (2010)
L-myo-inositol-1-phosphate synthase	Tobacco	Salt tolerence	Majee et al. (2004)
Serine-rich protein	Finger millet	Salt stress	Mahalakshmi et al. (2006)

3.9 SUMMARY AND CONCLUSION

Rice grows well under flooding conditions as water helps plant to leach salt from top layer of soil. It is crucial to improve rice tolerance under saline condition. *Porteresia* is a halophyte with remarkable salt stress tolerance.

P. coarctata has the potential to serve as an important source of salt-tolerance factors for rice, as indicated from the current trend of research. To find the novel salt stress tolerance related genes, comparative genomics is clearly turning out to be an important tool. Therefore, it is important to understand rice and its wild relative *Porteresia* at morphological, physiological, biochemical, and molecular level. With comparison to rice, *Porteresia* showed variations in traits which make it tolerant. The rhizomatous system and salt hairs of *Porteresia* play role for ion accumulation and regulations. The efficient osmoregulators such as pinitol can efficiently maintain osmotic balance and improve cell structure of plants. Its proteins, such as PcINO1 show remarkable difference in their structure and function when compared with rice. The salt directly affects ion homeostasis and overproduces ROS; *Porteresia* being a salt-tolerant species may have the effective antioxidant machinery for control of ROS in cells. Furthermore, different TFs were differentially regulated under salt and submergence stress. The new miRNA targets were also found in *Porteresia* under salt stress which may also enhance tolerance. Some stress-responsive genes from *Porteresia* are utilized to make salt-tolerant crop plant successfully. Therefore, the comparative physiological, biochemical, and molecular response of both the plants will help to identify salt tolerance related traits in plants. Being a wild relative of domesticated rice, it can be used for rice and other crop improvement program by breeding approach or transgenic approach.

KEYWORDS

- ABA biosynthesis
- adaxial surfaces
- apoplastic pathway
- C3 plant
- calcium-signaling pathway
- Calvin cycle
- carbohydrate metabolism
- cell expansion and metabolism
- environmental stresses
- growth regulation
- heat shock protein
- ion channel physiology
- lipid peroxidation
- MAPK signaling pathway
- membrane dynamics
- post-transcriptional level
- secondary metabolite biosynthesis
- transcriptomic analysis
- transport inhibitor response

REFERENCES

Anil, V. S.; Krishnamurthy, P.; Kuruvilla, S.; Sucharitha, K.; Thomas G; Mathew, M. K. Regulation of the Uptake and Distribution of Na⁺ in Shoots of Rice (*Oryza sativa*) Variety Pokkali: Role of Ca^{2+} in Salt Tolerance Response. *Physiol. Plant.* **2005**, *124*(4), 451–464.

Anonymous. Tenth Annual Report of M. S. Swaminathan Research Foundation, **2000**, 1999–2000.

Aranda-Sicilia, M. N.; Cagnac, O.; Chanroj, S.; Sze, H.; Rodriguez-Rosales, M. P.; Venema, K. Arabidopsis KEA2, a Homolog of Bacterial KefC, Encodes a K+ /H+ Antiporter with a Chloroplast Transit Peptide. *Biochim. Biophys. Acta* **2012**, *1818*, 2362–2371.

Bhattacharjee, S.; Mukherjee, A. K. Role of Free Radicals in Membrane Deterioration in Three Rice (*Oryza sativa* L.) Cultivars under NaCl Salinity at Early Germination Stage. *Indian J. Exp. Biol.* **1997**, *35*(12), 1365–1369.

Casano, L. M.; Martin, M.; Sabater, B. Sensitivity of Superoxide Dismutase Transcript Levels and Activities to Oxidative Stress is Lower in Mature-Senescent than in Young Barley Leaves. *Plant Physiol.* **1994**, *106*(3), 1033–1039.

Das-Chatterjee, A.; Goswami, L.; Maitra, S.; Dastidar, K. G.; Ray, S.; Majumder, A. L. Introgression of a Novel Salt-Tolerant L-Myo-Inositol 1-Phosphate Synthase from Porteresia Coarctata (Roxb.)Tateoka (PcINO1) Confers Salt Tolerance to Evolutionary Diverse Organisms. *FEBS Lett.* **2006**, *580*(16), 3980–3988.

Dionisio-Sese, M. L.; Tobita, S. Antioxidant Responses of Rice Seedlings to Salinity Stress. *Plant Sci.* **1998**, *135*(1), 1–9.

Elbein, A. D.; Pan, Y. T.; Pastuszak, I.; Carroll, D. New Insights on Trehalose: A Multifunctional Molecule. *Glycobiology* **2003**, *13*(4), 17R–27R.

El-Shabrawi, H.; Kumar, B.; Kaul, T.; Reddy, M. K.; Sneh, L.; Singla, P.; Sopory, S. K. Redox Homeostasis, Antioxidant Defense, and Methylglyoxal Detoxification as Markers for Salt Tolerance in Pokkali Rice. *Protoplasma* **2010**, *245*(1–4), 85–96.

FAO. *High Level Expert Forum—How to Feed the World in 2050;* Economic and Social Development Department, Food and Agricultural Organization of the United Nations: Rome, 2009.

Flowers, T. J.; Colmer, T. D. Salinity Tolerance in Halophytes. *New Phytol.* **2008**, *179*(4), 945–963.

Flowers, T. J.; Flowers, S. A.; Hajibagheri, M. A.; Yeo, A. R. Salt Tolerance in the Halophytic Wild Rice, Porteresia Coarctata. T. *New Phytol.* **1990**, *114*(4), 675–684.

Fukao, T.; Xu, K.; Ronald, P. C.; Bailey-Serres, J. A Variable Cluster of Ethylene Response Factor-Like Genes Regulates Metabolic and Developmental Acclimation Responses to Submergence in Rice. *Plant Cell* **2006**, *18*(8), 2021–2034.

Garcia, A. Salt Tolerance in the Halophytic Wild Rice, Porteresia Coarctata Tateoka. Ph.D. Thesis, University of Sussex, 1992.

Garg, A. K.; Kim, J. K.; Owens, T. G. Trehalose Accumulation in Rice Plants Confers High Tolerance Levels to Different Abiotic Stresses. *PNAS USA* **2002**, *99*(25), 15898–15903.

Garg, R.; Verma, M.; Agrawal, S.; Shankar, R.; Majee, M.; Jain, M. Deep Transcriptome Sequencing of Wild Halophyte Rice, *Porteresia coarctata*, Provides Novel Insights into the Salinity and Submergence Tolerance Factors. *DNA Res.* **2013**, *21*(1), 1–16.

Ghosh, S.; Bagchi, S.; Majumder, A. L. Chloroplast Fructose- 1,6-Bisphosphatase from Oryza Differs in Salt Tolerance Property from Porteresia Enzyme and is Protected by Osmolytes. *Plant Sci.* **2001**, *16*(6), 1171–1181.

Gong, Q.; Li, P.; Ma, S.; Indu Rupassara, S.; Bohnert, H. J. Salinity Stress Adaptation Competence in the Extremophile Thellungiella Halophila in Comparison with its Relative Arabidopsis Thaliana. *Plant J.* **2005,** *44*(5), 826–839.

Halliwell, B.; Gutteridge, J. M. C. *Free Radicals in Biology and Medicine;* Clarendan Press: Oxford, 1989; pp 446–489.

Hauser, F.; Horie, T. A Conserved Primary Salt Tolerance Mechanism Mediated by HKT Transporters: A Mechanism for Sodium Exclusion and Maintenance of High K/Na(+) Ratio in Leaves during Salinity Stress. *Plant Cell Environ. 33*(4), **2010,** 552–565.

Jagtap, T. G.; Bhosale, S.; Singh, C. Characterization of Porteresia Coarctata Beds along the Goa Coast, India. *Aquat. Bot.* **2006,** *84*, 37–44.

Jang, S. M.; Ishihara, A.; Back, K. Production of Coumaroylserotonin and Feruloylserotonin in Transgenic Rice Expressing Pepper Hydroxycinnamoyl-Coenzyme Aserotonin N-(Hydroxycinnamoyl) Transferase. *Plant Physiol.* **2004,** *135*(1), 346–356.

Jones-Rhoades, M. W.; Bartel, D. P.; Bartel, B. Micrornas and their Regulatory Roles in Plants. *Annu. Rev. Plant Biol.* **2006,** *57*, 19–53.

Jung, K. H.; Seo, Y. S.; Walia, H. The Submergence Tolerance Regulator Sub1A Mediates Stress-Responsive Expression of AP2/ERF Transcription Factors. *Plant Physiol.* **2010,** *152*(3), 1674–1692.

Khan, S. A.; Mulvaney, R. L.; Mulvaney, C. S. Accelerated Diffusion Methods for Inorganic-Nitrogen Analysis of Soil Extracts Water. *Soil Sci. Soc. Am. J.* **1997,** *61*(3), 936–942.

Kido, E. A.; Neto, J. R. F.; Silva, R. L.; Belarmino, L. C.; Neto, J. P. B.; Soares-Cavalcanti, N. M.; Pandolfi, V.; Silva, M. D.; Nepomuceno, A. L.; Benko-Iseppon, M. Expression Dynamics and Genome Distribution of Osmoprotectants in Soybean: Identifying Important Components to Face Abiotic Stress. *BMC Bioinf.* **2013,** *14*, S7.

Koiwa, H. Pathways and Genetic Determinants for Cell Wall-Based Osmotic Stress Tolerance in the Arabidopsis Thaliana Root System. In *Genes for Plant Abiotic Stress, Ch. 2;* Jenks, M. A., Wood, A. J., Eds.; Wiley-Blackwell: Oxford, UK, 2009.

Krishnamurthy, P.; Ranathunge, K.; Franke, R.; Prakash, H. S.; Schreiber, L.; Mathew, M. K. The Role of Root Apoplastic Transport Barriers in Salt Tolerance of Rice (*Oryza sativa* L.). *Planta* **2009,** *230*, 119–134.

Krishnamurthy, P.; Ranathunge, K.; Nayak, S.; Schreiber, L.; Mathew, M. K. Root Apoplastic Barriers Block Na$^+$ Transport to Shoots in Rice (*Oryza sativa* L.). *J. Exp. Bot.* **2011,** *62*(12), 4215–4228.

Latha, R.; Hosseini Salekdeh, G.; Bennett, J.; Swaminathan, M. S. Molecular Analysis of a Stress-Induced cDNA Encoding the Translation Initiation Factor, *eIF1*, from the Salt-Tolerant Wild Relative of Rice, *Porteresia coarctata. Funct. Plant Biol.* **2004,** *31*, 1035–1042.

Le Rudulier, D.; Bouillard, L. Glycine Betaine, an Osmotic Effector in *Klebsiella pneumoniae* and Other Members of the Enterobacteriaceae. *Appl. Environ. Microbiol.* **1983,** *46*(1), 152–159.

Liphschitz, N.; Shomer-Ilan, A.; Eshel, A.; Waisel, Y. Salt Glands on Leaves of Rhodes Grass (*Chloris gayana* Kth.). *Ann. Bot. (Lond.)* **1974,** *38* (2), 459–462.

Loewus, F. A.; Murthy, P. P. N. Myo-Inositol Metabolism in Plants. *Plant Sci.* **2000,** *150*(1), 1–19.

Mahalakshmi, S.; Christopher, G. S.; Reddy, T. P.; Rao, K. V.; Reddy, V. D. Isolation of a cDNA Clone (PcSrp) Encoding Serinerich-Protein from *Porteresia coarctata* T. and its Expression in Yeast and Finger Millet (*Eleusine coracana* L.) Affording Salt Tolerance. *Planta* **2006,** *224*(2), 347–359.

Majee, M.; Maitra, S.; Dastidar, K. G.; Pattnaik, S.; Chatterjee, A.; Hait, N. C.; Das, K. P.; Majumder, A. L. A Novel Salt-Tolerant L-Myo-Inositol-1-Phosphate Synthase from *Porteresia coarctata* (Roxb.) Tateoka, a Halophytic Wild Rice: Molecular Cloning, Bacterial Overexpression, Characterization, and Functional Introgression into Tobacco-Conferring Salt-Tolerance Phenotype. *J. Biol. Chem.* **2004,** *279*(27), 28539–28552.

Matysik, J., Bhalu, A. B.; Mohanty, P. Molecular Mechanisms of Quenching of Reactive Oxygen Species by Proline under Stress in Plants. *Curr. Sci.* **2002,** *82*(5), 525–532.

Mondal, T. K.; Ganie, S. A.; Debnath, A. B. Identification of Novel and Conserved MiRNAs from Extreme Halophyte, *Oryza coarctata*, a Wild Relative of Rice. *PLoS One* **2015,** *10*(10), e0140675.

Munns, R.; Tester, M. Mechanisms of Salinity Tolerance. *Annu. Rev. Plant Biol.* **2008,** *59*, 651–681.

Niroula, R. K.; Pucciariello, C.; Ho, V. T.; Novi, G.; Fukao, T.; Perata, P. SUB1A-Dependent and -Independent Mechanisms are Involved in the Flooding Tolerance of Wild Rice Species. *Plant J.* **2012,** *72*(2), 282–293.

Noctor, G.; Gomez, L.; Vanacker, H.; Foyer, C. H. Interactions between Biosynthesis, Compartmentation, and Transport in the Control of Glutathione Homeostasis and Signalling. *J. Exp. Bot.* **2002,** *53*(372), 1283–1304.

Ogawa, K. Glutathione Associated Regulation of Plant Growth and Stress Responses. *Antioxid. Redox Signaling* **2005,** *7*(7–8), 973–981.

Patra, B.; Ray, S.; Richter, A.; Majumder, A. L. Enhanced Salt Tolerance of Transgenic Tobacco Plants by Co-Expression of PciNO1 and MciMT1 is Accompanied by Increased Level of Myo-Inositol and Methylated Inositol. *Protoplasma* **2010,** *245*(1–4), 143–152.

Ray, P. K. S.; Islam, M. A. Genetic Analysis of Salinity Tolerance in Rice. *Bangladesh J. Agric. Res.* **2008,** *33*(4), 519–529.

Richharia, R. H.; Roy, J. K. Anatomical Studies in the Genus *Oryza*. I. Anatomy of *Oryza coarctata* in Relation to its Systematic Position in the Genus. *Oryza* **1965,** *2*(1), 1–9.

Saika, H.; Okamoto, M.; Miyoshi, K. Ethylenepromotes Submergence-Induced Expression of *OsABA8ox1*, a Gene that Encodes ABA 8'-Hydroxylase in Rice. *Plant Cell Physiol.* **2007,** *48*(2), 287–298.

Sengupta, S.; Majumder, A. L. Insight into the Salt Tolerance Factors of Wild Halophytic Rice, *Porteresia coarctata*: A Physiological and Proteomic Approach. *Planta* **2009,** *229*(4), 911–929.

Sengupta, S.; Majumder, A. L. *Porteresia coarctata* (Roxb.) Tateoka, a Wild Rice: A Potential Model for Studying Salt-Stress Biology in Rice. *Plant Cell Environ.* **2010,** *33*(4), 526–542.

Sengupta, S.; Patra, B.; Ray, S.; Majumder, A. L. Inositol Methyl Transferase from a Halophytic Wild Rice, *Porteresia coarctata* Roxb. (Tateoka): Regulation of Pinitol Synthesis under Abiotic Stress. *Plant Cell Environ.* **2008,** *31*(10), 1442–1459.

Slama, I.; Abdelly, C.; Bouchereau, A.; Flowers, T.; Savoure, A. Diversity, Distribution and Roles of Osmoprotective Compounds Accumulated in Halophytes under Abiotic Stress. *Ann. Bot. (Lond.)* **2015,** *115*(3), 433–447.

Smirnoff, N.; Cumbes, Q. J. Hydroxyl Radical Scavenging Activity of Compatible Solutes. *Phytochemistry* **1989,** *28*(4), 1057–1060.

Sreenivasulu, N.; Miranda, M.; Prakash, H. S.; Wobus, U.; Weschke, W. Transcriptome Changes in Foxtail Millet Genotypes at High Salinity: Identification and Characterization of a PHGPX Gene Specifically Up-Regulated by NaCl in a Salt-Tolerant Line. *J. Plant Physiol.* **2004,** *161*(4), 467–477.

Sunkar, R.; Li, Y. F.; Jagadeeswaran, G. Functions of Micrornas in Plant Stress Responses. *Trends Plant Sci.* **2012,** *17*(4), 196–203.

Suprasanna, P.; Nikalje, G. C.; Rai, A. N. Osmolyte Accumulation and Implications in Plant Abiotic Stress Tolerance. In *Osmolytes and Plants Acclimation to Changing Environment: Emerging Omics Technologies;* Iqbal, N., Nazar, R., Khan, N. A., Eds.; Springer: India, 2016; pp 1–12.

Suzuki, N.; Mittler, R. Reactive Oxygen Species and Temperature Stresses: A Delicate Balance between Signaling and Destruction. *Physiol. Plant.* **2006,** *126*, 45–51.

Tausz, M.; Sircelj, H.; Grill, D. The Glutathione System as a Stress Marker in Plant Ecophysiology: Is a Stress-Response Concept Valid. *J. Exp. Bot.* **2004,** *55*(404), 1955–1962.

Vaidyanathan, H.; Sivakumar, P.; Chakrabarty, R.; Thomas, G. Scavenging of Reactive Oxygen Species in NaCl-Stressed Rice (*Oryza sativa* L.) Differential Response in Salt-Tolerant and Sensitive Varieties. *Plant Sci.* **2003,** *165*(6), 1411–1418.

Verbruggen, N.; Hermans, C. Proline Accumulation in Plants: A Review. *Amino Acids* **2008,** *35*(4), 753–759.

Verma, D. K.; Mohan, M.; Yadav, V. K.; Asthir, B.; Soni, S. K. Inquisition of Some Physico-Chemical Characteristics of Newly Evolved Basmati Rice. *Environ. Ecol.* **2012,** *30*(1), 114–117.

Verma, D. K.; Mohan, M.; Asthir, B. Physicochemical and Cooking Characteristics of Some Promising Basmati Genotypes. *Asian J. Food Agro-Ind.* **2013,** *6*(2), 94–99.

Verma, D. K.; Mohan, M.; Prabhakar, P. K.; Srivastav, P. P. Physico-Chemical and Cooking Characteristics of Azad Basmati. *Int. Food Res. J.* **2015,** *22*(4), 1380–1389.

Xu, K.; Xu, X.; Fukao, T. Sub1A is an Ethylene Response- Factor-Like Gene that Confers Submergence Tolerance to Rice. *Nature* **2006,** *442*(7103), 705–708.

Yeo, A. R.; Flowers, T. J. The Physiology of Salinity Resistance in Rice (*Oryza sativa* L.) and a Pyramiding Approach to Breeding Varieties for Saline Soils. *Aust. J. Plant Physiol.* **1986,** *13*(1), 161–173.

Yeo, A. R.; Yeo, M. E.; Flowers, T. J. The Contribution of an Apoplastic Pathway to Sodium Uptake B.Y. Rice Roots in Saline Conditions. *J. Exp. Bot.* **1987,** *38*(7), 1141–1153.

Yeo, A. R.; Yeo, M. E.; Flowers, S. A.; Flowers, T. J. Screening of Rice (*Oryza sativa* L.) Genotypes for Physiological Characters Contributing to Salinity Resistance and their Relationship to Overall Performance. *Theor. Appl. Genet.* **1990,** *79*(3), 377–384.

Zeid, I. M. Trehalose as Osmoprotectant for Maize under Salinity-Induced Stress. *Res. J. Agric. Biol. Sci.* **2009,** *5*(5), 613–622.

Zhu JK Salt and Drought Stress Signal Transduction in Plants. *Annu. Rev. Plant Biol.* **2002,** *53*, 247–273.

CHAPTER 4

TECHNOLOGICAL DEVELOPMENT FOR ABIOTIC STRESS IN RICE

PANDIYAN MUTHURAMALINGAM[1],
SUBRAMANIAN RADHESH KRISHNAN[1,2],
DEEPAK KUMAR VERMA[3,*], MANIKANDAN RAMESH[1,4,*], and
PREM PRAKASH SRIVASTAV[3,5]

[1]*Department of Biotechnology, Alagappa University, Karaikudi, Tamil Nadu 630003, India, Tel: +91 4565 225215, Mob.: +91 9597771342, Fax: +91 4565 225202, E-mail: pandianmuthuramalingam@gmail.com*

[2]*Mob.: +91 9566422094, E-mail: radheshkrishnan.s@gmail.com*

[3]*Agricultural and Food Engineering Department, Indian Institute of Technology, Kharagpur, West Bengal 721302, India, Tel.: +91 3222281673, Mob.: +91 7407170259, Fax: +91 3222282224*

[4]*Mob.: +91 9442318200*

[5]*pps@agfe.iitkgp.ernet.in, Mob.: +91 9434043426*

Corresponding author. E-mail: deepak.verma@agfe.iitkgp.ernet.in; rajadkv@rediffmail.com; mrbiotech.alu@gmail.com

4.1 INTRODUCTION

The constant interaction between life and the environment is prerequisite for an efficient livelihood. The cultivation of cash crops has become immensely harsh due to various environmental stresses. As rice is the chief and staple crop for nearly half of the world's population (Verma et al., 2012, 2013, 2015), there occurs a definite need to feed the fastidiously proliferating population. A major portion of agricultural plots are affected by abiotic stresses (AbSs) such as salinity, heat, cold, drought, nutrient, ozone, heavy metals, ultraviolet light, visible light, chemical toxicity, and oxidative stress. These adverse environmental conditions pose serious threats to agriculture and hence are principal reason for reduction in the crop yield worldwide (Xiong and Ishitani, 2006). At the molecular, cellular, and whole plant levels various responses to these stresses are exhibited by plants (Hasegawa et al., 2000). In the earlier days, scientists researched in the biochemistry of AbS and attempted to explain the technological development in rice. In the recent past, the knowledge on plant AbS responses has advanced at a great pace with complete available information coupled with plant genomes and various genomics, proteomics, and metabolomic tools. Classical breeding techniques, biotechnology, and molecular biology have the potential to improve environmental stress tolerance of crops in which breeding of selected stress tolerant cultivars have already made some progress. The recent development in the understanding of molecular mechanisms in rice responses and tolerance to drought, salt, cold, and heat stresses are reviewed by active researchers in this field.

4.2 ABIOTIC STRESSES

4.2.1 HEAT STRESSES

Heat stress is a serious threat in many areas due to increased temperature and global warming at an alarming rate. It is often considered that rise in temperature beyond threshold level in plants induces stress that could cause an irreversible change in the plants growth and metabolism. Generally a change between 10–15°C is found to be ambient to cause this damage (Wahid et al., 2007). The stress has a complex mechanism with raise in the tempera- ture all along within a specific climatic zone with a time period during the photoperiodic session. Germination, flowering, and breaking of dormancy are critical steps in rice plant growth which can be manipulated by the application of certain temperature treatments depicted in Table 4.1 (Burdon, 1986). Morita et al. (2004) reported 34°C as threshold high temperature for rice plants at grain yield stage.

4.2.2 COLD STRESSES

Temperature affects the rate of maturity, development, and many other physiological processes in crop plants, which depend on the optimum

TABLE 4.1 Effect of Heat Stress Temperature on Different Growth Stage of Rice Plant.

Temperatures (°C)	Growth stage	Effect	References
29	Grain ripening	Reduction in grain filling	(Yoshida, 1981)
32	Booting	Decreasing in the number of pollen grains	(Shimazaki et al., 1964)
	Tillering	Reduction in height and tillering of plant	(Yoshida, 1978)
33·7	Anthesis	Poor anther dehiscence and sterility	(Jagadish et al., 2007)
34	Grain formation	Reduction in the Yield of rice plant	(Morita et al., 2004)
35	Flowering	Floret sterility	(Satake and Yoshida, 1978)
	Seedling	Poor growth of the seedling	(Yoshida, 1981)
40	Emergence	Delay and decrease in emergence	(Yoshida, 1978; Akman, 2009)

temperature that a plant can resist (Masaya and White, 1991). Crop plants are exposed to a rate lower stresses than its optimum temperature that alter considerably their lifecycle and metabolism. The plant development and agricultural yield is affected significantly by cold stress that is inclusive of freezing (defined as less than 0°C) and chilling injury (less than 20°C) (Kasuga et al., 1999; Lang et al., 2005). About 30–40% of rice yield are affected by cold stress annually (Andaya and Mackill, 2003). A decreased yield was observed in Australia (Godwin et al., 1994), the United States (Board et al., 1980), Korea, Nepal, Bangladesh, India and other countries (Kaneda and Beachell, 1974). It remains the chief factor along the temperate and arid regions. Other major economically important crops such as maize, soybean, and cotton are sensitive to chilling effect and hence a dearth is observed in their survival (Larcher, 1995). In the recent past, extensive research and analysis have been done in the development of cold stress-tolerant cultivars (Iba, 2002; Sakamoto and Murata, 2002; Chinnusamy et al., 2007). Repo et al. (2008) reported as a major factor which determine to the natural distribution of plants, yield potential, and phenology of economically important agricultural crops (Hayashi, 2001).

4.2.3 DROUGHT STRESSES

The third major limitation is the water availability and its direct effect on plant growth and production. The world food security is affected by the single most important factor that is unpredictable drought among AbSs that is limiting factor for rice plant growth and crop production and also acts as a catalyst to the great famines in the past. Moreover, the effects of successive droughts are more severe that are ensured by increased population pressures and fixed water supply in the world.

Paddy rice crops are one example for as voracious consumers of 5000 L of water to produce 1 kg of grain. At present, around 70% of the world's water is used for agriculture. It is expected that most of the Asian countries will be facing water shortage by 2025. Yet further concerns rise about the uncertainties over global warming. In rice plants, various biochemical and physiological responses are induced by drought stress. Recently at the transcriptional level, a number of genes have been described that respond to drought (Ingram and Bartels, 1996; Bray, 1997; Shinozaki and Yamaguchi-Shinozaki, 1996, 1997).

4.2.4 SALINITY STRESSES

Almost a third of the area farmed (some 380 million ha) is affected by salt that are associated with alkalinity and water logging. A raised water table (60 million ha), direct result of over-irrigation, brings underground salt (as NaCl) to the surface. Moons et al. (1995) showed that abscisic acid (ABA) accumulation upon salt stress was much more in a salt-tolerant rice cultivar as compared with a sensitive variety. The accumulation of a number of ABA-induced late embryogenesis abundant (LEA) dehydrin type proteins was also higher in the tolerant varieties. Whether the latter correlations are causal could not be established, since the various traits were not analyzed in segregating populations.

4.3 MORPHO-PHYSIOLOGICAL EFFECTS UNDER ABIOTIC STRESS (ABS) CONDITIONS

The AbS severely affects the plant physiological changes such as stomatal closure or opening, reduced photosynthesis, level of respiration, secondary metabolite biosynthesis (SMB), and transpiration. It leads to enhanced transpiration and reduced respiration. Morpho-physiological effects refer to number of shoot and root length changes (Zhou et al., 2007; Banerjee and Roychoudhury, 2015). Most importantly opening or closing of stomata is affected by AbS. From the environment, CO_2 passes to mesophyll cells for photosynthesis via stomata (Zhou et al., 2007). Stomatal closure mostly occurs in drought condition, it causes reduced photosynthesis due to reduced amount of CO_2 entry into the cell. Heat stress leads to stomatal opening because of cooling of leaf surface and increased transpiration.

Cell membrane damage and disruption cause changes in cell permeability and results in electrolyte (solutes) loss. The electrolytic leakage is a physiological parameter to measure the disruption to cellular membranes. Increased electrolytic leakage indicates reduced cellular tolerance against AbS (Cottee et al., 2007).

AbS condition increases the secondary metabolite (SM) biosynthesis that is generally SM, responsible for plant growth, stress tolerance and survival indirectly (Stamp, 2003). The major physiological effect of AbS combination is synthesis the osmolytes. These molecules are the responsible for maintaining the cell volume and fluid balance. The important osmolytes are proline, lysine, glycine, betaine, sucrose, fructose, myo-inosital, and mannitol (Burg and Ferraris, 2008). These osmolytes plays important role in AbS tolerance and avoidance mechanisms.

4.4 BIOCHEMICAL EFFECTS ON ABIOTIC STRESS

The reactive oxygen species (ROS) production occurs in response to AbS and biotic stress (BS) (Apel and Hirt, 2004). Hydrogen peroxide (H_2O_2), peroxide ions (O_2^{2-}) and superoxide (O_2^-) are the ROS molecules that are chemically reactive and can cause cellular membrane damage through oxidative stress to nucleic acids, proteins and fatty acids (Apel and Hirt, 2004). Biochemical changes triggered by stress may address the hormonal aberrations like decreased and increased level phytohormone such as gibberellic acid (GA), jasmonic acid (JA), cytokinin, salicylic acid (SA), ABA, ethylene (ET). These hormonal molecules trigger the expression of specific sets of stress responsible genes leading to AbS tolerance and avoidance. Biochemical alteration is an important aspect to study the molecular crosstalks of the rice as well as crop stress biology.

4.5 GENETIC BASIS OF ABIOTIC STRESS (ABS) TOLERANCE

Expression studies showed that AbS responsible genes can be clustered into three major categories: (i) Genes and transcription factors (TFs) involved in transcriptional control and signal transduction pathways (STPs); (ii) genes with protection of their own protein functions; and (iii) genes supporting through water, ion transport and uptake (Vierling, 1991; Ingram and Bartels, 1996; Smirnoff, 1998; Shinozaki and Yamaguchi-Shinozaki, 2000).

Rice plants adapt to AbS conditions such as drought, salinity, cold, heat, and oxidative stress circumstances by regulating exact set of these genes in response AbS signals that vary based on the factors such as the severity of environmental factors and AbS conditions (Table 4.2). Till date, genetic improvement of AbS resistance have involved manipulation of single or a few genes tangled in signaling and regulatory pathways or that encoding enzymes involved in these pathways (like antioxidants, osmolytes, ion transporters, water and osmoprotectants (Wang et al., 2003a). There are many interacting genes involved, and several efforts are mandate to improve crop AbS tolerance through genetic manipulation of one or a few of them is often associated with other, often unwanted, pleiotrophic and phenotypic (phenomic) changes (Hobo et al., 1999; Choi et al., 2000; Finkelstein and

TABLE 4.2 Potent Abiotic Stress Responsible Genes and their Properties/Engineering Improved Abiotic Stress Avoidance in Rice.

Gene/TF	Stress response	References
MYB	Drought, salinity	(Xiong et al., 2014)
DREB	Drought, cold, salinity	(Shinozaki and Yamaguchi-Shinozaki, 1996; Qin et al., 2007)
OsDREB1A	Cold, salinity, wound	(Dubouzet et al., 2003)
OsDREB1B	Cold	(Dubouzet et al., 2003)
OsDREB1C	Drought, Salinity, Cold, ABA, Wound	(Dubouzet et al., 2003)
OsDREB1G	Drought	(Chen et al., 2008)
OsDREB2A	Drought, salinity, cold, ABA	(Dubouzet et al., 2003; Matsukura et al., 2010; Mallikarjuna et al., 2011)
OsDREB1F	Drought, salinity, cold, ABA	(Wang et al., 2008)
OsDREB2B	Heat, cold	(Matsukura et al., 2010)
OsDREBL	Cold	(Chen et al., 2003)
CBF/DREB1 and DREB2	Drought, cold	(Stockinger et al., 1997; Liu et al., 1998)
MYC	Oxidative, drought, salinity	(Abe et al., 1997; Oryzabase database)
bZIP	Drought	(Tang et al., 2012)
NAC/CUC (Cup-shaped cotyledon)/NAC5	Drought, salinity	(Hu et al., 2006; Song et al., 2011; Hong et al., 2016; Khattab et al., 2014)
NAM (No apical meristem)	Drought, salinity	(Hu et al., 2006; Song et al., 2011)
ATAF (Arabidopsis transcription activation factor)	Drought, salinity	(Piao et al., 2001; Hu et al., 2006; Song et al., 2011)
NPK1 (Nicotiana protein kinase 1)	Drought	(Ning et al., 2008; Xiao et al., 2009)
β-carotene hydroxylase (BCH)	Drought	(Du et al., 2010)

TABLE 4.2 *(Continued)*

Gene/TF	Stress response	References
Ethylene response factor 3 (JERF3)	Drought, Osmotic stress	(Zhang et al., 2010)
Core-binding factor (CBF3)	Drought	(Xiao et al., 2009)
Salt Overly Sensitive 2 (SOS2) OsCIPK24, SOS2, OsSOS2, (Gene symbol-HKT6, CIPK24, HAK10, HAK7, HKT4, HKT1_ HAK16, NHX3, NHX1, SOS1, HKT8, CBL4)	Drought, salinity	(Martinez-Atienza et al., 2007; Xiao et al., 2009; Yang et al., 2012)
9-cis-epoxycarotenoid dioxygenase 2 (NCED2)	Drought	(Xiao et al., 2009)
RING domain-containing protein (RDCP1)	Drought	(Bae et al., 2011; Khattab et al., 2014)
C2H2-EAR zinc-finger protein (ZAT10)	Drought	(Xiao et al., 2009)
Na(+)/H(+) exchanger (NHX1)	Drought	(Xiao et al., 2009)
Salt- and drought-induced ring finger 1 (SDIR1)	Drought and Salt	(Gao et al., 2011)
Drought-responsive ethylene response factor (DREF1)	Drought	(Wan et al., 2011)
Heat shock proteins (Hsp17.0 and Hsp23.7)	Drought	(Zou et al., 2012)
OsiSAP8	Salt, drought, cold	(Kanneganti and Gupta, 2008)
OsCDPK	Salt, cold, drought	(Saijo et al., 2000)
Drought-induced lipid transfer protein (DIL)	Drought	(Guo et al., 2013)
DUF966-stress-repressive gene 2 (DSR2)	Drought, cold, salinity, heat, oxidative	(Luo et al., 2013)
NCED3	Drought, salinity	(Bang et al., 2013)
OsWRKY45	Drought, cold	(Tao et al., 2011)
P5CS	Drought, salinity	(Igarashi et al., 1997)

TABLE 4.2 *(Continued)*

Gene/TF	Stress response	References
OsHOS1	Cold	(Lourenço et al., 2013)
ZFP182	Drought, cold, salinity	(Huang et al., 2012)
ZFP245	Cold, Drought, Oxidative	(Huang et al., 2009)
ZFP252	Drought, salinity	(Xu et al., 2008)
OsCOIN	Cold, drought, salinity	(Liu et al., 2007)
OsCPK21	Salinity	(Asano et al., 2011)
OsLEA3–1	Drought	(Xiao et al., 2007)
OsCYP2	Salinity	(Ruan et al., 2011)
OsbZIP71	Drought, salinity, oxidative	(Liu et al., 2014)
SNAC1	Drought, salinity	(Hu et al., 2006)
ZFP182	Drought, salinity	(Huang et al., 2012)
ZFP252	Drought, salinity	(Xu et al., 2008)
OsSIKI	Drought, salinity	(Ouyang et al., 2010)
OsSIK2	Drought, salinity	(Chen et al., 2013)
OsPYL/RCA R5	Drought, salinity	(Kim et al., 2014)
OsLEA3–2	Drought, salinity	(Duan and Cai, 2012)
OsOAT (Ornithine δ-amino-transferase)	Drought, oxidative	(You et al., 2012)

Lynch, 2000; Lopez-Molina and Chua, 2000; Uno et al., 2000; Sakamoto and Murata, 2001; Kang et al., 2002; Yamanouchi et al., 2002 Wang et al., 2003b). These complex considerations, when coupled with the complexity of AbS and the plant-environmental crosstalks occurring at all levels of plant response to various AbS, delineates that the major task plant researchers are facing with engineering AbS-resistant rice is dauntingly complicated and extremely difficult.

4.6   ENGINEERING IMPROVED ABIOTIC STRESS TOLERANCE IN RICE

Several rice TFs belonging to different families are involved in AbS conditions. Numerous literature information revealed the gene expression analysis (Table 4.2), there are huge number of AbS-associated TFs that have been functionally characterized with transformational studies. Although, Table 4.2 comprises validated vital genes and their stress response in AbS, a theoretical diagram is represented for the overall understanding and complexity of AbS tolerance mechanisms (Fig. 4.1).

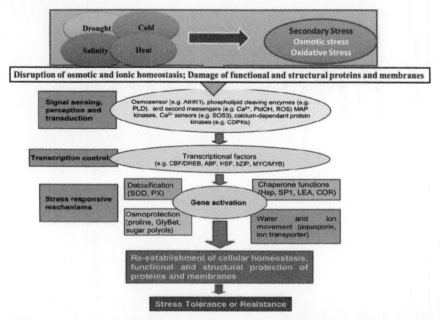

FIGURE 4.1   (See color insert.) The complexity of plant response to abiotic stress.

Source: Adapted from Wang et al. (2003a)

4.7 OMICS APPROACHES IN ABIOTIC STRESS OF RICE

Availability of whole transcript sequences and the recently evolving omics based research is vital approach to understand the molecular crosstalks under AbS conditions. The emerging high throughput omics related approaches in genomics, proteomics and metabolomics paved the way to identify and improve the crop production in rice. Combinatorial studies from omics related sources are focused in identifying the molecular system insights and biochemical properties that accelerates gene mining and its functional characters.

The era of genomics of rice stress biology involves transformation of pivotal players into rice pave the way to generate AbS tolerance in plants (Table 4.2). These genes had stress related traits that are required to efficiently target the various stress signaling pathways. The genome-wide transcription analysis in rice under several AbS has been studied intensively; some common examples includes salinity (Kawasaki et al., 2001), drought (Gorantla et al., 2007), heat (Yamakawa et al., 2007), cold (Cheng et al., 2007). However, under environmental conditions, rice is concurrently exposed to many stresses and, consequently, molecular mechanisms toward stress adaptation is different. Compared to those are openly exposed to single stress conditions due to the complex stress regulatory system is controlled by ABA mechanisms like ABA-dependent manner and ABA-independent (Rabbani et al., 2003; Mukherjee et al., 2006; Nakashima et al., 2007; Quan et al., 2010). Overexpression of genes quoted in Table 4.2 could induce the improved stress tolerance and altered phenotypic mechanisms.

Protein analysis is the direct high throughput approach to delineate the associated genes and its function. Though, the mentioned omics of an organism, especially genome and proteome not parallel to each other on a one to one analysis (Schroeder et al., 2001; Komatsu, 2007). Hence, the proteome- and metabolome-level analysis was performed for genomic study. The stress-induced protein functions have been outlined by proteomic analysis of specific tissues and organelles of rice, such as leaves (Salekdeh et al., 2002), embryo (Fukuda et al., 2003), root (Yan et al., 2005; Lee et al., 2009), anthers (Liu and Bennett, 2011), mitochondria (Chen et al., 2009) and cell suspension culture (Rao et al., 2010). Differential display (DD) following two-dimensional electrophoresis (2-DE), it has been developed for direct protein identification and the structure of protein identified by rice proteome database (http://gene64.dna.affrc.go.jp/RPD/main.html) or by matrix-assisted laser desorption/ionization-time of flight-time of flight (MALDI-TOF-TOF), Edman sequencing and MS, Swissprot and so forth. Particularly, the advancement of two dimensional-polyacrylamide gel electrophoresis (2D-PAGE) coupled with the application 5′-iodoacetamidofluorescein (5′IAF) (Baty et al; 2002; Cuddihy et al., 2008) or MS like two dimensional-fluorescense difference gel electrophoresis (2D-FDIGE) (Fu et al., 2008) supports to the identification of AbS-related proteins. Gel free proteomic analysis systems are also developed for differential proteome analysis like multidimensional protein identification methods (MudPIT) (Koller et al., 2002), isobaric tags for relative and absolute quantification (iTRAQ), isotope-coded affinity tags (ICAT) (Griffin et al., 2001). These techniques are used to identify modification of protein by means of the massive differences between stress treated and untreated rice proteomes for better AbS understanding.

In cellular process, secondary metabolites (SMs) are the end products and they reflect the systemic biological response to climatic changes (Royuela et al., 2000). Understanding the metabolic response to AbSs provides novel insights on biological response mechanisms that can hinder or regulate its survival. Recent trends in metabolomics studies are to derive the cellular status at exact time point by quantification of the total metabolites in the cellular system (Hollywood et al., 2006). These techniques are complemented with other techniques such as proteomics that reveal the whole cellular process. Various analytical technologies are available for the plant metabolomic study with the use of high throughput approaches such as Fourier-transform infrared (FT-IR) (Johnson et al., 2003; Okazaki and Saito, 2012), ultra high-resolution Fourier transform-ion cyclotron MS (Hirai et al., 2004), gas chromatography-mass spectrometry (GC-MS) (Kaplan et al., 2004), and nuclear magnetic resonance (NMR) (Kim et al., 2010). Rice metabolomics study speculates that quality of the metabolites resembles the type of metabolites that can promote seed germination (Shu at al., 2008), metabolic variance between treated and untreated rice plants (Wakasa et al., 2006), metabolomics profiling at different developmental stages (Tarpley et al., 2005), and natural metabolic variance between rice cultivars (Kusano et al., 2007). Few studies are available for AbS treatments such as drought, salt (Fumagalli et al., 2009), submergence (Barding et al., 2012), and oxidative stress (Ishikawa et al., 2010). These metabolomic techniques are used to understand the complex biochemical response and plant physiological performance of rice crops against AbSs.

4.8   ASSESSMENT AND FUTURE RESEARCH NEEDS

Technological developments of genome-scale expressed sequence tags (ESTs), genomic sequencing, and complementary DNA (cDNA) microarray like approaches have played candid role by potentiality prompting the isolation of key players that are essential to attain AbS tolerance. Therefore, it is vital to generate datasets from AbS experiments required to be merged so that the comparisons between different cellular and different genotype of rice can be chalked out. In functional genomics, data mining is an important approach for collecting the tagged mutant, overexpression and complementation tests, along with microarray analyses to reveal the classified relationships between vital signaling components and downstream effector genes.

Functional and integrative genomics approaches are essential to understand the molecular crosstalks in rice. The key players which intricate in AbS tolerance mechanisms have been well identified, categorized, and considered to find their transcriptional activity by using high throughput technologies such as EST libraries or whole genome sequencing. Many genome-wide approaches have been analyzed in rice to validate the AbS responsive transcripts by serial analysis of gene expression (Super SAGE), suppression subtractive hybridization (SSH), EST sequencing (Xiong et al., 2001; Molina et al., 2008; Mantri et al., 2007; Varshney et al., 2009). SSH is a powerful technology for polymerase chain reaction (PCR)-based amplification of differentially expressed cDNA in a particular PCR condition. SSH technique comprises the removal of double stranded DNA (dsDNA) formed by a control and stress-treated sample hybridization. Hence, eliminating cDNA of similar abundance, and retaining genomic sequences or differentially expressed (DE) transcripts. These techniques have been used worldwide to compare gene expression patterns in tissues and developmental stages in different conditions. Hitherto, identification of the DE transcripts (up and downregulated) in rice response to AbS at various stages especially in flowering stage of plants has not been used (Deokar et al., 2011). Studying the several AbS responses in rice by the utilization of transcript profiling (Mantri et al., 2007; Ashraf et al., 2009). NGS, computational omics approaches in rice AbS biology mechanisms that have not yet interpreted. With availability of rice genome sequences, in silico approaches have become easy to analyze and may pave the unparalleled way of rice AbS biology history for better understanding the molecular mechanisms.

4.9   CONCLUSION AND SUMMERY

This chapter briefly summarizes the "Technological Development for Abiotic Stress in Rice: A Critical Overview" and offers understanding about the rice development use of high throughput approaches. The role of AbS responsible genes and transcription factors, and their stress mechanisms, genetic-based development was mentioned. Rice omics approaches pave the way to better understanding about AbS tolerance mechanisms. With the availability of rice full transcriptome, in silico omics technologies are useful to develop the AbS-resistant rice crops.

ACKNOWLEDGMENT

PM acknowledges Alagappa University Research Fund (Ph.D./1215/AURF Fellowship/2015 dated 25/11/2015), Karaikudi, Tamil Nadu, India for providing research fellowship. The authors gratefully acknowledge the use of Bioinformatics Infrastructure Facility, Alagappa University funded by Department of Biotechnology, Ministry of Science and Technology, Government of India Grant (No. BT/BI/25/015/2012). Deepak Kumar Verma and Prem Prakash Srivastav are indebted to Department of Science and Technology, Ministry of Science and Technology, Govt. of India for an individual research fellowship (INSPIRE Fellowship Code No.: IF120725; Sanction Order No. DST/INSPIRE Fellowship/2012/686 & Date: 25/02/2013).

KEYWORDS

- Abiotic stress
- biochemical alterations and response
- breeding techniques
- crop stress biology
- Edman and EST sequencing
- Fourier-transform infrared
- gas chromatography-mass spectrometry
- genomic sequencing
- heat stress
- hybridizations
- MALDI-TOF
- metabolomics profiling
- microarray
- molecular crosstalk
- nuclear magnetic resonance
- phenotypic mechanisms
- proteomic analysis
- systemic biological
- transcription analysis
- world food security

REFERENCES

Abe, H.; Yamaguchi-Shinozaki, K.; Urao, T.; Iwasaki, T.; Hosokawa, D.; Shinizaki, K. Role of Arabidopsis MYC and MYB Homologs in Drought- and Abscisic Acid-Regulated Gene Expression. *Plant Cell* **1997**, *9,* 1859–1868.

Akman, Z. Comparison of High Temperature Tolerance in Maize, Rice and Sorghum Seeds, by Plant Growth Regulators. *J. Anim. Vet. Adv.* **2009**, *8,* 358–361.

Andaya, V. C.; Mackill, D. J. Mapping of QTLs Associated with Cold Tolerance During the Vegetative Stage in Rice. *J. Exp. Bot.* **2003**, *54,* 2579–2585.

Apel, K.; Hirt, H. Reactive oxygen species: metabolism, oxidative stress, and signal transduction. *Annu.Rev.PlantBiol.*, **2004**, *55,* 373–399.

Asano, T.; Hakata, M.; Nakamura, H.; Aoki, N.; Komatsu, S.; Ichikawa, H.; Hirochika, H.; Ohsugi, R. Functional Characterisation of OsCPK21, a Calcium-Dependent Protein Kinase that Confers Salt Tolerance in Rice. *Plant Mol. Biol.* **2011**, *75*(1), 179–191.

Ashraf, N.; Ghai, D.; Barman, P.; Basu, S.; Gangisetty, N.; Mandal, M. K.; Chakraborty, N.; Datta, A.; Chakraborty, S. Comparative Analyses of Genotype Dependent Expressed Sequence Tags and Stress-Responsive Transcriptome of Chickpea Wilt Illustrate Predicted and Unexpected Genes and Novel Regulators of Plant Immunity. *BMC Genomics* **2009**, *10,* 415.

Bae, H.; Kim, S. K.; Cho, S. K.; Kang, B. G.; Kim, W. T. Overexpression of OsRDCP1, a Rice RING Domain-Containing E3 Ubiquitin Ligase, Increased Tolerance to Drought Stress in Rice (*Oryza sativa* L.). *Plant Sci.* **2011**, *180,* 775–778.

Banerjee, A.; Roychoudhury, A. WRKY Proteins: Signaling and Regulation of Expression During Abiotic Stress Responses. *Sci. World J.* **2015**, *2015,* 807560.

Bang, S. W.; Park, S. H.; Jeong, J. S.; Kim, Y. S.; Jung, H.; Ha, S. H.; Kim, J. K. Characterization of the Stress-Inducible OsNCED3 Promoter in Different Transgenic Rice Organs and Over Three Homozygous Generations. *Planta* **2013**, *237*(1), 211–224.

Barding, G. A.; Béni, S.; Fukao, T.; Bailey-Serres, J.; Larive, C. K. Comparison of GC-MS and NMR for Metabolite Profiling of Rice Subjected to Submergence Stress. *J. Proteome Res.* **2012**, *12,* 898–909.

Baty, J. W.; Hampton, M. B.; Winterbourn, C. C. Detection of Oxidant Sensitive Thiol Proteins by Fluorescence Labeling and Two-Dimensional Electrophoresis. *Proteomics* **2002**, *2,* 1261–1266.

Board, J. E.; Peterson, M. L.; Ng, E. Floret Sterility in Rice in a Cool Environment. *Agron. J.* **1980**, *72,* 483–487.

Bray, E. A. Plant Responses to Water Deficit. *Trends Plant Sci.* **1997**, *2,* 48–54.

Burdon, R. H. Heat Shock and the Heat Shock Proteins. *Biochem. J.* **1986**, *240,* 313–324.

Burg, M. B.; Ferraris, J. D. Intracellular Organic Osmolytes: Function and Regulation. *J. Biol. Chem.*, **2008**, *283*(12), 7309–7313.

Chen, J.; Dong, Y.; Wang, Y.; Liu, Q.; Zhang, J.; Chen, S. An AP2/EREBP-type transcription-factor gene from rice is cold-inducible and encodes a nuclear-localized protein. *Theoretical and Applied Genetics*, **2003**, *107*, 972–979.

Chen, J. Q.; Meng, X. P.; Zhang, Y.; Xia, M.; Wang, X. P. Over-expression of OsDREB genes lead to enhanced drought tolerance in rice. *Biotechnology Letters*, **2008**, *30,* 2191–2198.

Chen, X.; Wang, Y.; Li, J.; Jiang, A.; Cheng, Y.; Zhang, W. Mitochondrial Proteome During Salt Stress-Induced Programmed Cell Death in Rice. *Plant Physiol. Biochem.* **2009,** *47,* 407–415.

Chen, L. J.; Wuriyanghan, H.; Zhang, Y. Q.; Duan, K. X.; Chen, H. W.; Li, Q. T.; et al. An S-Domain Receptor-Like Kinase, OsSIK2, Confers Abiotic Stress Tolerance and Delays Dark-Induced Leaf Senescence in Rice. *Plant Physiol.* **2013,** *163,* 1752–1765.

Cheng, C.; Yun, K. Y.; Ressom, H.; Mohanty, B.; Bajic, V.; Jia, Y.; Yun, S.; de los Reyes, B. An Early Response Regulatory Cluster Induced by Low Temperature and Hydrogen Peroxide in Seedlings of Chilling-Tolerant Japonica Rice. *BMC Genomics* **2007,** *8,* 175.

Chinnusamy, V.; Zhu, J.; Zhu, J. K. Cold Stress Regulation of Gene Expression in Plants. *Trends Plant Sci.* **2007,** *12,* 444–451.

Choi, H. I.; Hong, J. H.; Ha, J.; Kang, J. Y.; Kim, S. Y. ABFs, a Family of ABA-Responsive Element Binding Factors. *J. Biol. Chem.* **2000,** *275,* 1723–1730.

Cottee, N. S.; Tan, D. K. Y.; Bange, M. P.; Cheetham, J. A. In *Simple Electrolyte Leakage Protocols to Detect Cold Tolerance in Cotton Genotype*, Proceedings 4th World Cotton Research Conference, Lubbock, Texas, Sept 10–14, 2007, page 9.

Cuddihy, S. L.; Baty, J. W.; Brown, K. K.; Winterbourn, C. C.; Hampton, M. B. Proteomic Detection of Oxidized and Reduced Thiol Proteins in Cultured Cells. *Methods Mol. Biol.* **2008,** *519,* 363–375.

Deokar, A. A.; Kondawar, V.; Jain, P. K.; Karuppayil, S. M.; Raju, N. L.; Vadez, V.; Varshney, R. K.; Srinivasan, R. Comparative Analysis of Expressed Sequence Tags (ESTs) Between Drought-Tolerant And-Susceptible Genotypes of Chickpea Under Terminal Drought Stress. *BMC Plant Biol.* **2011,** *11,* 70.

Du, H.; Wang, N.; Cui, F.; Li, X.; Xiao, J.; Xiong, L. Characterization of the B-Carotene Hydroxylase Gene DSM2 Conferring Drought and Oxidative Stress Resistance by Increasing Xanthophylls and Abscisic Acid Synthesis in Rice. *Plant Physiol.* **2010,** *154,* 1304–1318.

Duan, J.; Cai, W. OsLEA3–2, an Abiotic Stress Induced Gene of Rice Plays a Key Role in Salt and Drought Tolerance. *PLoS One* **2012,** *7,* e45117.

Dubouzet, J. G.; Sakuma, Y.; Ito, Y.; Kasuga, M.; Dubouzet, E. G.; Miura, S.; Seki, M.; Shinozaki, K.; Yamaguchi-Shinozaki, K. OsDREB genes in rice, *Oryza sativa* L., encode transcription activators that function in drought-, high-salt-, and cold-responsive gene expression. *Plant J.* **2003,** *33,* 751–763.

Finkelstein, R. R.; Lynch, T. J. The Arabidopsis Abscisic Acid Response Gene ABI5 Encodes a Basic Leucine Zipper Transcription Factor. *Plant Cell* **2000,** *12,* 599–609.

Fu, C.; Hu, J.; Liu, T.; Ago, T.; Sadoshima, J.; Li, H. Quantitative Analysis of Redox-Sensitive Proteome with DIGE and ICAT. *J. Proteome Res.* **2008,** *7,* 3789–3802.

Fukuda, M.; Islam, N.; Woo, S. H.; Yamagishi, A.; Takaoka, M.; Hirano, H. Assessing Matrix Assisted Laser Desorption/Ionization-Time of Flight-Mass Spectrometry as a Means of Rapid Embryo Protein Identification in Rice. *Electrophoresis* **2003,** *24,* 1319–1329.

Fumagalli, E.; Baldoni, E.; Abbruscato, P.; Piffanelli, P.; Genga, A.; Lamanna, R.; Consonni, R. NMR Techniques Coupled with Multivariate Statistical Analysis: Tools to Analyse *Oryza sativa* Metabolic Content Under Stress Conditions. *J. Agron. Crop Sci.* **2009,** *195,* 77–88.

Gao, T.; Wu, Y.; Zhang, Y.; Liu, L.; Ning, Y.; Wang, D.; Tong, H.; Chen, S.; Chu, C.; Xie, Q. OsSDIR1 Overexpression Greatly Improves Drought Tolerance in Transgenic Rice. *Plant Mol. Biol.* **2011,** *76,* 145–156.

Godwin, D. C.; Meyer, W. S.; Singh, U. Simulation of the Effect of Chilling Injury and Nitrogen Supply on Floret Fertility and Yield in Rice. *Aust. J. Exp. Agric.* **1994**, *34,* 921–926.

Gorantla, M.; Babu, P.; Reddy, L. V.; Reddy, A.; Wusirika, R.; Bennetzen, J. L.; Reddy, A. R. Identification of Stress-Responsive Genes in an Indica Rice (*Oryza sativa* L.) Using ESTs Generated from Drought-Stressed Seedlings. *J. Exp. Bot.* **2007**, *58,* 253–265.

Griffin, T. J.; Sherman, J.; Aebersold, R. *Quantitative Proteomics (ICAT™);* John Wiley & Sons Ltd.: Hoboken, NJ, USA, 2001.

Guo, C.; Ge, X.; Ma, H. The rice OsDIL gene plays a role in drought tolerance at vegetative and reproductive stages. *Plant Mol. Biol.* **2013**, *82,* 239–253.

Hasegawa, P. M.; Bressan, R. A.; Zhu, J. K.; Bohnert, H. J. Plant Cellular and Molecular Responses to High Salinity. *Annu. Rev. Plant Physiol. Plant Mol. Biol.* **2000**, *51,* 463–499.

Hayashi, H. Plant Temperature Stress. In *Encyclopedia of Life Sciences;* John Wiley & Sons Ltd.: Chichester, 2001. DOI:10.1038/npg.els.0001320. http://www.els.net/

Hirai, M. Y.; Yano, M.; Goodenowe, D. B.; Kanaya, S.; Kimura, T.; Awazuhara, M.; Arita, M.; Fujiwara, T.; Saito, K. Integration of Transcriptomics and Metabolomics for Understanding of Global Responses to Nutritional Stresses in *Arabidopsis thaliana. Proc. Natl. Acad. Sci. U. S. A.* **2004**, *101,* 10205–10210.

Hobo, T.; Kowyama, Y.; Hattori, T. A bZIP Factor, TRAB1, Interacts with VP1 and Mediates Abscisic Acid-Induced Transcription. *Proc. Natl. Acad. Sci. U. S. A.* **1999**, *96,* 15348–15353.

Hollywood, K.; Brison, D. R.; Goodacre, R. Metabolomics: Current Technologies and Future Trends. *Proteomics* **2006**, *6,* 4716–4723.

Hong, Y.; Zhang, H.; Huang, L.; Li, D.; Song, F. Overexpression of a Stress-Responsive NAC Transcription Factor Gene ONAC022 Improves Drought and Salt Tolerance in Rice. *Front Plant Sci.* **2016**, *7,* 4.

Hu, H.; Dai, M.; Yao, J.; Xiao, B.; Li, X.; Zhang, Q.; Xiong, L. Overexpressing a NAM, ATAF, and CUC (NAC) Transcription Factor Enhances Drought Resistance and Salt Tolerance in Rice. *Proc. Natl. Acad. Sci.* **2006**, *103,* 12987–12992.

Huang, J.; Sun, S. J.; Xu, D. Q.; Yang, X.; Bao, Y. M.; Wang, Z. F.; Tang, H. J.; Zhang, H. Increased Tolerance of Rice to Cold, Drought and Oxidative Stresses Mediated by the Overexpression of a Gene that Encodes the Zinc Finger Protein ZFP245. *Biochem. Biophys. Res. Commun.* **2009**, *389,* 556–561.

Huang, J.; Sun, S.; Xu, D.; Lan, H.; Sun, H.; Wang, Z.; Bao, Y.; Wang, J.; Tang, H.; Zhang, H. A TFIIIa-Type Zinc Finger Protein Confers Multiple Abiotic Stress Tolerances in Transgenic Rice (*Oryza sativa* L.). *Plant Mol. Biol.* **2012**, *80*(3), 337–350.

Iba, K. Acclimative Response to Temperature Stress in Higher Plants: Approaches of Gene Engineering for Temperature Tolerance. *Annu. Rev. Plant Biol.* **2002**, *53,* 225–245.

Igarashi, Y.; Yoshiba, Y.; Sanada, Y.; Yamaguchi-Shinozaki, K.; Wada, K.; Shinozaki, K. Characterization of the Gene for Delta1-Pyrroline-5-Carboxylate Synthetase and Correlation Between the Expression of the Gene and Salt Tolerance in *Oryza sativa* L. *Plant Mol. Biol.* **1997**, *33*(5), 857–865.

Ingram, J.; Bartels, D. The Molecular Basis of Dehydration Tolerance in Plants. *Annu. Rev. Plant Physiol. Plant Mol. Biol.* **1996**, *47,* 377–403.

Ishikawa, T.; Takahara, K.; Hirabayashi, T.; Matsumura, H.; Fujisawa, S.; Terauchi, R.; Uchimiya, H.; Kawai-Yamada, M. Metabolome Analysis of Response to Oxidative Stress in Rice Suspension Cells Overexpressing Cell Death Suppressor Bax Inhibitor-1. *Plant Cell Physiol.* **2010**, *51,* 9–20.

Jagadish, S. V. K.; Craufurd, P. Q.; Wheeler, T. R. High Temperature Stress and Spikelet Fertility Inrice (*Oryza sativa* L.). *J. Exp. Bot.* **2007,** *58,* 1627–1635.

Johnson, H. E.; Broadhurst, D.; Goodacre, R.; Smith, A. R. Metabolic Fingerprinting of Salt-Stressed Tomatoes. *Phytochemistry* **2003,** *62,* 919–928.

Kaneda, C.; Beachell, H. M. Response of Indica-Japonica Rice to Low Temperatures. *Sabrao J.* **1974,** *6,* 17–32.

Kang, J. Y.; Choi, H. I.; Im, M. Y.; Kim, S. Y. Arabidopsis Basic Leucine Zipper Proteins that Mediate Stress-Responsive Abscisic Acid Signaling. *Plant Cell* **2002,** *14,* 343–357.

Kanneganti, V.; Gupta, A. K. Overexpression of OsiSAP8, a Member of Stress Associated Protein (SAP) Gene Family of Rice Confers Tolerance to Salt, Drought and Cold Stress in Transgenic Tobacco and Rice. *Plant Mol. Biol.* **2008,** *66*(5), 445–462.

Kaplan, F.; Kopka, J.; Haskell, D. W.; Zhao, W.; Schiller, K. C.; Gatzke, N.; Sung, D. Y.; Guy, C. L. Exploring the Temperature-Stress Metabolome of Arabidopsis. *Plant Physiol.* **2004,** *136,* 4159–4168.

Kasuga, M.; Liu, Q.; Miura, S.; Yamaguchi-Shinozaki, K.; Shinozaki, K. Improving Plant Drought, Salt and Freezing Tolerance by Gene Transfer of a Single Stress Inducible Transcriptional Factor. *Nat. Biotechnol.* **1999,** *17,* 287–291.

Kawasaki, S.; Borchert, C.; Deyholos, M.; Wang, H.; Brazille, S.; Kawai, K.; Galbraith, D.; Bohnert, H. J. Gene Expression Profiles During the Initial Phase of Salt Stress in Rice. *Plant Cell* **2001,** *13,* 889–906.

Khattab, H. I.; Emam, M. A.; Emam, M. M.; Helal, N. M.; Mohamed, M. R. Effect of Selenium and Silicon on Transcription Factors NAC5 and DREB2A Involved in Drought-Responsive Gene Expression in Rice. *Biol. Plant* **2014,** *58,* 265–273.

Kim, H. K.; Choi, Y. H.; Verpoorte, R. NMR-Based Metabolomic Analysis of Plants. *Nat. Protoc.* **2010,** *5,* 536–549.

Kim, H.; Lee, K.; Hwang, H.; Bhatnagar, N.; Kim, D.-Y.; Yoon, I. S.; et al. Over Expression of PYL5 in Rice Enhances Drought Tolerance, Inhibits Growth, and Modulates Gene Expression. *J. Exp. Bot.* **2014,** *65,* 453–464.

Koller, A.; Washburn, M. P.; Lange, B. M.; Andon, N. L.; Deciu, C.; Haynes, P. A.; Hays, L.; Schieltz, D.; Ulaszek, R.; Wei, J. Proteomic Survey of Metabolic Pathways in Rice. *Proc. Natl. Acad. Sci. U. S. A.* **2002,** *9,* 11969–11974.

Komatsu, S. Rice Proteomics: A Step Toward Functional Analysis of the Rice Genome. In *Rice Functional Genomics;* Upadhyaya, N. M., Ed.; Springer: New York, NY, USA, 2007, 61–89.

Kusano, M.; Fukushima, A.; Kobayashi, M.; Hayashi, N.; Jonsson, P.; Moritz, T.; Ebana, K.; Saito, K. Application of a Metabolomic Method Combining One-Dimensional and Two-Dimensional Gas Chromatography-Time-Of-Flight/Mass Spectrometry to Metabolic Phenotyping of Natural Variants in Rice. *J Chromatogr. B* **2007,** *855,* 71–79.

Lang, P.; Zhang, C. K.; Ebel, R. C.; Dane, F.; Dozier, W. A. Identification of Cold Acclimated Genes in Leaves of Citrus Unshiu by mRNA Differential Display. *Gene* **2005,** *359,* 111–118.

Larcher, W. Physiological Plant Ecology—Ecophysiology and Stress Physiology of Functional Groups, 3rd ed.; Springer: Berlin/Heidelberg/New York, 1995.

Lata, C.; Prasad, M. Role of DREBs in Regulation of Abiotic Stress Responses in Plants. *J. Exp. Bot.* **2011,** *62*(14), 4731–4748.

Lee, D. G.; Ahsan, N.; Lee, S. H.; Lee, J. J.; Bahk, J. D.; Kang, K. Y.; Lee, B. H. Chilling Stress-Induced Proteomic Changes in Rice Roots. *J. Plant Physiol.* **2009,** *166,* 1–11.

Liu, Q.; Kasuga, M.; Sakuma, Y.; Abe, H.; Miura, S.; Goda, H.; Shimada, Y.; Yoshida, S.; Shinozaki, K.; Yamaguchi-Shinozaki, K. Two transcription factors, DREB1 and DREB2, with an EREBP/AP2 DNA binding domain separate two cellular signal transduction pathways in drought- and low-temperature-responsive gene expression, respectively, in *Arabidopsis. Plant Cell* **1998**, *10*, 391–406.

Liu, J. X.; Bennett, J. Reversible and Irreversible Drought-Induced Changes in the Anther Proteome of Rice (*Oryza sativa* L.) Genotypes IR64 and Moroberekan. *Mol. Plant* **2011**, *4*, 59–69.

Liu, K.; Wang, L.; Xu, Y.; Chen, N.; Ma, Q.; Li, F.; Chong, K. Overexpression of OsCOIN, a Putative Cold Inducible Zinc Finger Protein, Increased Tolerance to Chilling, Salt and Drought, and Enhanced Proline Level in Rice. *Planta* **2007**, *226*, 1007–1016.

Liu, C.; Mao, B.; Ou, S.; Wang, W.; Liu, L.; Wu, Y.; et al. OsbZIP71, a bZIP Transcription Factor, Confers Salinity, and Drought Tolerance in Rice. *Plant Mol. Biol.* **2014**, *84*, 19–36.

Lopez-Molina, L.; Chua, N. H. A Null Mutation in a bZIP Factor Confers Aba-Insensitivity in Arabidopsis Thaliana. *Plant Cell Physiol.* **2000**, *41*, 541–547.

Lourenço, T.; Sapeta, H.; Figueiredo, D. D.; Rodrigues, M.; Cordeiro, A.; Abreu, I. A.; Saibo, N. J.; Oliveira, M. M. Isolation and Characterization of Rice (*Oryza sativa* L.) E3-Ubiquitin Ligase OsHOS1 Gene in the Modulation of Cold Stress Response. *Plant Mol. Biol.* **2013**, *83*(4–5), 351–363.

Luo, C.; Guo, C.; Wang, W.; Wang, L.; Chen, L. Overexpression of a New Stress-Repressive Gene OsDSR2 Encoding a Protein with a DUF966 Domain Increases Salt and Simulated Drought Stress Sensitivities and Reduces Aba Sensitivity in Rice. *Plant Cell Rep.* **2013**, *33*, 323–336.

Mallikarjuna, G.; Mallikarjuna, K.; Reddy, M. K.; Kaul, T. Expression of OsDREB2A Transcription Factor Confers Enhanced Dehydration and Salt Stress Tolerance in Rice (*Oryza sativa* L.). *Biotechnol. Lett.* **2011**, *33*, 1689–1697.

Mantri, N. L.; Ford, R.; Coram, T. E.; Pang, E. C. Transcriptional Profiling of Chickpea Genes Differentially Regulated in Response to High-Salinity, Cold and Drought. *BMC Genomics* **2007**, *8*, 303.

Martinez-Atienza, J.; Jiang, X.; Garciadeblas, B.; Mendoza, I.; Zhu, J. K.; Pardo, J. M.; Quintero, F. J. Conservation of the Salt Overly Sensitive Pathway in Rice. *Plant Physiol.* **2007**, *143*(2), 1001–1012.

Masaya, P.; White, J. W. Adaptation to Photoperiod and Temperature. In *Common Beans Research for Crop Improvement;* van Schoonhoven, A., Voysest, O. Eds.; CAB International and CIAT: CA, 1991; pp 455–500.

Matsukura, S.; Mizoi, J.; Yoshida, T.; Todaka, D.; Ito, Y.; Maruyama, K.; Shinozaki, K.; Yamaguchi-Shinozaki, K. Comprehensive Analysis of Rice DREB2-Type Genes that Encode Transcription Factors Involved in the Expression of Abiotic Stress-Responsive Genes. *Mol. Genet. Genomics* **2010**, *283*(2), 185–196.

Molina, C.; Rotter, B.; Horres, R.; Udupa, S. M.; Besser, B.; Bellarmino, L.; Baum, M.; Matsumura, H.; Terauchi, R.; Kahl, G. SuperSAGE: the Drought Stress Responsive Transcriptome of Chickpea Roots. *BMC Genomics* **2008**, *9*, 553.

Moons, A.; Bauw, G.; Prinsen, E.; et al. Molecular and Physiological Response to Abscisic Acid and Salts in Roots of Salt-Sensitive and Salt-Tolerant Indica Rice Varieties. *Plant Physiol.* **1995**, *107*, 177–186.

Morita, S.; Siratsuchi, H.; Takanashi, J.; Fujita, K. Effect of High Temperature on Ripening in Rice Plant. Analysis of the Effect of High Night and High Day Temperature Applied to the Panicle in Other Parts of the Plant. *Jpn. J. Crop. Sci.* **2004**, *73*, 77–83.

Mukherjee, K.; Choudhury, A.; Gupta, B.; Gupta, S.; Sengupta, D. An Abre-Binding Factor, OSBZ8, is Highly Expressed in Salt Tolerant Cultivars than in Salt Sensitive Cultivars of Indica Rice. *BMC Plant Biol.* **2006**, *6,* 18.

Nakashima, K.; Tran, L. S. P.; van Nguyen, D.; Fujita, M.; Maruyama, K.; Todaka, D.; Ito, Y.; Hayashi, N.; Shinozaki, K.; Yamaguchi, S. K. Functional Analysis of a NAC-Type Transcription Factor OsNAC6 Involved in Abiotic and Biotic Stress-Responsive Gene Expression in Rice. *Plant J.* **2007**, *51,* 617–630.

Ning, J.; Liu, S.; Hu, H.; Xiong, L. Systematic Analysis of NPK1-Like Genes in Rice Reveals a Stress-Inducible Gene Cluster Co-Localized with a Quantitative Trait Locus of Drought Resistance. *Mol. Genet. Genomics* **2008**, *280,* 535–546.

Okazaki, Y.; Saito, K. Recent Advances of Metabolomics in Plant Biotechnology. *Plant Biotechnol. Rep* **2012**, *6,* 1–15.

Ouyang, S. Q.; Liu, Y. F.; Liu, P.; Lei, G.; He, S. J.; Ma, B.; et al. Receptor-Like Kinase OsSIK1 Improves Drought and Salt Stress Tolerance in Rice (*Oryza sativa*) Plants. *Plant J.* **2010**, *62,* 316–329.

Piao, H. L.; Lim, J. H.; Kim, S. J.; Cheong, G. W.; Hwang, I. Constitutive Over-Expression of AtGSK1 Induces NaCl Stress Responses in the Absence of NaCl Stress and Results in Enhanced NaCl Tolerance in Arabidopsis. *Plant J.* **2001**, *27,* 305–314.

Qin, F.; Kakimoto, M.; Sakuma, Y.; Maruyama, K.; Osakabe, Y.; Tran, L. S.; Shinozaki, K.; Yamaguchi-Shinozaki, K. Regulation and functional analysis of ZmDREB2A in response to drought and heat stresses in *Zea mays* L, *The Plant Journal*, **2007**, *50,* 54–69.

Quan, R.; Hu, S.; Zhang, Z.; Zhang, H.; Zhang, Z.; Huang, R. Overexpression of an ERF Transcription Factor TSRF1 Improves Rice Drought Tolerance. *Plant Biotechnol. J.* **2010**, *8,* 476–488.

Rabbani, M. A.; Maruyama, K.; Abe, H.; Khan, M. A.; Katsura, K.; Ito, Y.; Yoshiwara, K.; Seki, M.; Shinozaki, K.; Yamaguchi-Shinozaki, K. Monitoring Expression Profiles of Rice Genes Under Cold, Drought, and High-Salinity Stresses and Abscisic Acid Application Using cDNA Microarray and RNA Gel-Blot Analyses. *Plant Physiol.* **2003**, *133,* 1755–1767.

Rao, S. R.; Ford, K. L.; Cassin, A. M.; Roessner, U.; Patterson, J. H.; Bacic, A. Proteomic and Metabolic Profiling of Rice Suspension Culture Cells as a Model to Study Abscisic Acid Signalling Response Pathways in Plants. *J. Proteome Res.* **2010**, *9,* 6623–6634.

Repo, T.; Mononen, K.; Alvila, L.; Pakkanen, T. T.; Hanninen, H. Cold Acclimation of Pedunculate Oak (*Quercus robur* L.) at its Northernmost Distribution Range. *Environ. Exp. Bot.* **2008**, *63,* 59–70.

Royuela, M.; Gonzalez, A.; Gonzalez, E. M.; Arrese-Igor, C.; Aparicio-Tejo, P. M.; Gonzalez-Murua, C. Physiological Consequences of Continuous, Sublethal Imazethapyr Supply to Pea Plants. *J. Plant Physiol.* **2000**, *157,* 345–354.

Ruan, S. L.; Ma, H. S.; Wang, S. H.; Fu, Y. P.; Xin, Y.; Liu, W. Z.; Wang, F.; Tong, J. X.; Wang, S. Z.; Chen, H. Z. Proteomic Identification of OsCYP2, a Rice Cyclophilin that Confers Salt Tolerance in Rice (*Oryza sativa* L.) Seedlings when Overexpressed. *BMC Plant Biol.* **2011**, *11*(34), 1–15.

Saijo, Y.; Hata, S.; Kyozuka, J.; Shimamoto, K.; Izui, K. Overexpression of a Single Ca^{2+}-Dependent Protein Kinase Confers Both Cold and Salt/Drought Tolerance on Rice Plants. *Plant J.* **2000**, *23,* 319–327.

Sakamoto, A.; Murata, N. The Use of Bacterial Choline Oxidase, a Glycinebetaine-Synthesizing Enzyme, to Create Stressresistant Transgenic Plants. *Plant Physiol.* **2001,** *125,* 180–188.

Sakamoto, A.; Murata, N. The Role of Glycine Betaine in the Protection of Plants from Stress: Clues from Transgenic Plants. *Plant Cell Environ.* **2002,** *25,* 163–171.

Salekdeh, G. H.; Siopongco, J.; Wade, L. J.; Ghareyazie, B.; Bennett, J. Proteomic Analysis of Rice Leaves During Drought Stress and Recovery. *Proteomics* **2002,** *2,* 1131–1145.

Satake, T.; Yoshida, S. High Temperature-Inducedsterility in Indica Rices at Flowering. *Jpn. J. Crop Sci.* **1978,** *47,* 6–17.

Schroeder, J. I.; Kwak, J. M.; Allen, G. J. Guard Cell Abscisic Acid Signalling and Engineering Drought Hardiness in Plants. *Nature* **2001,** *410,* 327–330.

Shimazaki, Y.; Satake, T.; Watanabe, K.; Ito, N. Effect of Day- and Night-Temperature Accompanied by Shading Treatment During the Booting Stage Upon the Induction of Sterile Spikelets in Rice Plants. (Studies of cool Weather Injuries of Rice Plants in Northern Part of Japan. IV.) [In Japanese, with English summary]. *Res. Bull. Hokkaido Natl. Agric. Exp. Sta.* **1964,** *83,* 10–16.

Shinozaki, K.; Yamaguchi-Shinozaki, K. Molecular Responses to Drought and Cold Stress. *Curr. Opin. Biotechnol.* **1996,** *7,* 61–167.

Shinozaki K, Yamaguchi-Shinozaki K. Gene Expression and Signal Transduction in Water Stress Response. *Plant Physiol.* **1997,** *115,* 327–334.

Shinozaki, K.; Yamaguchi-Shinozaki, K. Molecular Responses to Dehydration and Low Temperature: Differences and Cross Talk Between Two Stress Signalling Pathways. *Curr. Opin. Plant Biol.* **2000,** *3,* 217–233.

Shu, X. L.; Frank, T.; Shu, Q. Y.; Engel, K. H. Metabolite Profiling of Germinating Rice Seeds. *J. Agric. Food Chem.* **2008,** *56,* 11612–11620.

Smirnoff, N. Plant Resistance to Environmental Stress. *Curr. Opin. Biotechnol.* **1998,** *9,* 214–219.

Song, S. Y.; Chen, Y.; Chen, J.; Dai, X. Y.; Zhang, W. H. Physiological Mechanisms Underlying OsNAC5-Dependent Tolerance of Rice Plants to Abiotic Stress. *Planta* **2011,** *234,* 331–345.

Stamp, N. Out of the Quagmire of Plant Defense Hypotheses. *Q. Rev. Biol.* **2003,** *78*(1), 23–55.

Stockinger, E. J.; Gilmour, S. J.; Thomashow, M. F. Arabidopsis thaliana CBF1 encodes an AP2 domain-containing transcriptional activator that binds to the C-repeat/DRE, a cis-acting DNA regulatory element that stimulates transcription in response to low temperature and water deficit. Proc. Natl. Acad. Sci. USA **1997,** *94,* 1035–1040.

Tang, N.; Zhang, H.; Li, X.; Xiao, J.; Xiong, L. Constitutive Activation of Transcription Factor OsbZIP46 Improves Drought Tolerance in Rice. *Plant Physiol.* **2012,** *158,* 1755–1768.

Tao, Z.; Kou, Y.; Liu, H.; Li, X.; Xiao, J.; Wang, S. OsWRKY45 Alleles Play Different Roles in Abscisic Acid Signalling and Salt Stress Tolerance but Similar Roles in Drought and Cold Tolerance in Rice. *J. Exp. Bot.* **2011,** *62*(14), 4863–4874.

Tarpley, L.; Duran, A.; Kebrom, T.; Sumner, L. Biomarker Metabolites Capturing the Metabolite Variance Present in a Rice Plant Developmental Period. *BMC Plant Biol.* **2005,** *5,* 8.

Uno, Y.; Furihata, T.; Abe, H.; Yoshida, R.; Shinozaki, K.; Yamaguchi-Shinozaki, K. Novel Arabidopsis bZIP Transcription Factors Involved in an Abscisic-Acid-Dependent Signal

Transduction Pathway Under Drought and High Salinity Conditions. *Proc. Natl. Acad. Sci. U. S. A.* **2000,** *97,* 11632–11637.

Varshney, R. K.; Hiremath, P. J.; Lekha, P.; Kashiwagi, J.; Balaji, J.; Deokar, A. A.; Vadez, V.; Xiao, Y.; Srinivasan, R.; Gaur, P. M.; Siddique, K. H.; Town, C. D.; Hoisington, D. A. A Comprehensive Resource of Drought- and Salinity Responsive ESTs for Gene Discovery and Marker Development in Chickpea (*Cicer arietinum* L.). *BMC Genomics* **2009,** *15,* 523.

Verma, D. K.; Mohan, M.; Yadav, V. K.; Asthir, B.; Soni, S. K. Inquisition of Some Physico-Chemical Characteristics of Newly Evolved Basmati Rice. *Environ. Eco.* **2012,** *30*(1), 114–117.

Verma, D. K.; Mohan, M.; Asthir, B. Physicochemical and Cooking Characteristics of Some Promising Basmati Genotypes. *Asian J. Food Agro-Ind.* **2013,** *6*(2), 94–99.

Verma, D. K.; Mohan, M.; Prabhakar, P. K. Physico-Chemical and Cooking Characteristics of Azad Basmati. *Int. Food Res. J.* **2015,** *22*(4), 1380–1389.

Vierling, E. The Roles of Heat Shock Proteins in Plants. *Annu. Rev. Plant Physiol. Plant Mol. Biol.* **1991,** *42,* 579–620.

Wahid, A.; Gelani, S.; Ashraf, M.; Foolad, M. R. Heat Tolerance in Plants: an Overview. *Environ. Exp. Bot.* **2007,** *61,* 199–223.

Wakasa, K.; Hasegawa, H.; Nemoto, H.; Matsuda, F.; Miyazawa, H.; Tozawa, Y.; Morino, K.; Komatsu, A.; Yamada, T.; Terakawa, T. High-Level Tryptophan Accumulation in Seeds of Transgenic Rice and Its Limited Effects on Agronomic Traits and Seed Metabolite Profile. *J. Exp. Bot.* **2006,** *57,* 3069–3078.

Wan, L.; Zhang, J.; Zhang, H.; Zhang, Z.; Quan, R.; Zhou, S.; Huang, R. Transcriptional Activation of OsDERF1 in OsERF3 and OsAP2–39 Negatively Modulates Ethylene Synthesis and Drought Tolerance in Rice. *PLoS One* **2011,** *6,* e25216.

Wang, Q.; Guan, Y.; Wu, Y.; Chen, H.; Chen, F.; Chu, C. Overexpression of a rice OsDREB1F gene increases salt, drought, and low temperature tolerance in both Arabidopsis and rice, *Plant Molecular Biology,* **2008,** *67,* 589–602.

Wang, R. C.; Okamoto, M.; Xing, X. J.; Crawford, N. M. Microarray Analysis of the Nitrate Response in Arabidopsis Roots and Shoots Reveals Over 1000 Rapidly Responding Genes and New Linkages to Glucose, Trehalose-6-Phosphate, Iron, and Sulfate Metabolism. *Plant Physiol.* **2003a,** *132,* 556–567

Wang, W.; Vinocur, B.; Altman, A. Plant Responses to Drought, Salinity and Extreme Temperatures: Towards Genetic Engineering for Stress Tolerance. *Planta* **2003b,** *218,* 1–14.

Xiao, B.; Huang, Y.; Tang, N.; Xiong, L. Over-Expression of a Lea Gene in Rice Improves Drought Resistance Under the Field Conditions. *Theor. Appl. Genet.* **2007,** *115,* 35–46.

Xiao, B. Z.; Chen, X.; Xiang, C. B.; Tang, N.; Zhang, Q. F.; Xiong, L. Z. Evaluation of Seven Function-Known Candidate Genes for their Effects on Improving Drought Resistance of Transgenic Rice Under Field Conditions. *Mol. Plant* **2009,** *2,* 73–83.

Xiong, L.; Ishitani, M. Stress Signal Transduction: Components, Pathways and Network Integration. In *Abiotic Stress Tolerance in Plants Toward the Improvement of Global Environment and Food;* Rai, A. K., Takabe, T., Eds.; Springer: Netherland, 2006; pp 3–30.

Xiong, L.; Lee, M. W.; Qi, M.; Yang, Y. Identification of Defense-Related Rice Genes by Suppression Subtractive Hybridization and Differential Screening. *Mol. Plant-Microbe Interact.* **2001,** *14*(5), 685–692.

Xiong, H.; Li, J.; Liu, P.; Duan, J.; Zhao, Y.; Guo, X.; Li, Y.; Zhang, H.; Ali, J.; Li, Z. Overexpression of OsMYB48-1, a Novel MYB-Related Transcription Factor, Enhances Drought and Salinity Tolerance in Rice. *PLoS One* **2014,** *9*(3), e92913.

Xu, D. Q.; Huang, J.; Guo, S. Q.; Yang, X.; Bao, Y. M.; Tang, H. J.; Zhang, H. S. Overexpression of a TFIIIa-Type Zinc Finger Protein Gene ZFP252 Enhances Drought and Salt Tolerance in Rice (*Oryza sativa* L.). *FEBS Lett.* **2008,** *582,* 1037–1043.

Yamakawa, H.; Hirose, T.; Kuroda, M.; Yamaguchi, T. Comprehensive Expression Profiling of Rice Grain Filling-Related Genes Under High Temperature Using DNA Microarray. *Plant Physiol.* **2007,** *144,* 258–277.

Yamanouchi, U.; Yano, M.; Lin, H.; Ashikari, M.; Yamada, K. A Rice Spotted Leaf Gene, Spl7, Encodes a Heat Stress Transcription Factor Protein. *Proc. Natl. Acad. Sci. U.S.A.* **2002,** *99,* 7530–7535.

Yan, S.; Tang, Z.; Su, W.; Sun, W. Proteomic Analysis of Salt Stress-Responsive Proteins in Rice Root. *Proteomics* **2005,** *5,* 235–244.

Yang, C.; Zhao, N.; Xu, C.; Liu, B.; Shi, D. Regulation of Ion Homeostasis in Rice Subjected to Salt and Alkali Stresses. *Aust. J. Crop Sci.* **2012,** *6,* 724–731.

Yoshida, S. Tropical Climate and its Influence on Rice. IRRI Research Paper Series 20; IRRI: Los Baños, The Philippines, 1978.

Yoshida, S. *Fundamentals of Rice Crop Science;* IRRI: Los Baños, The Philippines, 1981.

You, J.; Hu, H.; Xiong, L. An Ornithine δ-Aminotransferase Gene OsOAT Confers Drought and Oxidative Stress Tolerance in Rice. *Plant Sci.* **2012,** *197,* 59–69.

Zhang, H.; Liu, W.; Wan, L.; Li, F.; Dai, L.; Li, D.; Zhang, Z.; Huang, R. Functional Analyses of Ethylene Response Factor JERF3 with the Aim of Improving Tolerance to Drought and Osmotic Stress in Transgenic Rice. *Transgenic Res.* **2010,** *19,* 809–818.

Zhou, Y., Lam, H. M.; Zhang, J. Inhibition of Photosynthesis and Energy Dissipation Induced by Water and High Light Stresses in Rice. *J. Exp. Bot.* **2007,** *58,* 1207–1217.

Zou, J.; Liu, C.; Liu, A.; Zou, D.; Chen, X. Overexpression of OsHSP17.0 and OsHSP23.7 Enhances Drought and Salt Tolerance in Rice. *J. Plant Physiol.* **2012,** *169,* 628–635.

PART II
Biochemical Trends and Advances in Rice Research

PART II

CHAPTER 5

ASSESSMENT OF AROMATIC CONTENT AND IN-VITRO RESPONSES IN TRADITIONAL INDIAN RICE VARIETIES

SRINIVASAN BALAMURUGAN[1],
SHANMUGARAJ BALA MURUGAN[1,2],
INCHAKALODY P. VARGHESE[1,3], ASHWINI MALLA[1,4],
KANTILAL V. WAKTE[5], SARIKA MATHURE[5,6],
ALTAFHUSAIN B. NADAF[5,7], and
RAMALINGAM SATHISHKUMAR[1,8,*]

[1]*Plant Genetic Engineering Laboratory, Department of Biotechnology, Bharathiar University, Coimbatore, Tamil Nadu 641046, India, Mob.: +919566878007, E-mail: bala.svm@gmail.com*

[2]*balagene3030@gmail.com, Mob.: +919003608162*

[3]*vinchakalody@gmail.com*

[4]*am08.loyola@gmail.com, Mob.: +919963978146*

[5]*Cytogenetics Laboratory, Department of Botany, Savitribai Phule Pune University, Pune, Maharashtra 411007, India, Mob.: +918888888385, E-mail: kanti999ster@gmail.com*

[6]*sarika.mathure@gmail.com, Mob.: +917774000575*

[7]*abnadaf@unipune.ernet.in; abnadaf@unipune.ac.in, Tel.: +912025601439, Mob.: +917588269987, 9822878306, Fax: +912025601439*

[8]*Mob.: +919360151669, Fax: +914222422387*

Corresponding author. E-mail: rsathish@buc.edu.in

5.1 INTRODUCTION

Rice is the staple food for more than half of the world's population, which provides 27% of caloric need and 20% of the dietary protein intake (Fresco, 2005; Verma and Srivastav, 2017). About 90% of rice production and consumption were reported in Asian countries (Dutta et al., 2012). Giri and Vijayalakshmi (2000) reported that by the year 2050 the world population is estimated to be approximately 11 billion people out of which 90% of the people reside in the developing countries of the South. As rice is considered as the staple food for the tropical and subtropical region people, the increase in rice yield is a must especially in Asia (Cha-um et al., 2009). In India, there are lots of indigenous traditional rice varieties which are economically valuable due to properties like aroma, medicinal properties, grain yield, short duration, and so forth. The fragrant or aromatic rice are the special varieties that emit delicious fragrance when cooked (Itani et al., 2004). Due to its pleasant aroma, it is preferred over non-aromatic rice varieties (Joshi and Behera, 2006; Nadaf et al., 2006). Aroma content of the rice is the primary selection criteria employed for its assessment of quality (Reinke et al., 1991; Verma and Srivastav, 2016). Most of the aromatic rice varieties are cultivated in the regions of India, Thailand, and Pakistan. Basmati rice in India is considered to be one of the most widely used aromatic rice across the world due to its exquisite aroma. In India, there are many indigenous aromatic rice varieties apart from Basmati, which are small, medium grained and good in fragrance (Nadaf et al., 2006). Gandhakasala and Jeerakasala are fine white rice varieties grown in South India which are believed to contain fragrance and also having the high grain yield of 2743 and 2179 kg/ ha respectively. Gandhakasala and Jeerakasala are traditionally believed to be aromatic rice varieties (George et al., 2005). Due to difficulty in cultivation and lack of scientific data, these valuable rice genotypes are in the verge of extinction though it has many special traits, especially aroma, medicinal properties, and so forth. Buttery et al. (1983a) reported that there are more than 114 volatile compounds responsible for aroma in the aromatic rice varieties. Among these volatile compounds, 2-acetyl-1-pyrroline (2AP) is the principal compound which imparts delicious popcorn-like flavor to the rice varieties which is mostly controlled by the loss function of betaine aldehyde dehydrogenase gene (*badh2* of *fgr*) on rice chromosome 8. The non-aromatic rice also contains 2AP, but 10–100 times lower than that of the aromatic variety (Buttery et al., 1983a; Laksanalamai and Ilangantileke, 1993). A recessive gene (*fgr* gene) located on exon 7 of the Chromosome

8 is proved to be responsible for the rice fragrance. Therefore, the 2AP analysis, by both quantitative and qualitative analysis, is essential to identify the rice variety as aromatic or non-aromatic in order to reveal its economic importance. In vitro callus induction and plant regeneration are the essential prerequisite for crop improvement, conservation and genetic transformation studies (Bhaskaran and Smith, 1988). The effective callus formation and regeneration is considered as variety specific (Saharan et al., 2004) and the efficient regeneration in *indica* rice varieties is still a major problem as rice is considered as recalcitrant (Sathishkumar and Manoharan, 1996) which limits the genetic transformation studies towards those rice varieties. Although a copious literature reporting the in vitro regeneration from different explants exists (Karthikeyan et al., 2009), very little is reported about the efficient and high-frequency regeneration particularly for these rare Indian rice varieties. Establishment of the in vitro system is necessary for both crop improvement and to conserve these valuable natural resources from extinction. In this report, the 2AP content was analyzed by sensory analysis, histochemical staining, quantified by gas chromatography (GC) method and an efficient methodology for the in vitro regeneration is optimized which can be applied for the improvement of these rare cultivars with medicinal applications and also ensures the protection of these valuable rice varieties from the verge of extinction.

5.2 MATERIALS AND METHODS

Mature seeds of *indica* rice cvs. Gandhakasala and Jeerakasala were procured from Wayanad, India. Mature seeds were dehusked and surface sterilized consecutively for 3 minutes with 70% ethanol followed by 0.1% mercuric chloride treatment for 2 minutes. After several rounds of washing in sterile water these are finally blotted dry. These surface sterilized seeds were used as explants for in vitro experiments.

5.2.1 SENSORY EVALUATION

Sensory aroma was analyzed according to the method described by Sood and Siddiq (1978) by a trained panel, comprising of 15 university students and the ranking was done according to the method described by Therakulkait et al. (2009) with little modification using 0–3 point scale (0—no aroma; 1—subtle aroma; 2—clear aroma; 3—strong aroma).

5.2.2   QUANTIFICATION OF 2-ACETYL-1-PYRROLINE

Quantification of 2AP using HS-SPME/GC-FID in rice varieties was performed by standard addition approach (Nadaf, 2009, Personal Communication). Extraction of aroma volatiles was performed in 4 mL screw cap vials (15×45 mm) with PTFE silicon septa (Chromatography research supplies, Louiseville, KY, USA). One gram rice containing 300 µL of odour-free water was pre-incubated at 80°C for 30 min followed by 20 min adsorption with 1 cm long fiber coated with Carboxen/Divinyl-benzene/Poly-dimethyl-siloxane (CAR/DVB/PDMS). GC (Shimadzu 17 A, Japan) with BP-20 capillary column (30 m×0.32 µm) and flame ionization detector (FID) was used to separate and to analyze the head-space volatiles from rice. For this, splitless injections were carried out at 250°C using SPME manual holder assembly (57330-U) equipped with commercially available SPME fiber (Supleco, Bellefonte, PA, USA). The GC oven program was, 1 min hold at 50°C, ramped to 100°C at the rate of 4°C/min and further ramped to 240°C at the rate of 50°C/min with a final hold of 2 min.

5.2.3   HISTOCHEMICAL STAINING OF 2-ACETYL-1-PYRROLINE

The rice seeds were soaked in the water overnight then dehusked manually and thin longitudinal sections were taken using the razor blade. The thin sections were treated in alcohol gradients (10, 20, 30, 50, 70, 90% and absolute alcohol) and transferred to the 2,4-dinitrophenylhy-drazine (DNPH) reagent. The sections were incubated in the hot air oven at 600 C for 30 minutes. After the incubation, the sections were photographed using the microscope (Olympus) at 400X magnification (Nadaf et al., 2006).

5.2.4   IN VITRO STUDIES

5.2.4.1   CALLUS INDUCTION

For callus induction, surface sterilized seeds were inoculated in the Murashige and Skoog basal medium supplemented with 0.8% agar with various concentrations of 2,4 Dichlorophenoxyaceticacid (2,4-D 1.0, 1.5,

2.0, 2.5, 3.0, 3.5 and 4.0 mg · L^{-1}) were tested for callus induction. The cultures were maintained at $25 + 2°C$ and $55 + 5\%$ relative humidity under continuous light (1500 lx) provided by cool day light fluorescent tubes for 16 h/day.

Twenty-five seeds were inoculated in each petri dish as triplicates and percentage of callus induction efficiency was calculated by using the formula given below.

$$\text{Percentage of Callus Induction } (\%) = \frac{\text{Number of Calli induced}}{\text{Number of Seeds Inoculated}} \times 100$$

5.2.4.2   PLANT REGENERATION AND HARDENING

The calli were sub-cultured every 2 weeks in the same media for callus proliferation. After 4 weeks, the callus in the plates was allowed to grow under dark condition till the induction of embryogenic callus. The mature seed-derived callus were transferred to the MS medium supplemented with various hormonal combinations, (a) NAA (0.5 mg · L^{-1}) with Kinetin (KIN 2 mg · L^{-1}), (b) NAA (2.5 mg · L^{-1}) with BAP (0.5 mg · L^{-1}), (c) Kinetin (2 mg · L^{-1}) with IAA (1 mg · L^{-1}), (d) BAP (2 mg · L^{-1}), Tryptone (1 g · L^{-1}) with Kinetin (2 mg · L^{-1}), (e) IAA (1 mg · L^{-1}) with Kinetin (4 mg · L^{-1}) and (f) NAA (0.5 mg · L^{-1}) with BAP (0.5 mg · L^{-1}) and cultures were maintained in the condition mentioned above. After the shoot and root initiation from callus, the shoots were separated and subcultured. The regenerated plantlets with a height of 6 cm were transferred to the pot containing a mixture of sterile soil and coir pith (1:1) was maintained at 20°C in growth chamber and was gradually hardened by controlling humidity and temperature then finally hardened in the soil. In plant regeneration, 10 embryogenic calli explants per treatment were used as triplicate.

5.2.5   STATISTICAL ANALYSIS

Each experiment was repeated three times. The data obtained were subjected to one-way analysis of variance (ANOVA) followed by Duncan's multiple range test (DMRT) at $p < 0.05$ using SPSS V13.0.

5.3 RESULTS

5.3.1 SENSORY ANALYSIS

The sensory evaluation proved that Basmati rice (2.33) is good in fragrance as expected. Among the traditional varieties, Gandhakasala have high aroma rank (1.80) than Jeerakasala (0.86) and negative control ADT 43 (0.41) which showed negligible quantity of aroma (Fig. 5.1).

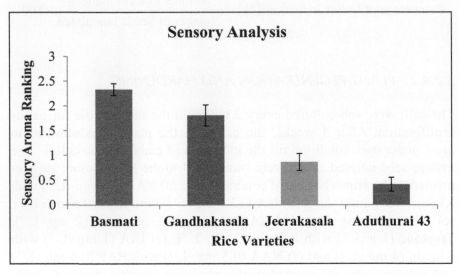

FIGURE 5.1 **(See color insert.)** Sensory evaluation of aroma in the four rice varieties. Data represented the mean value ± standard error of 15 panelists as triplicates.

5.3.2 QUANTIFICATION OF 2-ACETYL-1-PYRROLINE

The quantification of the 2AP was done by GC technique in four rice varieties and the results are represented in Figure 5.2. Gas chromatogram showed many peaks which reveal the presence of more volatile compounds in the rice seeds (Data not shown). 2AP was eluted between the retention time (RT) of 6.98–7.02 minutes. The peak area was measured and 2AP content was quantified which were expressed in parts per billion (ppb) in which positive control Basmati 370 showed highest followed by Gandhakasala where Jeerakasala and Aduthurai 43 showed very low 2AP content.

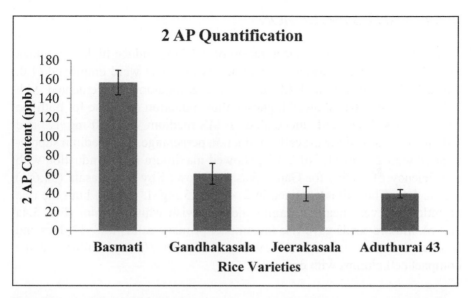

FIGURE 5.2 (See color insert.) Quantification of 2AP in four rice varieties. Data represented the mean ± standard error from three experiments.

5.3.3 HISTOCHEMICAL STAINING OF 2-ACETYL-1-PYRROLINE

Microscopic study of transverse section of rice seeds stained with DNPH revealed that Basmati 370, and Gandhakasala have more orange red spots (2-acetylphenyl hydrazone) when observed under 400X, indicating the presence of 2AP (Fig. 5.3). This has confirmed the presence of 2AP in rice seeds.

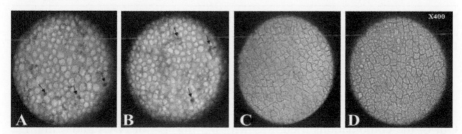

FIGURE 5.3 (See color insert.) Histochemical staining of 2AP in rice using DNPH reagent. Arrow heads indicate the presence of 2AP. (A) Basmati 370, (B) Gandhakasala, (C) Jeerakasala, and (D) Aduthurai 43.

5.3.4 CALLUS INDUCTION

To identify the optimum concentration of 2,4-D to induce high percentage of callus formation, various concentrations of 2,4-D were employed (1.0, 1.5, 2.0, 2.5, 3.0, 3.5, and 4.0 mg · L^{-1}). MS medium supplemented with 2.5 mg · L^{-1} of 2,4-D showed highest callus induction response for both the rice varieties. Rice seeds inoculated on MS medium with 1.0 mg · L^{-1} and 4.0 mg · L^{-1} showed lowest callus induction percentage. MS medium supplemented with 2.5 mg · L^{-1} of 2,4-D, showed maximum callus induction with a efficiency of 70.6% for Gandhakasala followed by Jeerakasala (54.6%) (Table 5.1). The callus obtained in 2, 4-D (2.5 mg · L^{-1}) from both the rice varieties was very healthy, fragile and yellowish white in color (Fig. 5.4a and b). Mature seeds gave rise to compact primary callus in 8–12 days and friable secondary callus in 16–19 days and after 30 days callus had the compact cell clumps with embryos.

TABLE 5.1 Effect of 2,4-D Concentration on Callus Induction from the Rice Seeds Var. Gandhakasala and Jeerakasala.

S. N.	Concentration of 2,4-D (mg · L^{-1})	Percentage of callus induction (%)	
		Gandhakasala	Jeerakasala
1	1.0	37.3±3.5[d]	29.3±1.3[e]
2	1.5	42.6±1.3[c]	33.3±1.3[d]
3	2.0	57.3±3.5[b]	41.3±4.8[b]
4	2.5	70.6±2.6[a]	54.6±2.6[a]
5	3.0	62.6±4.6[b]	48.0±1.6[b]
6	3.5	57.3±3.5[b]	42.6±3.5[b]
7	4.0	52.0±2.3[b]	37.3±1.3[c]

Note: Data represented as the mean value ± SE (n=3). Mean followed by same letters in a column are not significantly different at p<0.05 level as determined by ANOVA followed by Duncan's multiple range test.

5.3.5 PLANT REGENERATION

Out of the combinations tried, MS with 0.5 mg · L^{-1} of NAA and 2 mg · L^{-1} of Kinetin showed better plant regeneration for both the rice varieties Gandhakasala and Jeerakasala after 30 days of inoculation (Fig. 5.4c and d). In the above concentrations, both the root and shoot regenerated at the same time. Even though MS supplemented with NAA (2.5 mg · L^{-1}) and BAP (0.5 mg · L^{-1})

gave green callus, but there was no regeneration of shoot or root. Other tested combination like MS with BAP (2.0 mg·L⁻¹) + KIN (2.0 mg·L⁻¹) + Tryptone (1 g·L⁻¹) induced rapid proliferation of callus but no root or shoot differentiation was observed. The necrosis was observed in the callus inoculated in MS+ NAA (0.5 mg·L⁻¹) with BAP (0.5 mg·L⁻¹) and dried slowly with rooting. However in this study, mature seed-derived callus was found to be compact for the efficient regeneration. The regenerated plantlets after attaining a height of 6 cm were transferred to the pot, which contained mixture of sterilized soil, sand and coir pith finally hardened to grow in soil (Fig. 5.4e and f).

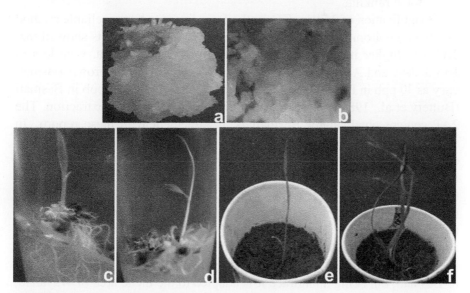

FIGURE 5.4   (See color insert.) Stages of regeneration of rice varieties Gandhakasala and Jeerakasala from seed derived embryogenic calli. (a and b) 4 weeks old Callus of var. Gandhakasala and Jeerakasala; (c and d) Development of shooting and rooting; (e and f) Hardened plants of Gandhakasala and Jeerakasala.

5.4   DISCUSSION

Among the 114 volatile compounds detected, 2AP is the principal compound responsible for the fragrance in rice (Buttery et al., 1983a). Due to the pleasant aroma, aromatic rice varieties command higher price and demand in the local and international market. In the present study, qualitative and quantitative analyses of 2AP were performed in the rare rice varieties Gandhakasala and Jeerakasala in order to reveal its economic importance. Additionally an efficient in vitro system was optimized for these varieties for its trait development. Fragrance content in the rice is the primary selection method employed for its assessment of quality (Reinke et al., 1991). Nagaraju et al. (1975) and Sood and Siddiq (1978) reported that sensory evaluation technique is a rapid method for determination of fragrance. This method has some draw backs when large volume of samples has to be analyzed. The detection sense is not unique between all the panelists and the solution consists of KOH or I^2-KI which can cause damage to the nasal passages (Cordeiro et al., 2002). From the results obtained, it is proved that traditional aromatic variety Gandhakasala is a promising variety due to its high aroma ranking.

Quantification of 2AP by GC method was found to be the reliable method to assess the aromatic content in the rice varieties. The results showed that 2AP was higher in Basamti 370, moderate in Gandhakasala, very less in Jeerakalasa and Aduthurai 43. Reports on 2AP content of brown Basmati vary as 30 ppb in Pusa Basmati (Nadaf et al., 2006) and 170 ppb in Basmati (Buttery et al., 1983b) by steam distillation and continuous extraction. The 2AP content obtained for Basmati is in agreement with these reports. In study, it was observed that 2AP content of Jeerakasala was equivalent to that of non-scented cultivar Aduthurai 43. The aroma of scented rice samples could have reduced during their storage prior to the analysis rendering mild scented Jeerakasala close to non-scented rice. Such effect on scented rice resulting in reduction in 2AP content by drying methods and storage conditions was revealed by Wongpornchai et al. (2004) in Khao Dawk Mali 105 and confirmed by Borompichaichartkul et al. (2007) using Thai Jasmine rice. Gandhakasala rice possessed moderate content of 2AP as compared to Basmati, but being a landrace it is adapted to the niche area of Wayanad, South India, it could be utilised for breeding and development of aromatic rice. Histochemical staining of 2AP in rice seeds revealed that the fragrant rice varieties contain the considerable amount of 2AP which has been visualized in the form of red spots, while the non-fragrant varieties contain minimal amount of red spots indicating the lower amount of 2AP. The intensity of red spots is directly proportional to the amount of 2AP present in the rice seeds. Nadaf et al. (2006) reported the principle behind in the DNPH staining that, 2,4-dinitrophenylhydrazine reacts with active methyl ketone group of 2AP to form orange red color compound 2-acetylphenyl hydrazone, which can be visually assessed. Nadaf et al. (2006) have also reported that aleurone layer of the rice seed contains more 2AP than the endosperm.

The callus induction is considered as the most important initial step for plant genetic transformation. Callus induction frequency was highly varied between the rice varieties and hormonal combinations tested. 2,4-D is the hormone, which is commonly used for the callus induction medium (Raina, 1989; Zhenyu et al., 1999; Saharan et al., 2004). Higher concentration of 2,4-D turned the callus brown. Zaidi et al. (2006) reported in-vitro response in indica rice is genotype specific. The callus induction in rice is considered as the important step and 2,4-D is the major growth factor required for the callus induction (Saharan et al., 2004). Noor et al. (2005) reported the effective callus initiation occurs in 2 mg $\cdot$ L^{-1} (2,4-D) in N$_6$ medium. Karthikeyan et al. (2009) also found that Linsmaier and Skoog's medium supplemented with 2.5 mg $\cdot$ L^{-1} of 2,4-D for ADT 43. In this study, we have optimized the hormonal combination for efficient callus induction for the rare India rice varieties Gandhakasala and Jeerakasala, which would be very useful for the genetic improvement studies. The efficient plant regeneration is the ultimate prerequisite for genetic manipulation. Hence, it is necessary to find out the optimum media and hormonal concentration for the efficient regeneration system. Gandhakasala showed fast response in the regeneration medium than Jeerakasala in the above said hormonal combinations. Zaidi et al. (2006) found that parameters such as carbon source, growth regulators and agar concentration play a key role in the plant regeneration. The selection of the explants for the plant regeneration is another critical factor for the regeneration of the plants. Dong et al. (1996) used immature derived callus and Hiei et al. (1994), Toki (1997) and Rashid et al. (2001) used callus derived for scutellum of mature seeds.

5.5 CONCLUSION

In conclusion, 2AP analysis revealed that Gandhakasala as aromatic rice variety for the first time which implied its economic importance. Besides, an efficient in vitro system was optimized for high-frequency callus induction and plant regeneration through seed-derived embryogenic callus which is the essential prerequisite for the future genetic manipulation of these rare Indian traditional rice varieties. Taken together, this report will provide further insight into the conservation of the rare traditional rice varieties from the verge of extinction.

5.6 SUMMARY

Gandhakasala and Jeerakasala are the rare Indian traditional rice varieties grown in South India, which are known for their fragrance and higher yield, but are difficult to grow. Hence, this study was undertaken to develop the efficient and reliable protocol for in vitro callus induction, plant regeneration and also to assess the aromatic compound 2AP and also to develop the efficient and reliable protocol for in vitro callus induction, plant regeneration in two rare Indian rice varieties, Gandhakasala and Jeerakasala. 2AP content was extracted and quantified using HS-SPME-GC-FID; qualitatively assessed by sensory evaluation, histochemical staining in Gandhakasala and Jeerakasala seeds with Basmati 370 and Aduthurai 43 as control. Different concentrations of auxin and different combinations of both auxins and cytokinins were fortified with MS medium in order to optimize the callus induction medium and plant regeneration medium respectively. The GC results revealed that 2AP content was found to be higher in Basmati 370 (156.7 ppb) followed by Gandhakasala (60.52 ppb), Aduthurai 43 (39.58 ppb), and Jeerakasala (39.53 ppb). Sensory analysis and histochemical staining proved qualitatively that Gandhakasala has significant aroma like Basmati 370 whereas Jeerakasala and Aduthurai 43 does not. The maximum callus induction was obtained in Murashige and Skoog (MS) medium fortified with 2,4-D (2.5 mg $\cdot$ L^{-1}) for both Gandhakasala (70.60%) and Jeerakasala (54.6%). Plant regeneration from the seed-derived callus was obtained in the MS media with Kinetin (2.0 mg $\cdot$ L^{-1}) and NAA (0.5 mg $\cdot$ L^{-1}) for both Gandhakasala and Jeerakasala. This study proved that the Gandhakasala is an aromatic rice variety for the first time by both the means of quantitative and qualitative analysis. In addition, in vitro conditions were also optimized for these varieties which will be useful for the crop improvement by biotechnology means having many pharmaceutical applications.

KEYWORDS

- 2-acetyl-1-pyrroline
- aromatic rice varieties
- betaine aldehyde dehydrogenase
- callus formation and induction
- crop improvement
- extraction of aroma volatiles

- flame ionization detector
- gas chromatography
- genetic transformation
- high frequency plant regeneration
- recessive gene
- seed-derived callus
- volatile compounds

REFERENCES

Bhaskaran, S.; Smith, R. H. Enhanced Somatic Embryogenesis in *Sorghum biocolor* L. from Shoot Tip Culture. *In Vitro Cell. Dev. Biol. Plant* **1988**, *24*, 65–70.

Borompichaichartkul, C.; Wiset, L.; Tulayatun, V.; Tuntratean, S.; Thetsupamorn, T.; Impaprasert, R.; Waedalor, I. Comparative Study of Effects of Drying Methods and Storage Conditions on Aroma and Quality Attributes of Thai Jasmine Rice. *Drying Technol.* **2007**, *25*, 1185–1192.

Buttery, R. G.; Juliano, B. O.; Ling, L. C. Identification of Rice Aroma Compound 2-Acetyl-1-Pyrroline in Pandan Leaves. *Chem. Ind.* (London), **1983a**, 478.

Buttery, R. G.; Ling, L. C.; Juliano, B. O.; Turnbaugh, J. G. Cooked Rice Aroma and 2-Acetyl-1-Pyrroline. *J. Agric. Food Chem.* **1983b**, *31*, 823–826.

Cha-um, S.; Srianan, B.; Pichkum, A.; Kridmanee, C. An Efficient Procedure for Embryogenic Callus Induction and Double Haploid Plant Regeneration Through Anther Culture of Thai Aromatic Rice (*Oryza sativa* L. Subsp. Indica). *In Vitro Cell. Dev. Biol. Plant* **2009**, *45*, 171–179.

Cordeiro, G. M.; Christopher, M. J.; Henry, R. J.; Reinke, R. F. Identification of Microsatellite Markers for Fragrance in Rice by Analysis of the Rice Genome Sequence. *Mol. Breed.* **2002**, *9*, 245–250.

Dong, J.; Teng, W.; Buchholz, W. G.; Hall, T. C. Agrobacterium Mediated Transformation of Javanica Rice. *Mol. Breed.* **1996**, *2*, 267–276.

Dutta, A. K.; Gope, P. S.; Banik, S.; Makhnoon, S.; Siddiquee, M. A.; Kabir, Y. Antioxidant Properties of Ten High Yielding Rice Varieties of Bangladesh. *Asian Pac. J. Trop. Biomed.* **2012**, *2*, S99–S10.

Fresco, L. Rice is Life. *J. Food Compos. Anal.* **2005**, *18*, 249–253.

George, S. P.; Bastian, D.; Radhakrishnan, S. V.; Aipe, K. C. Evaluation of Aromatic Rice Varieties in Wayanad, Kerala. *J. Trop. Agric.* **2005**, *43*, 67–69.

Giri, C. C.; Vijayalaxmi, G. Production of Transgenic Rice with Agronomically Useful Genes: An Assessment. *Biotechnol. Adv.* **2000**, *18*, 653–683.

Hiei, Y. S.; Ohta, S.; Komari, T.; Kumanshiro, T. Efficient Transformation of Rice (*Oryza sativa* L.) Mediated by *Agrobacterium* and Sequence Analysis of the Boundries of the T-DNA. *Plant J.* **1994**, *6*, 271–282.

Itani, T.; Tamaki, M.; Hayata, Y.; Fushimi, T.; Hashizume, K. Variation of 2-Acetyl-1-Pyrroline Concentration in Rice Grains Collected in the Same Region in Japan and Factors Affecting Its Concentration. *Plant Prod. Sci.* **2004**, *7*, 178–183.

Joshi, R. K.; Behera, L. Identification and Differentiation of Indigenous Non-Basmati Aromatic Rice Genotypes of India Using Microsatellite Markers. *Afr. J. Biotechnol.* **2006**, *6*, 348–354.

Karthikeyan, A.; Karuthapandiyan, S.; Ramesh, M. High Frequency Plant Regeneration from Embryogenic Callus of a Popular *Indica* Rice (*Oriza sativa* L.). *Physiol. Mol. Biol. Plants* **2009**, *15*, 371–375.

Laksanalamai, V.; Ilangantileke, S. Comparison of Aroma Compound (2-Acetyl-1-Pyrroline) in Leaves from Pandan (*Pandanus amaryllifolius*) and Thai Fragrant Rice (Khao Daawk Mali-105). *Cereal Chem.* **1993**, *70*, 381–384.

Nadaf, A. B.; Wakte, K. V.; Krishnan, S. Histochemical and Biochemical Analysis of Major Aroma Compound (2-Acetyl-1-Pyrroline) in Basmati and other Scented Rice (*Oryza sativa* L). *Curr. Sci. India*, **2006**, *91*, 1533–1536.

Nagaraju, M.; Chaudhary, D.; Balkrishna Rao, M. J. A Simple Technique to Identify Scent in Rice and Inheritance Pattern of Scent. *Curr. Sci. India* **1975**, *44*, 599.

Noor, A.; Rashid, H.; Chaudhry, Z.; Mirza, B. High Frequency Regeneration from Scutellum Derived Calli of Basmati Rice cv. Basmati 385 and Super Basmati. *Pak. J. Bot.* **2005**, *37*, 673–684.

Raina, S. K. Tissue Culture in Rice Improvement: Status and Potential. *Adv. Agron.* **1989**, *42*, 339–398.

Rashid, H. S.; Bokhari, Y. A.; Quraishi, A. Callus Induction, Regeneration and Hygromycin Selection of Rice (Super Basmati). *OnLine J. Biol. Sci.* **2001**, *1*, 1145–1146.

Reinke, R. F.; Welsch, L. A.; Reece, J. E.; Lewin, L. G.; Blakeney, A. B. Procedure for Quality Selection of Aromatic Rice Varieties. *Int. Rice Res. Newsl.* **1991**, *16*, 10–11.

Saharan, V.; Yadav, R. C.; Yadav, N. R.; Chapagain, B. P. High Frequency Plant Regeneration from Desiccated Calli of Indica Rice (*Oryza sativa* L.). *Afr. J. Biotechnol.* **2004**, *3*, 256–259.

Sathishkumar, R.; Manoharan, K. Lipid Changes Due to Growth-Factor Supplements in Callus and Plasma Membrane-Enriched Fraction of Rice Cultures. *Phytochemistry* **1996**, *43*, 1171–1174.

Sood, B. C.; Siddiq, E. A. A Rapid Technique for Scent Determination in Rice. *Indian J. Genet. Plant Breed.* **1978**, *38*, 268–271.

Therakulkait, C.; Kaewka, K.; Cadwallader, K. R. Effect of Preparation Conditions on Composition and Sensory Aroma Characteristics of Acid Hydrolyzed Rice Bran Protein Concentrate. *J. Cereal Sci.* **2009**, *50*, 56–60.

Toki, S. Rapid and Efficient *Agrobacterium*-Mediated Transformation in Rice. *Plant Mol. Biol. Rep.* **1997**, *15*, 16–21.

Verma, D. K.; Srivastav, P. P. Extraction Technology for Rice Volatile Aroma Compounds In *Food Engineering Emerging Issues, Modeling, and Applications,* (as part of book series on Innovations in Agricultural and Biological Engineering)*;* Meghwal, M., Goyal, M. R., Eds.; Apple Academic Press: USA, 2016; Vol. 2

Verma, D. K.; Srivastav, P. P. Proximate Composition, Mineral Content and Fatty Acids Analyses of Aromatic and Non-Aromatic Indian Rice. *Rice Sci.* **2017**, *24*(1), 21–31.

Wongpornchai, S.; Dumri, K.; Jongkaewwattana, S.; Siri, B. Effects of Drying Methods and Storage Time on the Aroma and Milling Quality of Rice (*Oryza sativa* L.) cv. Khao Dawk Mali 105. *Food Chem.* **2004**, *87*, 407–414.

Zaidi, M. A.; Narayanan, M.; Sardana, R.; Taga, I.; Postek, S.; Johns, R.; McNulty, M.; Mottiar, Y.; Mao, J.; Loit, E.; Altosaar, U. Optimizing Tissue Culture Media for Efficient Transformation of Different Indica Rice Genotypes. *Agron Res.* **2006**, *4*, 563–572.

Zhenyu, G.; Gaozy, H. D.; Huang, D. N. Some Factors Influencing Callus Formation and Plant Regeneration in *Indica* Rice Varieties. *Plant Physiol.* **1999**, *35*, 113–115.

CHAPTER 6

BIOCHEMICAL EVALUATION OF IRRIGATED FLOODED TRANSPLANTED AND AEROBIC RICE (*ORYZA SATIVA L.*): A REVIEW

MANISHA KUMARI[1], BAVITA ASTHIR[1,2,*], DEEPAK KUMAR VERMA[3], and VISHAL SINGH[4]

[1]*Department of Biochemistry, Collage of Basic Sciences and Humanities, Punjab Agriculture University, Ludhiana, Punjab 141004, India, E-mail: manishabindra@gmail.com*

[2]*Mobile: +919216292388, Tel.: +911612562967*

[3]*Department of Agricultural and Food Engineering, Indian Institute of Technology, Kharagpur, West Bengal 721302, India, Mob.: +917407170260, Tel.: +913222281673, Fax: +913222282224, E-mail: deepak.verma@agfe.iitkgp.ernet.in, rajadkv@rediffmail.com*

[4]*Department of Agricultural Engineering, M. S. Swaminathan School of Agriculture, Centurion University of Technology and Management, Odisha, India, Mobile: +918093872582; +918348521736, E-mail: vishalsinghiitkgp87@gmail.com; vishalsinghiitkgp@cutm.ac.in*

Corresponding author. E-mail: b.asthir@rediffmail.com; basthir@pau.edu

6.1 INTRODUCTION

Rice (*Oryza sativa* L.) is the staple food of more than 60% of the world's population. It belongs to genus *Oryza* of Gramineae family. About 90% of all rice grown in the world is produced and consumed in the Asian region. Rice is grown under diverse conditions but submerged in water is the most common method used worldwide. Cultivated rice is generally considered a semi-aquatic annual grass. It has a main stem and a number of tillers. Each productive tiller bears a terminal flowering head or panicle. Plant height varies by variety to variety and environmental conditions, ranging from approximately 0.4 to 5 m in some floating areas. The growth duration of the rice is become 3–6 months, depending on the variety and the environment under which it is grown. During this time, rice completes two distinct growth phases: vegetative and reproductive, depicted in Figure 6.1 described by Brouwer et al. (1989). The vegetative phase is sub-divided into germination, early seedling growth, and tillering; the reproductive phase is subdivided into the time before and after anthesis. The time after anthesis is better known as the ripening period (Rahman and Zhang, 2013). Potential grain yield is primarily determined before anthesis. It develops after pollination and fertilization are completed. Rice grain contains about carbohydrates (80%), protein (7.12%), sugars (0.12%), dietary fiber (1.3%), water (11.62%) and traces of thiamine, riboflavin, niacin, pantothenic acid, vitamin B_6, folate, calcium, iron, magnesium, manganese, phosphorous, potassium, and zinc (Verma and Srivastav, 2017). Rice has a potential to improve nutrition, boost food security, foster rural development, and support sustainable land acre (Patel et al., 2010).

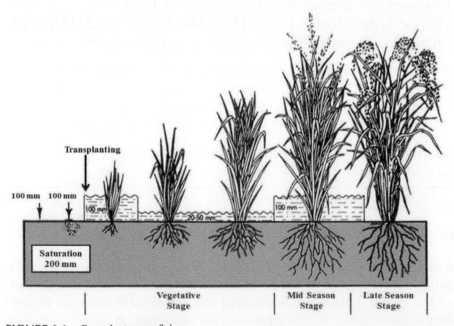

FIGURE 6.1   Growth stages of rice.

Rice serves as an important food crop for the majority of world population nevertheless; the demand for rice production is still rising because of the continuous increase in population (Verma et al., 2012, 2013, 2015). Rice productivity has to be increased by 3% per annum for feeding the future population and fulfillment of present consumption demand (Ke et al., 2009; Pyngrope et al., 2013). Rice is particularly sensitive to water stress as 50% of the world rice production is affected more or less by drought (Bouman et al., 2006a; Yang et al., 2008). Due to intensive agriculture approaches and high water consumption in farming besides industries and other purposes, ground water level is reducing rapidly causes scarcity of qualitative water. Area of rice cultivation and its production affected as per availability of water (e.g., in low rainfall regions the area of rice cultivation and production both are stressed). Water scarcity at vegetative stage significantly reduces tiller number per plant and at flowering stage great reduction in yield from the reduction in fertile panicle and filled percentage as compared to other stages of stress (Zulkarnain et al., 2013).

Due to less efficiency to avoid transpiration losses, rice crop is more sensible to water availability rather than other cereals (Sikuku et al., 2010). Worldwide rice productivity is being threatened by increased endeavors of water deficit stress. Under water-limited conditions, plant stops growing quickly due to drying of soil, so decline in root and shoot length, dry weight, chlorophyll content along with production of reactive oxygen species (ROS) was observed (Farooq et al., 2009). One of the best approaches for stress tolerance is to identify the cultivars having strong antioxidant system so as to survive and perform better under water deficit conditions. That is why the less water resisting varieties of rice has been identified and promoted in rain fed and less irrigated regions for attaining sufficient food for them (Farooq et al., 2011). Some of water stress favored rice varieties which adaptable in that environment, are practiced in particular water scary regions and achieved good yield (Kato and Katsura, 2014).

6.2   AEROBIC AND TRANSPLANTED RICE

Rice is a semi-aquatic crop that requires eminent amount of water for its proper growth and development. It serves as an important food crop for the majority of world population (Sikuku et al., 2010). Global rice consumption will rise from 439 million t in 2010–496 million t in 2020 and further increase to 555 million t in 2035 (Seck et al., 2012). Demand of food for entire world is accomplish up to maximum extent because of green revolution but some other scientific solution should be find out to feed the uncontrolled and rapid growing population. Rice system consumes 400,000 million L of water that is about 66% of irrigation water which resulted in sharp decline in underground water during the last two decades so, threatens the sustainability of traditional rice production (Lal et al., 2013). Due to decline in water table, an argent need has been emerged to intensive and serious study about response of paddy crop in the terms of environmental prospects because several environment disturbances are appearing in a serious form and threatening the bio-lives, crop diversity and agricultural climate in different way like drought and flood, and so forth, due to excess intervention of humans in ecology of nature in the form of exploitation of ground water, intensive use of chemical fertilizer and hybrid seeds, and so forth, Sustainable and progressive farming depends on preservation of ground water and rise in water table to arrange secure irrigation facility throughout the year, so it is necessary to utilize the ground water with high efficiency for good production of rice (Lal et al., 2013).

Conventional rice is called anaerobic rice because oxygen is absent due to flooding by irrigation and it grows well in standing water with water holding capacity (WHC) of about 40–50% throughout a growing season. Developed and good varieties of aerobic rice are cultivated in non-submerged and un-puddle but aerated soil structures as well as it's sown in non-saturated soil (WHC varies between 70–100%) also (Kreye et al., 2009; Vanitha and Mohandass, 2014). In case of Aerobic rice crop, appropriate soil moisture is required rather than standing water in other variety, so farmers can prevent to use of excess water and can increase the water efficiency up to 30% (Farooq et al., 2011), therefore, substantial water savings are possible from aerobic rice. Anaerobic rice fields differ physically from aerobic rice by a layer of standing water, leveling of fields and growth of plants in nursery. Morphologically, these differences mainly account in development of roots and their length; as in aerobic seminal roots develops first and have deep, extensive root system though coleoptiles develops first in transplanted (Ahmadi et al., 2007). The over exploitation of water by anaerobic type threatens the sustainability of our ecosystem. Therefore, the traditional practice of rice cultivation needs to be re-examined as water is becoming increasingly scarce (Zulkarnain et al., 2009). Present scenario of environment and limited water sources emphasize on adoption of less water resist rice variety without hampering the crop yield for making rice cultivation in such regions more viable and more ecofriendly. So there is urgent need to develop low water requiring rice varieties (aerobic rice) to sustain rice productivity and to conserve natural ground water reserve.

Germination, plant establishment, biomass formation, and grain filling are mainly dependent on water and less water results in drastic reduction in yield by interrupting floret initiation (which cause spikelet sterility) and grain filling during vegetative growth, flowering, and terminal period of rice cultivation, respectively. Free water level from 3 weeks before anthesis until maturity is closely related to grain yield in rain fed lowland rice (Fukai et al., 2008). Grain quality and filling is directly associated with biological aging process of plants. Generally, duration of grain filling influenced with plant senescence and reveals that scary of water causes the short duration of grain filling period and early biological aging but increases remobilization of assimilates (reserved in the stems and sheaths of rice and contribute grain weight) from the straw to the grains. Mostajeran and Rahimi-Eichi (2009) revealed that plant growth can get affected due to drying of soil in lack of standing water but grain production may not be hamper because of drought-tolerant rice variety. Water scarcity at vegetative stage may reduce yields less than terminal stage because of recovery growth in the later growing season; however it demands extra farm labor that raises concerns about unpredictable rainfall (Sarvestani et al., 2008). A number of management strategies to improve the efficiency of utilization of water include avoiding runoff, mitigation of losses from soil and plants, less stomatal conductance, transpiration inhibitors, correction in there imbalanced use, and integrated water management. To meet this challenge, aerobic rice will be a best way to overcome the problems like drought and global warming.

In the present situation where natural resources and environment health is in constraint like water resource and global warming respectively, the cultivation of aerobic rice is an appropriate step for sustainable source of food. During prolonged period of water scarcity, less availability of water affect the concentration of metabolites and causing hindrance in amino acids and carbohydrate metabolism. The accumulation of soluble sugars and activity of carbohydrate metabolizing enzymes is strongly correlated to the acquisition of water with holding capacity of rice. To obtain a high seed content having good quality with less water consumption, most of the supplied water would have to be translocated to the grain before maturity (Condon et al., 2004).

Different sugars involved in synthesis of several compounds and production of energy besides of stabilization of membranes by operating as compatible osmolytes which include proline, polyamines (PAs) and polyols (Mostajeran and Rahimi-Eichi, 2009). In water scarcity, for example the diversion of carbon to polyol biosynthesis that are non-reducing sugars may also store excess carbon that decreases the activity of carbon metabolizing enzymes such as invertase and sucrose synthase (SUSY). Water soluble carbohydrates can accumulate in rice stems and during later phase of grain filling, these become an important source of assimilates for grain yield in rice. Changes in carbohydrates are of particular interest because of the direct relationship with physiological processes such as photosynthesis, tolerance of assimilates and respiration (Jain, 2013). The determination pattern of carbohydrate metabolizing enzymes and their role in adaptation may provide the base for effective strategy leading to better aerobic rice variety.

To enhance the quality and productivity of crops, traditional and conventional breeding has been taken place from past 20 years by choosing the most suitable characteristics in the aspect of yield or technologies. Breeder has been focus on one or combination of more than one characteristic which associated directly or indirectly to improve the plant survival rate and economic yield and selection of better cultivars. Therefore, utility of different good characteristics has been emphasized which favored water productivity for better plant survival, good grain yield and amount of dry matter in water stress or in drought condition (Condon et al., 2004; Tuberosa, 2012). Ultimately, each and every characteristics and its ability enhance the plant breeding programs on individual basis or after mixing with others (Farooq et al., 2010). But still no real attempts are there to understand the methods for yield enhancement especially in relationship to less water consumption. Plant breeders must produce varieties having better yield along with less absorption of water. Rice crop is very sensitive to irrigation that is why step to reduce of water application may hamper the yield, so by considering the reduction in amount of water availability, some intensive study and development of some technologies which facilitates the rice production with good yield with minimum irrigation. Water use in irrigated rice is high because the crop is grown under "lowland" conditions: the soil is puddled (wet-land preparation to create a muddy layer) and, after the transplanting of rice seedlings, the field is kept flooded with 3–5 cm depth of water until some 10 days before harvest. Evaporation rate has been very high in rice field because of plenty of water applied in field which remains standing and due to sunny days, it evaporates rapidly besides seepage and percolation out of the root zone (Sandhu et al., 2012). Moreover, the puddling of rice fields requires an extra amount of water when compared with that of other grain crops such as maize or wheat. Transplanted system not only leads to wastage of water but also causes environmental degradation and reduces fertilizer use efficiency (FUE). It leads to destruction of soil aggregates and reduction in macro pore volume and to a large increase in micro pore space which subsequently reduce the yields of post rice crops.

Many water saving procedure and technologies such as alternate wetting and drying, direct wet seeding, and aerobic system are in practice for rice cropping in low land areas. Implementation of these scientific technologies, it is possible to adaptation of less water resist variety and aerobic system rather than traditional method of flood irrigation of rice cultivation besides developing new high yield and aerobic favored varieties (Farooq et al., 2011). Study reveals that nearly 60% of rice cultivars grown by fallowing traditional procedure. Several other observations reveals about study of genetic changes in different characteristic which probably can survive in water stressed condition includes thicker and deeper roots, deeper root penetration and membrane stability, and so forth, (Gowda et al., 2011). Aerobic rice varieties are more prominent and more capable to grow in water stress regions without reduction in yield. Duration of crop is evaluate with help of genotype and environment aspects and also determines the ability of crops to complete the growth cycle in shorter time period because short duration variety of rice becoming mature in less time to avoid the water deficient period and proven efficient and productive in water stress areas (Farooq et al., 2011).

Medium height and drought resist rice varieties are selected foe aerobic rice culture. In case of water and soil related stresses, the root characteristics of rice cultivar have a significant role and deep and thick root system is suitable to access the water from more depth and can survive in water stress soil. Rice cultivar of deep and thick root system have a higher yield in upland regions because it's able to withdraw water from soil in which water is available at higher depth (Sandhu et al., 2012).

Crop of water scary condition preserve the water for their survival by minimizing the water losses through transpiration besides fixation of carbon dioxide (CO_2) to supply the energy to crops. Farooq et al. (2011) stated that transpiration loss is reduced when crop varieties have short height and small leaf area is suitable for water stress areas but in result production is also reduced. Such plants withstand drought therefore grow well under aerobic conditions although their growth rate and biomass are relatively low (Farooq et al., 2011). Loss of chlorophyll contents was generally accepted to be the main determinant of reduced photosynthesis under water stress (Surendar et al., 2013). Limitation of photosynthesis under water stress was caused by reduced a photosynthetic pigment which leads to production of ROS (Mafakheri et al., 2010).

Breeders should have to focus and short out the constraint of anaerobic adaption, survival capability in water deficit condition and low humidity ration by releasing some suitable and capable variety of rice to assure the food availability in upland or rain fed areas. Rice crop is habituated in aerobic condition or in irrigation facilitated areas (e.g., low land or land well connected with water body) and in presence of sufficient water, this crop creates a superficial root zone which consume more water besides heavy non stomatal water losses from leaves (Lafitte and Bennet, 2003). Development and growth of roots in controlled conditions is better in case of aerobic rice verities results establishment of seedlings in lesser time and grow properly in water stress land (Konwar, 2009). Martin et al. (2007) revealed that root length, biomass and production are proportionally related with each other that is why a crop variety having longer and deeper roots structure are appropriate to cultivate in aerobic condition.

Aerobic rice cultivation was developed worldwide to reduce water consumption and produce rice with less water (Bernier et al., 2008; Qin et al., 2010). Aerobic rice undergoes oxidative stress due to less water availability which leads to increased ROS production and these adapt to water stress by enhancing water stress tolerance associated traits (root thickness, length density, pulling force, penetration ability, depth of rooting and osmotic adjustments), altering cellular metabolism and evoking various defense mechanisms (Gowda et al., 2011; Kato and Katsura, 2014). However, this ended up in causing lower plant growth rate, poor grain establishment and hence, less yield (Kato and Okami, 2011).

Crops which are selected from previous generations to produce moisture and lodging resisting genotypes are preferred for cultivation in aerobic system. Aerobic genotypes with better antioxidant defense system potentially contribute to water-saving rice cultivation under water-deficit scenarios. So, one of the best approach for stress tolerance is to identify the cultivars having strong antioxidant system so as to survive and perform better under water deficit conditions. Therefore, the identification of rice genotypes consuming less water is considered a promising move towards sustainable crop productivity in water-scarce areas (Kato and Katsura, 2014). Breeders tried to develop good yield rice verities suitable for water stress areas but unfortunately these varieties are not up to the mark in terms of yield and biomass that is why further research has been required to search out new varieties which provide better yield under aerobic rice system in water deficit areas. Different important and desired physiological and biochemical characteristics are identified as suitability of aerobic system to develop a high yielding and water resist aerobic rice varieties. For producing good yield and water stress breed, aerobic rice breeding nurseries should be managed for preparing qualitative and maximum effective rice breeds of desired characteristics rather than by imposing aerobic culture (Bouman et al., 2006b).

Hence, gradually shifting from traditional irrigated, conventional system of rice production to aerobic rice will be especially beneficial to overcome the decline in water table. Keeping in view the biochemical effects of growing aerobic rice, the chapter entitled "A Review on Biochemical Evaluation of Irrigated Flooded Transplanted and Aerobic Rice (*O. sativa* L.)" address a review on biochemical evaluation of irrigated and aerobic rice such as antioxidant activity and defense mechanism; polyamines and carbohydrate metabolism; protein and their sub-fractions; and its effect on yield components that suggest ways for development of high yielding aerobic rice.

6.3 WATER USE EFFICIENCY

Measuring index of capacity of cropping system to convert water into plant biomass or grain is known as "water use efficiency" (WUE), includes entire rainfall and water stored in soil within the growing period of crops (GRDC, 2009). WUE indicates the ratio of water consumed during plant metabolism and total water losses through transpiration (Bacon, 2004). It can be determined in different terms like harvested yield per unit of soil water used in total duration of crop season; total dry matter produced above the ground per unit of soil water used within a crop season and total produced crop per unit of total irrigation applied (in form of precipitation and artificial water supply) (Lambers et al., 2008). Grain Research and Development Corporation, Australia recommended "water use efficiency" (WUE), based on several factors like soil ability to preserve the water; crop ability to withdraw the water present in soil and precipitation and crop characteristics to change the biomass in grain (GRDC, 2009). The first time WUE concept was discussed in 19th century for betterment of irrigation systems. The main point which has impact on research and scientific approaches to study about WUE is discussed in Table 6.1 and for intensive study of WUE, Solomon (1983), Bos and Wolters (1989) and Clemmens and Solomon (1997) may also have to concern.

To feed the exploding population with rapid growth of urbanization and industrialization, increasing of agricultural production is essential but agriculture land facing water threats due to less rainfall and excessive exploitation of ground water. As per data, 117 cm annual rainfall recorded in India, the gross water utilization is nearly 145 cm/ha of gross irrigated area, if 70% of the total precipitation is available for crop's consumptive use (CCU) (Geetalakshmi et al., 2009). In India, efficiency of mostly irrigation systems varies between 25–45%. Traditional rice cultivation consumes large quantity of labor and water (according to estimation nearly 5,000 L of water is needed to produce 1 kg of rice). Water needed for irrigation in 2025 is estimated up to 910 km^3 as compared to 810 in 2010. Major constraints to rice production that India faces are water, land, labor, high quality germplasm, and other inputs such as fertilizers, pesticides, and insecticides (Uphoff et al., 2005). This clearly indicates the need for water resource development, conservation and optimum use. Rice farming is ongoing but subject to rapid change as rain fed area in the country, account for 60% of the cultivated area. Farmers are also seeking alternate methods of cultivation for growing rice to combat this water scarce situation.

TABLE 6.1 Brief Overview on the Key Studies of WUE Concepts Evolved Over 19th Century in Chronological Order.

Years	Remarkable Development	Reference
1920	Efficient water irrigation can be expressed by establishing the co-relation between the area under mature crop and the quantity of water used for its irrigation. It is also defined as "duty of water"	Brown (1920)
1928	The limits of improvements are ruled by economics and also used a significant term permissible waste	Fortier (1928)
1932	Ratio of transpiration of irrigation water by plants of an irrigation area and total water applied in the farm through rivers, canals or any other water body within the same duration of time, defined as "Irrigation efficiency"	Israelsen (1932)
1942	Uniformity coefficient is defined as the ratio of depth of lower proportion of application to applied depth of across the field in average	Christiansen (1942)
1960	Water storage efficiency defined as the ratio of stored water in root zone at the time of irrigation and water need to store in root zone before irrigation	Hansen (1960)
1960	Efficiency of seasonal application to extended application for single irrigation applied to the overall irrigation season	Hall (1960)
1965	Appropriate and efficient irrigation is directly related to the concept of water storage efficiency and can be measured by observing the amount of water applied to agricultural field fulfill or exceeded than actual needs.	Hart and Reynolds (1965)
1972	There are several physical, economical, and political factors which influenced the water application efficiency. Different characteristics like soil infiltration, dimensions of stream, measurement of storage of available water, sprinkler spacing and its nozzle size, pressure and wind condition as well as value of water and crop, cost of land preparation, labor and equipment besides water laws and geographical status.	Willardson (1972)

TABLE 6.1   *(Continued)*

Years	Remarkable Development	Reference
1978	Principle of application efficiency and distribution uniformity depends on the mean of lower quarter of determined applied depths for measurement of efficiency and adequacy both.	Merriam and Keller, (1978)
1983	Yield related efficiency measures assess the exact utilization of the applied water to the crop and estimate the significance of irrigation system and management decisions. Uniformity and efficiency of applied irrigation measures the expected yield from un-fluctuated irrigation of presumed crops.	Solomon, (1983)
1986	Adequacy of irrigation as the percentage of the root zone throughout a field that is restored to field capacity during irrigation. Therefore, obtaining an adequacy level of 100% will result in percolation losses because of the non-uniformity of application.	Whittlesey et al. (1986)
1988	Surface irrigation efficiency which incorporates the application efficiency characteristics, is determined with use of stress or excess efficiency.	Blair and Smerdon, (1988)
1997	Irrigation sagacity, which is a measure of prudent water use, was recommended by Solomon and Burt. Sagacious uses are defined as being either beneficial, or non-beneficial but reasonable. Non sagacious uses are those without economic, practical or other justification and are considered non-beneficial and unreasonable. The list the steps to establish whether a non-beneficial use of water is reasonable. The essential difference between the traditional irrigation efficiency and irrigation sagacity is the inclusion of the non-beneficial, though reasonable use, of water. For example, evaporation from channel, some soil evaporation, and deep percolation due to preferential flow are non-beneficial but reasonable losses.	Solomon and Burt (1997)
1998	Adjustment of both input and outcomes to appropriate prices is an economic aspect known as Allocative efficiency.	Omezzine and Zaibet, (1998)

To meet the major challenge of increasing rice production with less water use, aerobic rice cultivation has been developed as a novel water-saving technique (Bernier et al., 2008; Kato and Katsura, 2014). Under aerobic conditions, rice crop can survive in water deficit soil as well as moisten soil is sufficient to grow it rather than flooded water in traditional cultivation. Numbers of aerobic condition suited rice verities are developed but their methodology of aerobic adaptation are quite unclear (Wang et al., 2002). Aerobic cultivation system of rice act as resources conservation technology because it utilize minimum water with high efficiency besides reduction in labor cost up to 50% with help of mechanization (Bouman et al., 2002). Different factors like artificial ground water recharge, effective rain water utilization and proper management of salinity in coastal regions should be focused for proper and sustainable development and appropriate utilization of ground water.

6.4   ANTIOXIDANT ACTIVITY AND DEFENSE MECHANISM

Plants have evolved several defense mechanisms to prevent or alleviate the damage from antioxidant species which include scavenging the antioxidant species by natural antioxidants such as ascorbate and α-tocopherol and use of an enzymatic antioxidants such as superoxide dismutase (SODs), ascorbate peroxidase (APx), glutathione reductase (GR), catalase, and peroxidase which acts in tandem (Sharma et al., 2012). Several isozymes are also present in plants and their relative compositions change during exposure to different regime of water. Therefore, the study of these enzymes may help in developing high yielding aerobic rice varieties. Tohidi-Moghadan et al. (2009) revealed several interlinks between water deficit and endogenous extent of water-soluble antioxidants. Water stress and temperature and other environmental stresses influenced the most of all physiological and physicochemical aspects of plants and hamper the yield. Earlier work correlated the association of enzymes having antioxidant characteristics and other biochemical attributes available in plants which accelerate the yield has been considered for short duration experiments in laboratory (Liang et al., 2003) but no relevant results were obtained to develop an aerobic rice variety having high yield. Limited reports were available on physio-morpho-biochemical traits involved in common practice grown rice and aerobic rice. Large gaps in yield and grain quality traits between aerobic and flood irrigated systems could thus overweighed by these studies.

Generally, soil is going to dry in situation of water deficit condition due to that crop growth is also retard and decline in root and shoot length, dry weight, chlorophyll content and production of ROS was observed (Farooq et al., 2011; Sandhu et al., 2012). Water stress tolerance is a complex trait associated with several physiological and biochemical attributes. Plants being sessile have evolved many adaptations to counteract water deficit. In a genetic sense, these adaptations or the mechanisms of drought tolerance can be classified into four categories: avoidance (development and physiological traits), tolerance (physiological and biochemical adaptations), escape and recovery (Bhushan et al., 2007). These mechanisms are inter-related as there is no fixed line of demarcation in between them. Tolerance is the ability of cells to metabolize even at low leaf water status (Toker et al., 2007; Miro and Ismail, 2013).

In water stressed plants, stress indicates stomatal closure and water saving tactic of plants for their survival because water can be preserved for plants by reducing the transpiration losses via regulating the stomatal opening but it's constrain the availability of CO_2. This process hampers the photosynthesis and ROS production of plants as well as oxidation characteristics of cells due to water deficiency. ROS are common components of biochemical changes in the chloroplasts, mitochondria or in peroxisomes, when plants are subjected to water stress conditions (Damanik et al., 2010). Accumulation of ROS activates antioxidative defence mechanism in plants. Under water stress, production of hydroxyl radical in chloroplast and mitochondria and accumulation of these materials catalyze the harmful reactions caused diminishing of thylakoidal membranes and photosynthetic gadgetry besides the mitochondrial electron transport chains (Zlatev and Lidon, 2012).

ROS in plant cells produces from different organelles (e.g. chloroplast, mitochondria and peroxisomes etc.) with high oxidizing metabolic process and rate of electron flow (Sharma et al., 2012). The ROS formation increased with 1–2% consumption of oxygen. Generally, oxygen available in atmosphere can enhance the level of ROS through deduction in single electron causing evaluation of superoxide radical which generates hydrogen peroxide after addition of proton (H^+) (Bhattacharjee, 2005). In same sequence, other generated superoxide radical's form per hydroxy radical after receiving proton shown by reaction in Figure 6.2 described by Apel and Hirt (2004). Further reactions occurred when transition metals like copper and iron will be present and according to an example of Fenton or Haber-Weiss mechanism, hydroxyl radical evolved which known as most reactive chemical element in entire biological world (Del Rio et al., 2006). Per-oxy-nitrite ($OONO^-$) can be formed due to reaction of superoxide radical (O_2^-) with nitric oxide (NO) and singlet oxygen is treated as another ROS (Fig. 6.3), there is no plus of any extra electron to superoxide radical rather an electron moves to higher energy orbital. During photo-excitation of chlorophyll and its reaction with Oxygen also generate Superoxide radical ($^1O_2$).

The various reactions showing production of ROS from oxygen and their scavenging have been listed in below Figure 6.2 and Figure 6.3.

High amount of superoxide radical (O_2^-) and hydrogen peroxide (H_2O_2) can causes the death of cells whereas low amount of it shows favorable mechanism and adjustable characteristics with oxidative and abiotic constraint (Nakagami et al., 2005).

Creation of ROS is proportional to its ability of free movement and interaction between different cellular compartments. Hydrogen peroxide (H_2O_2) have a characteristic to move from the place of its generation and also able to move across the biological membranes while superoxide radical (O_2^-), singlet oxygen ($^1O_2$) and hydroxyl radical (OH) are incapable to move. In case of hydrogen peroxide (H_2O_2), biological membranes are less permeable and its movement is takes place via specific aquaporin. This is another procedure of balancing the local concentrations, thus the biological effect of H_2O_2 (Yang et al., 2012).

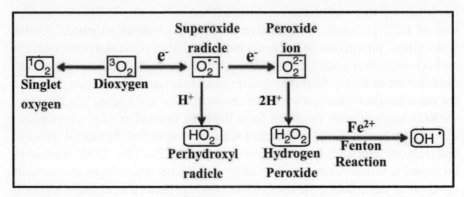

FIGURE 6.2   Generation of ROS by energy transfer (Adapted from Apel and Hirt, 2004).

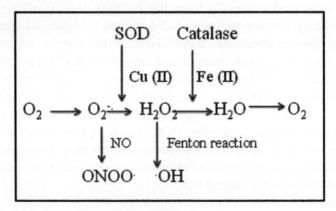

FIGURE 6.3   Sources and scavenging of ROS

6.4.1 REACTIVE OXYGEN SPECIES (ROS)

Within the condition of biotic and abiotic stresses, the ROS have a vital role in destruction of cells (Candan and Tarhan, 2003). The enhanced production of ROS can damage biological membranes (lipid peroxidation), nucleic acids, oxidizing proteins and other classes of macromolecules. Stress injury of cells and organized structures within the cell through membrane were noticed due to effect of peroxidation of essential membrane lipids in intracellular organelles which produced after reaction between ROS and unsaturated fatty acids (Sharma et al., 2012). The characteristics of ROS like behavior as damaging, protective or signaling factor proportional to the balance between creation and use of ROS at appropriate time and location are important to understand. It is dispersed in very short periphery before the reaction with cellular molecules that is why the responding characteristics of cell firmly influenced the reactive rate of ROS. Plants resist the effects of over-produced ROS by a complex antioxidant defense system consisting of enzymatic and non-enzymatic antioxidants where enzymatic components are considered to be the most efficient mechanisms against oxidative stress (Sharma et al., 2012).

Production of H_2O_2 is abruptly takes place with the plant body but during biotic and abiotic stresses, abrupt increasing in the production rate of hydrogen peroxide (H_2O_2) was noticed. H_2O_2 has been proposed as a candidate biomarker for biochemical stress. Changes in H_2O_2 homeostasis trigger genetic programs that promote stress acclimation or induce programme cell death. H_2O_2 have an vital role in plant because of it work as signaling molecule within the plant system besides behave as protective substance during water stress (Hossain et al., 2013), whereas its higher concentration deteriorates the plants with age and destruct the plants cells. Increasing quantity of hydrogen peroxide has a impact on oxidative damage and status of lipid peroxidation (Al-Ghamdi, 2009; Nounjan and Theerakulpisut, 2012). To prevent the plants from oxidative stress, different characteristics of hydrogen peroxide like its generation, spreading and disappearing should be linked but the mechanism can be changed genotypically (genetic makeup) along with multiplying favored condition in the various stage of aerobic or transplanted rice. The enzyme SODs converts superoxide (O^{2-}) to H_2O_2 whereas catalase (CAT) and peroxidases catabolizes H_2O_2 into H_2O and O_2. Guaiacol peroxidase (GP_x) degraded the hydrogen peroxide with help of guaiacol (phenolic compound), treated as electron donor whereas uses of APX minimized the form of ascorbate to prevent the destruction of cells by H_2O_2 (Fig.6.4). H_2O_2 causes retardation in synthesis of antioxidative enzymes besides a vital key under stress condition (Basu et al., 2010).

A study observed the hydrogen peroxide (H_2O_2) and hydroxyl radical (OH•) is produced in high rate during last stage of drought period in juvenile been plants which reveals the circumstances of oxidative stress in cells (Zlatev, 2005). Hydrogen peroxide has an important role to produce lignin by peroxidase-mediated oxidative polymerization of alcohols and many enzymatic systems which contribute in formation of H_2O_2 at the upper layer of plant cells (Lutje et al., 2000). Water scary and water logging stresses on *Vigna sinensis* plants promote the antioxidant metabolites like phenolic compounds, hydrogen peroxides and ascorbic acid (El-Enany et al., 2013).

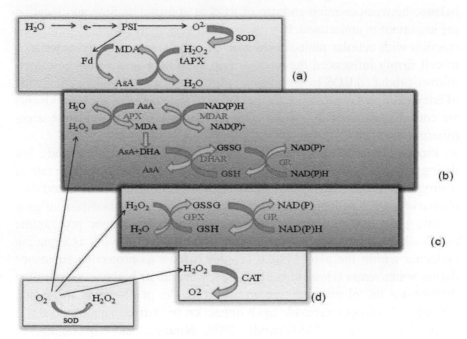

FIGURE 6.4 (See color insert.) ROS scavenging pathways in plants. The water-water cycle (b), the ascorbate glutathione cycle (c), the glutathione-peroxidase (GPx) cycle (d) Catalase (CAT), superoxide dismutase (SODs) acts as the first line of defense converting O^{2-} into H_2O_2. In the contrast to CAT (d), APx and GPx require an ascorbate (ASA) and/ or a glutathione (GSH) regenerating cycle (a-c). This cycle uses electron directly from the photosynthetic apparatus (a) or NADPH (b, c) as reducing power. [DHA: Dehydroascorbate; DHAr: DHA reductase; Fd: Ferrodoxin; GR: Glutathione reductase; GSSG: Oxidised glutathione; MDA: Monodehydro-ascorbate; MDAR: MDA reductase; PSI: Photosystem I; and tAPx: Thylakoid-bound APx]

Different process like rearrangement of components of membrane lipid, alter in photosynthetic apparatus, mitochondrial electron transport chain regulation and stress-induced ROS scavenging system happens under stress (Fig.6.4). Stability of the membrane is accomplished by reducing the leakage of solutes from the cell. The cell water content is maintained by accumulating compatible solutes like fructan, glycinebetine, polyamines, polyols, proline and trehalose that are nontoxic and do not interfere with cellular activities (Asthir et al., 2010). Proline and other amino acids stabilize the cellular proteins destruction which cause by stress. Each and every cellular constituent have different antioxidant enzymes and locally present antioxidants. However, when this local antioxidant capacity cannot cope with ROS production then H_2O_2 can leak into the cytosol and diffuse to other compartments (for example, during stress or temporarily reduced antioxidants levels due to development signals).

Different antioxidants including low molecular weight also like ascorbate and tocopherol in spite of antioxidative enzymes such as AP_x, GP_x, CAT and SODs, regulates the creation and diffusion of ROS effectively (Suzuki and Mittler, 2006; Turhan et al., 2008) but impact of oxidative destruction appears in form of oxidation of cellular lipids and proteins, damage of photosynthetic pigments and deactivation of photosynthetic enzymes because of exceeding of accumulation of ROS over the elimination strength of antioxidant system due to water stress (Soleimanzadeh, 2012). ROS antioxidant defense mechanism has been described under following headings:

6.4.1.1 MEMBRANE STABILITY PARAMETERS

Photosynthetic pigments are important regulators of photosynthesis but its damaging in plants due to stress. Several researches have revealed that in the condition of stress, chlorophyll content is reduced because of its unfavorable impact on membrane firmness (Bijanzadeh and Emam, 2010; Mafakheri et al., 2010; Din et al., 2011). Reduction of total chlorophyll concentration causes losses in chlorophyll "a" in case of susceptible rice cultivation (Singh and Dubey, 1995) and Tewari et al. (2002) studied tomato plants and found that the variation in chlorophyll content has been taken place when chlorophyll a/b ratio in resistant tomato reduced in situation of water stress.

In case of Zea mays, crop production in water stress condition, higher reduction of chlorophyll content were noticed in mesophyll cells and lesser but significant loss from cells of bundle sheath and plants degradation with age signifies the retardation of photosynthesis due to reduction of chlorophyll. Sikuku et al. (2010) observed that lamellar content of the light harvesting chlorophyll a/b protein is causing the losses of chlorophyll in plants which is grown in water deficit areas and also concluded that many factors can be associated with evaluation of Photosynthetic capacity of a crop or plant out of which constituents of photosynthetic pigment. Generation of chlorophyll by process of photosynthesis have a very vital role in growth and health of plants and in reduction in amount of chlorophyll were observed due to salt deficiency also restrict the growth and yield of some of plants species (Chaves et al., 2009; Ashraf and Harris, 2013). Jamil et al. (2013) also stated that lesser quantity of salt causes retardation of photosynthesis in various plants. Potential of photosynthesis is directly proportional to the concentration of its pigments. Reduced quantity of chlorophyll due to water deficiency in crop field results oxidation retardation and also can be signifies the degradation in pigment photooxidation and chlorophyll besides decreasing the numbers of green leaves (Munawarti et al., 2013).

Lipid peroxidation is a biochemical is act as an indicator for free radical mediated injuries. Zlatev and Lidon, (2012) concluded that polyunsaturated fatty acids (a constituent of membrane lipids) are susceptible to peroxidation. Small hydrocarbons divide as ketones, and malonoaldehyde, and so forth, during lipid peroxidation. After crossing the threshold ROS level, lipids per-oxidation of cellular and organelle membranes begins (Garg and Manchanda, 2009; Sharma et al. 2012).

The overall process of lipid peroxidation involved three different steps viz. initiation, progression and termination depicted in Figure 6.5. Initiation step involves abstraction of a hydrogen atom by transition metal complexes (Fe and Cu), O_2^-, OH and H_2O_2, in an unsaturated fatty acyl chain of a polyunsaturated fatty acid (PUFA) residue, mainly by OH· ROO· increases by adding of oxygen at carbon centered lipid radical with fatty acid and further after a extracting hydrogen atom from chain of PUFA results formation of different reactive species (including malondialdehyde and lipid epoxides) after breakage of lipid hydro peroxide. A single initiation event thus has the potential to generate multiple peroxide molecules by a chain reaction. Lipid peroxidation causes the reduction in membrane fluidity, ascending the leakage of membrane which damages the membrane proteins, deactivating of receptors and enzymes and ion channels by allowing the materials to pass rather than through certain defined channels (El-beltagi and Mohamed, 2013).

In general, the plants exposed to various abiotic stresses exhibit an increase in lipid peroxidation due to generation of ROS (Gill and Tuteja, 2010). Different characteristics of the process like branching of the chain reactions permit many steps of regulation. Structure of the membranes: composition and organization of lipids inside the bilayer, the degree of polyunsaturated fatty acid, mobility of lipids within the bilayer, localization of per oxidative process in a particular membrane and the preventive antioxidant system (ROS scavenging and lipid peroxidation product detoxification) are the regulated properties of preventing lipid peroxidation.

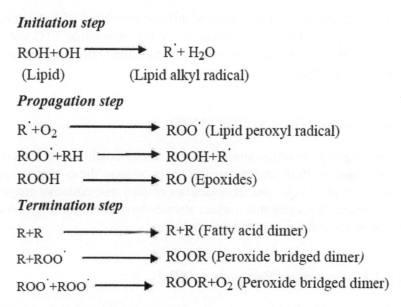

Initiation step

$$ROH + OH \longrightarrow R^\bullet + H_2O$$
(Lipid)   (Lipid alkyl radical)

Propagation step

$$R^\bullet + O_2 \longrightarrow ROO^\bullet \text{ (Lipid peroxyl radical)}$$

$$ROO^\bullet + RH \longrightarrow ROOH + R^\bullet$$

$$ROOH \longrightarrow RO \text{ (Epoxides)}$$

Termination step

$$R + R \longrightarrow R + R \text{ (Fatty acid dimer)}$$

$$R + ROO^\bullet \longrightarrow ROOR \text{ (Peroxide bridged dimer)}$$

$$ROO^\bullet + ROO^\bullet \longrightarrow ROOR + O_2 \text{ (Peroxide bridged dimer)}$$

FIGURE 6.5   Lipid peroxidation reactions.

The concept of lipid peroxidation treated as destructive process were altered within these last ten years, its reflects that lipid hydro peroxides and oxygenated products of lipid degradation in spite of lipid peroxidation initiators like ROS can involve in the signal transduction cascade (Zlatev and Lidon, 2012).

Triphenyl tetrazolium chloride (TTC) cell viability assay is based on the principles of tetrazolium salt reduction to formazan by dehydrogenase respiratory enzymes (Ibrahim and Quick, 2001; Mikuła et al., 2006; Asthir et al., 2002). Therefore, the TTC test evaluates the mitochondrial electron transport chain, and thus, it represents respirational activity. This assay has a long history of use as an estimate of plant cell and tissue viability under stress (Chen et al., 1982) suggested TTC assays could be used to evaluate stress tolerance in plants. TTC assay was evaluated in barley to detect genotypic differences in heat tolerance by Asthir et al. (2002). Due to dehydrogenase process in the cells of living organism the colorless TTC is minimized to red triphenyl formazan. Germination ability of seeds of various agricultural produces like maize, wheat and paddy also has been determined with help of TTC test (Marrero-Ponce et al., 2007). In present, TTC is one of the most simple and practicing test for finding out the viability of seeds of different agricultural crops and plants as well as germination and viability of maize seeds has been determined with help of this test correlated with the conclusion of conductivity test (Govender et al., 2008), meanwhile TTC test also used in viability determination of tissue (Ibrahim and Quick, 2001) and cell suspension (Mikuła et al., 2006) in plants.

6.4.1.2 ENZYMATIC ANTIOXIDANTS

Exposure of plants to unfavourable environmental conditions can increase the production of ROS and to protect them against these toxic oxygen intermediates, plant cells and its organelles employ antioxidative enzymes. Different research reveals that various stresses have been prevented due to induction of the cellular antioxidant machinery.

6.4.1.2.1 Ascorbate Peroxidase (APx)

Ascorbate peroxidase (APx) is involved in scavenging of H_2O_2 in water-water and ascorbate-glutathione (ASA-GSH) cycles and utilizes ascorbate (ASA) as an electron donor (Fig. 6.4). Asorbate peroxidase utilizes the reducing power of ascorbic acid to eliminate potentially harmful H_2O_2. APx has a higher affinity for H_2O_2 than CAT and GPx (guaiacol peroxidase) therefore it may have a more crucial role in the management of ROS during stress (Asthir et al., 2002).

6.4.1.2.2 Catalase (CAT)

Catalase (CAT) catalyzes the removal of H_2O_2 produced by dismutation of superoxide radical (Fig. 6.4). One molecule of Catalase changes nearly six millions of molecules of hydrogen peroxide into water and oxygen in one minute. Catalase (CAT) supports in elimination of hydrogen peroxide (H_2O_2) which produced during peroxisomes by oxidases participated in β-oxidation of fatty acids, photorespiration and purine catabolism (Garg and Manchanda, 2009; Ahmadizadeh et al., 2013).

6.4.1.2.3 Glutathione Reductase (GR)

Glutathione reductase (GR) is found in both prokaryotes and eukaryotes (Romero-Puertas et al., 2006). GR is also associated in ASA-GSH cycle and perform an important role defense system in counter of ROS with the help of sustainable status of reduced GSH. GR catalyzes the nicotinamide adenine dinucleotide phosphate (NADPH) dependent reaction of disulphide bond of GSSG (oxidized glutathione) and is thus important for maintaining the GSH pool (Fig. 6.4). Different enzymes like APx along with CAT, GPx and SODs are known as one of the main enzymes within the mechanism of antioxidative defense which direct evaluate the cellular concentration of O^{2-} and H_2O_2 (Gill and Tuteja, 2010; Sharma et al., 2012).

6.4.1.2.4 Guaiacol Peroxidase (GPx)

Guaiacol peroxidase (GPx) eliminates toxic H_2O_2 by reducing it into water and guaiacol is oxidized to the quinone. The peroxidases with broad specificities had been observed in cell wall where they utilized hydrogen peroxide to produce phenoxy compounds which polymerized to generate components like lignin (Patel et al., 2010). GPx preferred guaiacol and pyragallol (aromatic electron donors) which oxidize ascorbate with rate of 1% approximately that of guaiacol.

6.4.1.2.5 Superoxide Dismutases (SODs)

Superoxide dismutases (SODs) establish the first line of defense against ROS (Fig. 6.4). SODs are mainly involved in dismutation of superoxide radical to H_2O_2 and oxygen to prevent the main enzymes and organelles which participated in various metabolic steps of cells (Mittler, 2002; Gur et al., 2010). Each of three SODs isozymes (Cu Zn SOD, Fe SOD, Mn SOD) are independently controlled as per degree of oxidative stress acknowledged in the respective sub cellular components (Del Rio et al., 2006).

6.4.1.2.6 Case Study on Enzymatic Antioxidants

Antioxidative enzymes (APx, GPx, CAT, SODs and GR) have highest activity at anthesis and milky dough stage in leaves and grains of rice respectively under normal and water stress conditions (Liang et al., 2003; Qin et al., 2010). Similar observations were reported by Almeselmani et al. (2006) and Kumar et al. (2012) in wheat under normal, late sowing and high temperature conditions. Protective enzymes such as CAT, POD, SODs and

GPx play a main role in eliminating ROS and these enzymes were enhanced gradually from 3–5 days after heading and reached to the peak on the 14th day then began to decline. The activities of CAT, peroxidases, SODs and GPx were found to be higher in stress tolerant cultivars as compared to susceptible ones. It has also been reported that submergence reduces the activity of these enzymes, however after exposure to air their activity again increases (Das et al., 2005).

Different cultivars having higher antioxidant characteristics and lower level of membrane lipid peroxidation, reflects drought resistance characteristics also. In condition of higher stress, enzyme activities were degraded due to excess collection of free radicals and these enzymes are also reduced in heavy stress that led to the degradation of these enzymes in maize cultivars (Zheng et al., 2010). Qin et al. 2010) reported that activity of CAT and APx get decreased in non-flooded cultivation as compared to flooded one however, SODs activity increased significantly under non-flooded conditions in rice leaves. It was further supported by Liang et al. (2003) that these enzymes activities get decreased in rice leaves under zero-mulching condition. Reduction in SODs activity related genes under the heat stress treated tobacco plants was observed by Gur et al. (2010).

Under stress, catalase activities retarded and inactivation of enzyme protein also noticed because of ROS, catalase photo-inactivation or reduction in enzyme synthesis or change in assembly of enzyme subunits and it also caused reduction in the capacity of the leaves to decompose hydrogen peroxide (H_2O_2) (Sharma et al., 2014).

In the crop like wheat, excess collection of hydrogen peroxide (H_2O_2) causes heat shock which can be prevent with increasing of catalase (CAT) activities in plants (Kumar et al., 2012). Induction in APx activity was reported in response to chilling, drought, mild water stress, Cu toxicity, ozone toxicity and UV-B radiation. The activity of GPx varied considerably depending upon plant species and stress condition (Patel and Hemantaranjan, 2012). Rich glutathione reductase (GR) activities help in *de-novo* synthesis of enzyme protein because this enzyme treated as an internal agent to counter the oxidative destruction in rice crop. In the stage of sub-lethal stress, level of per-oxidase activity has been work as biomarker to estimate the stress intensity (Verma and Dubey, 2003). Increased activity of antioxidant enzymes related with drought tolerance characteristics of sorghum crop during germination stage however in tolerant cultivars significant increase was observed by Singh and Sharma (2013).

Intensive study reveals that enhancement in enzymatic antioxidants under stress prevent the membrane damage in case of different drought tolerant verities. Commonly acclimation to drought was causing low ROS level through antioxidant system (Zaefyzadeh et al., 2009), simultaneously a series of adaptive responses to water stress takes place including the induction of preventive enzymes like SODs, APx and CAT due to enhancement of hydrogen peroxide (H_2O_2) content under mild water deficit condition, leading to promote draught resistance. Study reveals that the activities of protective enzymes accelerated more in case of different drought resistance cultivars rather than less-resistive cultivars (Zheng et al., 2010). Rostami and Rahemi (2013) reported that CAT, APx and GPx activity decreased significantly however SODs activity increased under drought stress in susceptible cultivars as compared to tolerant cultivars of caperfig varieties which suggest the possibility of evaluating antioxidant enzymes activities as a physiological marker in drought tolerance screening of caprifig and related species.

6.4.1.3   NON-ENZYMATIC ANTIOXIDANTS

6.4.1.3.1   Ascorbic acid and tocopherol

Ascorbic acid and α-tocopherol are readily vulnerable to O_2 attack. Furthermore, the oxidized forms of these antioxidants may be highly unstable under physiological conditions: for instance, DHA at pH values higher than pH 6.0 decomposes to yield tartrate and oxalate (Foyer and Noctor, 2011). Therefore, the sizes of both the ascorbate pool and α-tocopherol pool would inevitably become smaller in plant cells when exposed to the condition of an excessive O_2 formation, unless biosynthesis of those antioxidants keeps pace with their degradation.

6.4.1.3.1.1   Ascorbic Acid

Ascorbic acid is one of the most important, water soluble and abundant antioxidant used to reduce the destruction of plants due to ROS (Farooq et al., 2013). Ascorbic acid is available in photosynthetic cells, meristems and other plant tissues. The matured leaves of plant having maximum quantity of chlorophyll, also contains highest concentration of ascorbic acid. Generally, present ascorbate in leaves remains in reduced form within the normal physiological conditions (Smirnoff, 2000; Eduardo et al., 2011). Because of antioxidant properties, it is react with super-oxide, hydrogen peroxide (H_2O_2) or tocopheroxyl radical to produce monodehydroascorbic acid or dehydroascorbic acid (Fig. 6.6).

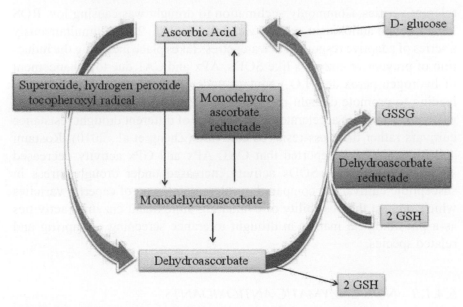

FIGURE 6.6 **(See color insert.)** Synthesis and degradation pathway of L-Ascorbic acid in plant tissues. [GSH: Reduced glutathione; GSSG: Oxidized glutathione].

Reduced forms are again recycled to ascorbic acid with help of mono-dehydro ascorbate reductase and de-hydro ascorbate reductase using reducing equivalents from NAD (P) H or glutathione, respectively (Fig.6.7). Completely oxidized dehydroascorbic acid has been lost if it is not reduced as well as having short half-life that is why re-generation of ascorbate is highly essential besides of that ascorbate treated as most powerful ROS scavenger due to its capability to transfer electrons during enzymatic reaction (Foyer and Noctor, 2011).

In addition to the importance of ascorbate in the ascorbate-glutathione cycle, it also plays a vital role to prevention of enzymatic activities which have prosthetic transition metal ions. Besides, ascorbate associated cell division, cell wall elongation and other developmental process for proper growth of plants (Dolatabadian and Jouneghani, 2009).

While plants resist in stress, the ascorbate pool gradually reduced and this the capability to preserve the ascorbate pool accounts for genotype difference in leaves of some particular chosen rice genotypes (Frei et al., 2008). Bartoli et al. (2004) reported a decrease in ascorbic acid content in wheat leaves exposed to drought conditions. Shao and Chu (2005) also have shown reduction in ascorbic acid content in maize during drought stress, results its key role in deciding the oxidative response. Ascorbic acid catalyzes the different enzymatic functions besides the reduction in destruction occurred by oxidative process via synergic activities with some of other antioxidants (Pourcel et al., 2007). Similarly, mung bean showed a lesser concentration of ascorbic acid when induced by various ROS generating substances (Kanwal et al., 2013).

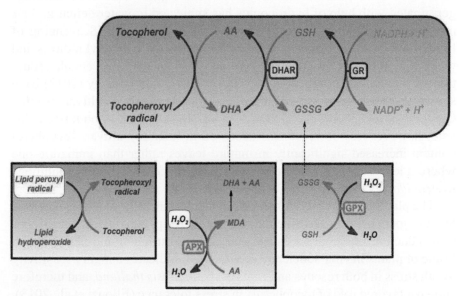

FIGURE 6.7   (See color insert.) Ascorbic acid and tocopherol role as antioxidants. [AA: Ascorbic acid; APx: Ascorbate peroxidase, DHA: Dehydroascrobic acid; GPx: Guaiacol peroxidase; GSH: Reduced glutathione; GSSG: Oxidized glutathione; and MDA: Malondialdehyde].

6.4.1.3.1.2   Tocopherol

Tocopherol has been available in most of all plants and generally stored in in photosynthetic and non-photosynthetic tissues (Wu et al., 2004). Tocopherols are lipid soluble antioxidant, known for potential scavengers of ROS and lipid radicals (Hollander-Czytko et al., 2005). Because of α-tocopherol contains three methyl groups, it has the maximum antioxidative activity in comparison of other isomers like β-, γ- and δ-form of tocopherols (Falk and Munne-Bosch, 2010). It is treated as an influent free radical trap due to its preventive characteristics of chain propagation steps in lipid autooxidation (Fig. 6.7). It had been expected that one molecule of α-tocopherol can scavenge upto 120 singlet oxygen by resonance energy transfers (Munne-Bosch and Falk, 2004). Oxidative stress activated the expression of genes responsible for the synthesis of tocopherols in higher plants and so increase in tocopherol content had also been reported (Kumar et al., 2013).

Quantity of endogenous α-tocopherol also affected due to its deterioration and recycling under stress. α-tocopherol levels decreases due to increment of ROS in chloroplasts and in the situation of higher stress. Rice seedlings germinated with help of hydroponics has subjected to water deficit in 30% polyethylene glycol indicates a reduction of α-tocopherol. Scavenging of lipid peroxyl radical by α-tocopherol produces α-tocopheroxyl radicals and when it is not recycled back by ascorbate then causing irreversible reduction of α-tocopherol (Szarka et al., 2012). Turan and Tripathy (2012) have shown that tocopherol accumulation in response to salt is cultivar specific. Contents of tocopherol declined in response to salt stress in both rice cultivars, however less decrease was observed in tolerant cultivar. Tocopherol content increased significantly in mature leaves rather than immature one whereas leaves age have no impact under regulated or salt stress conditions in *Arabidopsis thaliana* (Cela et al., 2011).

The plants having stress resistance capacity, generally showing increased level of tocopherol meanwhile the most sensitive plants reflects net tocopherol reduction under stress causes oxidative destruction and cell damage. Some of plants having deficiency of vitamin-E indicates a higher sensitivity to salt stress in both rosettes and roots of *Arabidopsis thaliana,* and therefore shown a favored role of tocopherols in stress tolerance (Ellouzi et al., 2013).

6.4.1.3.2   Osmoprotectants

Osmoprotectants has been proved as most important factors to protect plant cells from dehydration by lowering the cellular osmotic potential and protect the macromolecular structures against the destabilizing effect of water stress. They are non-toxic and stored to significant levels without disturbing plant metabolism. Some of the common osmolytes are amino acids (e.g. proline), higher polyamines (putrescine, spermidine, spermine), polyols, quaternary ammonium compounds (glycine betaine) and sugars (mannitol, trehalose, sucrose) (Krasensky and Jonak, 2012). Collective amount of polyols for example manitol, sorbitol, inositol and their derivatives are interlink with drought and salinity stress tolerance in several plants due to associated with any of the membranes, protein complexes and enzymes to secure them by ROS (Dinakar and Bartels 2013). Fructans are one of the polymers of fructose and treated as main storage carbohydrate in many plant varieties. In plants, fructane accumulate within vacuoles and are thought to be involved in abiotic stress tolerance by stabilizing both the quaternary structure of proteins and the highly ordered structure of membranes (Ende, 2013). Osmotic stress takes place due to water deficiency can be resolved by osmotic adjustment via. collection of cellular solutes like proline (Farooq et al., 2013).

6.4.1.3.3 Proline

Proline is a potent non-enzymatic antioxidant and potential inhibitor of programme cell death. Proline is required by the microbes, animals and plants to mitigate the adverse effects of ROS (Ashraf and Foolad, 2007). It's have versatile action like stabilization of proteins, membranes and subcellular structures, and protecting cellular functions by scavenging ROS under osmotic stress condition (Trovato et al., 2008). Salt, drought and metal stress caused accumulation of proline which may be due to increased synthesis or decreased degradation (Cecchini et al., 2011).

Proline increase proportionally faster under stress than others amino acid in plant and been suggested as evaluating parameter for selecting drought resistant varieties (Jaleel et al., 2009). Content of proline is proportional to the different characteristics of plants like age of plant and leaves and position and parts of leaves. Gradually increasing of proline quantity due to draught was more harmful during flowering period in the comparison of vegetative stage. Proline content increased nearly ten times in drought stress during vegetative period of plants probably leading to balance osmotic capability which resulted in drought stress elimination in chickpea (Mafakheri et al., 2010). It is preserved in cells as an osmoprotectant behalf of different abiotic conditions and hydrogen peroxide treatments (Ozden et al., 2009). However, there are reports that heat stress caused a decline in the level of proline accumulation in wheat (Sridevi et al., 2009). Proline may act as both as a metabolic substrate to sustain the needs of rapidly dividing cells and, in turn as a feedback signal to fine-tune the developmental processes by supporting the energetic need of rapidly dividing or elongating cells (Lehmann et al., 2010).

6.5 POLYAMINES CATABOLISM

Polyamines (PAs) are low molecular-weight organic compound with two or more primary amino ($-NH_2$) groups (Fig. 6.8) which play their crucial roles in growth, development and stress responses of all living cells. The biological activity of polyamines is attributed to the cationic nature of these molecules (Bassard et al., 2010; Gupta et al., 2013). Several mechanisms are employed to achieve homeostasis of intracellular polyamine levels (Fig. 6.9). Spermidine (SPD) is formed from putrescine (PUT) as well as spermine (SPM) from Spermidine by successive summation of aminopropyl groups formulated from decarboxylated S-adenosylmethionine which is produced due to activity of SAMDC (Mattoo et al., 2010). Diverse climatic and environmental constrains of plants countered with help of polyamines. It confers plant tolerance to abiotic resistance through treating as direct ROS scavengers, binding to antioxidant enzyme to scavenge free radicals. It means that polyamines may modulate directly or indirectly plant defense response to stress through regulation of H_2O_2 production (Shi and Chan, 2014).

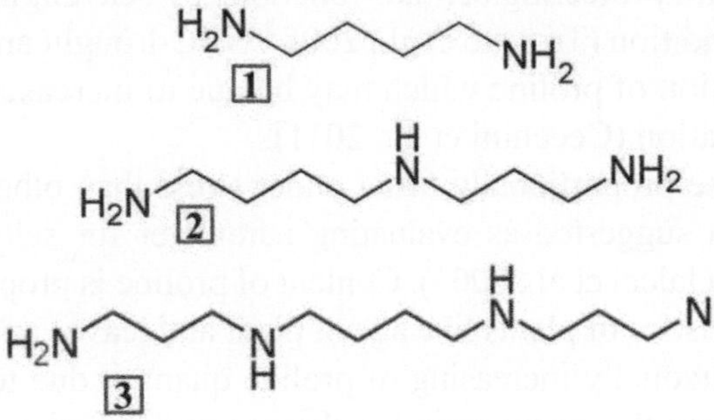

FIGURE 6.8   Structure of some important polyamines (PAs). 1) putrescine ($C_4H_{12}N_2$), 2) spermidine ($C_7H_{19}N_3$), and 3) spermine ($C_{10}H_{26}N_4$).

Polyamines levels are modulated by conjugation either with small molecules. Polyamines are reported to increase in response to various stresses. Putrescine implicated in direct removal of free radicals, thereby directly reducing oxidative damage, or may act indirectly by elevating contents of antioxidants through reduced TBARS content (Asthir et al., 2010, Pottosin et al., 2014). Decreasing the stress induced ROS production has been noticed due to PAs through induction of antioxidant enzymes and enhancement in glutathione levels

Net photosynthetic rate (NPR) and WUE in leaves of rice subjected to drought stress for seven days were significantly improved by spraying of plants with PUT, SPD and SPM solutions, while amongst the PAs, SPM was the most effective (Farooq et al., 2009). Spraying of SPD or SPM to panicle at early grain-filling stage significantly enhance the activities of sucrose synthetase, adenine diphosphoglucose pyrophosphorylase and soluble starch synthetase in inferior spikelets, resulting in improvement of grain-filling rate, seed-setting rate and grain weight (Tan et al., 2009).

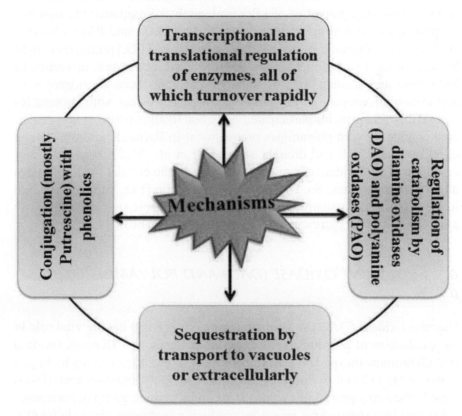

FIGURE 6.9   (See color insert.) Mechanisms involved in homeostasis of intracellular polyamine.

A good level of free PUT in initial stage of stress as well as insoluble conjugated PUT helps the plant to generate stress habitant characteristics. Higher level of free SPD or SPM and insoluble conjugated PUT in spite of initially collected PAs in water deficit condition are beneficial physiological traits of rice in development of drought habitant capability (Yang et al., 2007). Do et al., (2013) reported that SPD and SPM content increased significantly under long term drought stress conditions in rice as compared to PUT. No correlations between polyamine contents and drought tolerance were reported at genetic level.

The rise in PAs was higher in tolerant wheat genotypes suggesting the special role of SPD and SPM synthesis under water deficit conditions in plants. PAs having good antioxidative influence because of its different properties like anionic and cationic binding characteristics in radical scavenging, inhibiting properties of LP, metal-catalyzed oxidative reaction and by production of hydrogen peroxide (H_2O_2) by DAO and PAO (Grzesiak et al., 2013). Polyamine-derived hydrogen peroxide (H_2O_2) production might be work as dual purpose under abiotic stress conditions cause dissection of ROS level and accumulation of osmolytes also has been considered after application of exogenous polyamines, which consistent with proteomics caused different proteins participated in carbon fixation pathway were mediated commonly with polyamines pretreatment in Bermuda grass (*Cynodon dactylon*) under salt and drought stress (Shi et al., 2013). Any change or deviation of antioxidative enzymes may perturb the cellular redox of plants under abiotic stresses. So, PUT has diminished the H_2O_2 impact that could interfere with antioxidative enzymes activity by their significant increase after application of putrescine in *Salvinia natans* (Mandal et al., 2013).

6.5.1 DIAMINE OXIDASE (DAO) AND POLYAMINE OXIDASE (PAO)

Diamine oxidase (DAO) and polyamine oxidase (PAO) having vital role in the catabolism of polyamines inside of the plant tissues. Diamine Oxidase (DAO) promote the oxidative deamination of putrescine producing hydrogen peroxide (H_2O_2) and pyrroline and another side polyamine oxidase (PAO) cleaves the aminopropyl side chains at secondary amino group of polyamine substrates like spermidine, spermine producing hydrogen peroxide (H_2O_2), 1,3-diaminopropane and 1Δ-pyrroline, respectively (Moschou et al., 2012). These enzymes have a role in peroxidase-mediated lignin formation (Asthir et al., 2002) in barley. It has been found that DAO, PAO are involved in providing resistance to barley against *Blumeria graminis* (Asthir et al., 2002; Gemperlova, 2006). Different products of diamine oxidase after reaction like pyrroline, which catabolized further to α-aminobutyric acid and treated as main key of signal transduction pathways during stress response of several plants (Moschou et al., 2012).

Asthir et al. (2010) reported that temperature is directly proportional to activity rate of peroxidase (POX) and superoxide dismutase (SODs) activities and in other hand reversely proportional to diamine oxidase (DAO) and polyamine oxidase (PAO) activities were reduced in roots, shoots and developing grains. PUT application within the condition of high temperature increased the activity of diamine oxidase (DAO), polyamine oxidase (PAO), POX and SODs along with increased content of ascorbate and tocophereol in grains. Filippou et al. (2013) indicated that increased DAO activity carry out PUT oxidation which results in higher accumulation of gamma-amino butyric acid (about 39%) under salt stress which in turn help in providing tolerance to leaves of *Medicago truncatula* plants.

PUT is catabolized by cell wall-located diamine oxidase (DAO) and this catabolism generate hydrogen peroxide (H_2O_2), resulting in peroxidative damage of the plasma membrane and accelerated the leakiness of cell membrane to ions. Gupta et al. (2013) reported that substantial increase in PAs as well as higher accumulation of PAOs and polyamine (PA) biosynthesizing arginine decarboxylase were observed during heat stress in rice. PUT oxidation by DAOs is also increased during the early stages of drought treatment in Arabidopsis as reported by Alcazar et al. (2011). It also supports the model which predicts that a PA recycling-loop occurring under drought stress serves as amplification of ROS signaling by recurrent generation of H_2O_2 that may contribute to ROS-mediated drought stress responses. The creation of ROS was strongly linked in sensitive genotypes to catabolism of polyamines (PAs) by polyamine oxidase (PAO) and reduced polyamine oxidase (PAO) activity coincided with storage of proline. Within water deficit regions, the polyamine oxidase (PAO) activities of drought resistance seedlings was poor than in sensitive seedlings due to higher accumulation catabolism of polyamines (PAs) in tolerant seedlings (Lotfi et al., 2010). Catabolism of polyamines (PAs) catabolism considered an important factor in the control of stomatal aperture. Stomatal closure induced by ethylene in *Arabidopsis* involves induction of DAO and PAO activity, respectively, as a source of H_2O_2 under stress conditions (Hou et al., 2013). It has also been demonstrated that the activities of DAO contribute to ethylene induced H_2O_2 production under stress conditions.

6.6   CARBOHYDRATE METABOLISM

Carbohydrates perform various roles in plants, such as the energy storage (like starch, glycogen and sucrose) and its transportation as associates of structural components like cellulose in plants and chitin in animals, and so forth, Changes in carbohydrate content are of particular importance because of their direct proportionality with physiological processes like photosynthesis, translocation and respiration. Carbohydrates such as sugars (glucose, fructose, sucrose) and starch accumulates under a variety of environmental stresses, playing a leading role in osmoregulation, osmotic adjustment, carbon storage and radical scavenging (Jain, 2013). Sugars, known to act as signaling molecules regulating a variety of genes in different physiological pathways, also induce extracellular invertase (Rolland, 2002). Sugars regulate the expression of some genes involved in germination of seeds (Gill et al., 2002).

Carbohydrates are synthesized in source leaves and translocated to sink tissues in most species in the form of sucrose to maintain heterotrophic metabolism and multiplication in spite of stocked as sucrose or starch. Change in relations of source and sink associated with plants growth and multiplication as well as plants (like sessile life forms) developed controlling mechanisms which enable a fluctuated response with respect to assimilate bifurcation to particular need of the habitat like biotic and abiotic stress factors. That is why metabolism, partitioning and sensing is very important for all the stages of plant lifecycle (Chen et al., 2012).

6.6.1 SUCROSE

Sucrose known as a carbon and energy source, requires the cleavage of the α1-β2-glycosidic bond, which can be activated by two different enzymes like invertases and sucrose synthase cleavage. The reversible cleavage by the glycosyltransferase sucrose synthase to UDP-glucose and fructose conserves the bond energy. By contrast, β-fructosidase, β-fructofuranosidase and invertases catalyze the irreversible hydrolyses to glucose and fructose (Lakshmanan et al., 2013). On the basis of their pH optima, invertases have been categorized as acid invertase and neutral invertase (Chandra and Tyagi, 2004). Acid invertase and neutral invertases are defined on basis of pH required for maximum activity. Acid invertase is situated in cytoplasm and cell wall of intact tissue and alkaline invertase is located exclusively in the cytoplasm. Sucrose cleavage at the assimilate entry site allow building up and maintain sink strength by accumulating sugars (Ruana et al., 2010).

Sucrose phosphate synthase (SPS) is an important enzyme in source-sink relationships. The reaction catalyzed by sucrose phosphate synthase is not irreversible, so, this is the first committed reaction in sucrose synthesis pathway (Fig. 6.10). It can be induced by illumination both in the photosynthetic and non-photosynthetic tissues and its activity can be regulated by osmotic stress, light and darkness according to Lee et al. (2005). In photosynthetic tissue, sucrose phosphate synthase is regulated at different steps, including allosteric activation by glucose-6-phosphate and inhibition by inorganic phosphate (which permits sucrose synthesis to proceed at times when substrate is sufficient) and deactivated due to protein phosphorylation. In leaves, changes in SPS activity often correlate with change in the rate of sucrose synthesis and its export. Sucrose synthase is homologous to sucrose phosphate synthase, which catalyzes the penultimate step of sucrose.

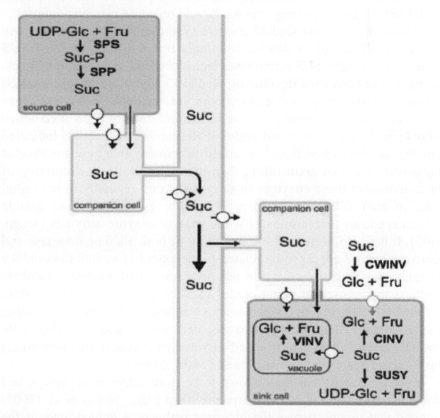

FIGURE 6.10 Overview of the life and death of a sucrose molecule. Following synthesis in the source, sucrose is transported to the sink, where is can be utilized or stored. Sucrose transport depends on sucrose transporters, as indicated by black circles with arrows. The light grey transporter sign represents hexose transporters. [Suc: sucrose; Fru: fructose; Glc: glucose; UDP-Glc: UDP-glucose; SPP: sucrose phosphatase; SPS: sucrose phosphate synthase; SUSY: sucrose synthase; CWINV: cell wall invertase; VINV: Vacuolar invertase, CINV: cytosolic/plastidic/mitochondrial invertases].

Sucrose remains predominant soluble sugar in the seed with hexoses for example glucose and fructose available at significantly lower levels.

Substantial enhancement of sucrolytic enzymes was regulated at the time of maximum seed filling stage (10–20 days after flowering). The activity pattern of sucrose synthase highly paralleled the phase of quick seed filling and therefore, can be interlinked with sink strength. Asthir et al. (2010) have suggested that conversion of sucrose may depend upon the processes occurring in the tissue at that time. As in actively growing and storing sink tissue, sucrose synthase (cleavage) activity may predominate and in other particularly those undergoing expansion, acid invertase may predominate.

Fallahi et al. (2008) studied sucrose synthase (SUSY) in developing seeds and siliques of *Arabidopsis thaliana* reveals diverse role of SUS during development. SUS activity was highest in developing whole siliques, young rosette leaves and developing seed at 13 days after flowering which indicate that SUS may involve in temporary starch deposition during early stages of seed development, while in the later stages SUS metabolize sucrose in the cotyledons and embryo. Research evidence has indicated that the activities of SUS and invertase in growing rice grains peaked at the earlier stages of grain filling then thereafter declined. Promoting of the activities of these enzymes in growing grains generally in the initial stage of grain filling may be causes from a starting period of soluble protein synthesis that terminates while peak of enzyme activities (Jenga, 2003). It has also been found that activity of both alkaline invertase and sucrose synthase appear to be related to the extent of assimilation uptake by numerous sink tissues while the involvement of sucrose phosphate synthase in sucrose importing organs remains equivocal. Acid invertase is found to be most active where sucrose exited the transport path rather than at storage. There is high sucrose content with high sucrose phosphate synthase and low acid invertase and sucrose synthase activity in the wheat during water scarcity (Magneschi and Perata, 2009).

Invertase activity has been shown to be most active in regions of cell elongation and lower in areas of mature plant parts. Wobus et al. (2005) reported that in sink tissues the invertase pathway is guided towards the development and cell elongation thus the invertase activity looks like essential to early seed growth, which is further supported by Sitthiwong et al. (2005) in developing fruit of soybean or elongating stem of bean and in wheat grain by Baraka (2008). Thus, the function of invertases in these tissues is to hydrolyze sucrose under condition where there is a high demand for hexoses.

The modulation of SPS activity could be linked to changes in the carbo-hydrate pool size brought about by a decreasing photosynthesis rate during drought. Increase in SPS activity during the first steps of the drought process is related to an increase in soluble sugar content in *Triticum durum* (including fructans) (Fresneau et al., 2007). Kaushal et al. (2013) measured levels of various carbohydrates and activities of key enzymes of sucrose metabolism in pollen grains of chickpea and cereals under stressed conditions and found that invertase, a sucrose-cleaving enzyme, was also inhibited along with SPS and SS which leads to pollen sterility. Invertase activity has been shown to be affected by elevated temperature in a variety of plants. Pattanagul and Thitisakasakul (2008) also reported the sensitivity of invertase to environmental stress in rice and they found reduction in acid invertase activity under stress conditions which leads to reduction in growth. Studies also reveal the role of sucrose metabolizing enzymes in growth of seedling under metal stress conditions.

Yativ et al. (2010) has studied about goof genetic changes in sugar content and in the ration of available sucrose, glucose and fructose in watermelon fruits. Cell wall surrounded invertase activity was high and stagnant within entire duration of fruit development in regulated lines and in genotypes storing low sucrose contents. Acid invertase activity was significantly lower in genotypes accumulating high levels of sucrose in the comparison of low-sucrose-accumulating genotypes. Reguera et al. (2013) reported that in the transgenic rice plants, photosynthesis was maintained at control levels during stress and their flag leaf showed increased sucrose phosphate synthase activity and reduced sucrose synthase and invertase activities, leading to increased sucrose contents. The increase in sugar content was not correlated with increases in glycolytic metabolites, reflecting that the sugar accumulation simultaneously constitute part of the osmotic adjustment response of wild-type plants to dehydration brought about through water deficit situation.

Several observation and study has been taken place about carbohydrate collection during various abiotic stresses in the temperate grasses and cereals from the Gramineae family where long term carbohydrate accumulation takes place at the time of reproductive development (Parvaiz and Satyawati, 2008; Chen et al., 2012). The accumulation of sugars in response to water scarcity is also quite well documented (Mostajeran and Rahimi-Eichi, 2009). Soluble sugars allow plants to maximize sufficient carbohydrate storage under standard conditions however, under stress they regulate internal osmolarity and stabilize the membranes and protein, thus protect them against different biotic and abiotic stresses (Andersen et al., 2011). Sugars substitutes for water in maintaining hydrophilic structures in their hydrated orientation, facilitates vitrification and also protects the cell during desiccation by formation of an intracellular glass which prevents cellular collapse and functional damage (Parvaiz and Satyawati, 2008). Osmotic adjusted cultivars undergo enhancement in sugar content resulted in better yield under stress conditions. Amount of glucose, fructose and sucrose in flag leaves and food grains of sensitive crops decreased due to water stress. These changes in flag leaves and food grain were occurred due to sharp reduction in cell wall invertase and soluble invertase activities, whereas a transient enhance in the activity of invertases of flag leaves has been noticed in resistive cultivar, however invertase activity reduced in leaves and grains of resistant cultivar due to substantial increase in sucrose synthase activity as seed development proceeds (Saeedipour and Moradi, 2011). Guan-fu et al. (2010) observed that water deficit stress increases sucrose content and decreases pollen grains in panicle and spikelets of rice genotypes. Water deficit induced spikelet sterility due to destruct antioxidant enzyme activities, decreasing of carbohydrate in spikelets and reducing pistil water strength during flowering of paddy.

Water stress during grain development causes large yield losses in cereals including wheat. This reduction is mainly accounted by a reduction in starch accumulation, since in general over 65% of cereal dry weight is accounted by starch (Veesar et al., 2007).

6.6.2 STARCH

Starch is the dominating substance and energy store in several main foods, including the seeds of cereal and vegetables like potato tuber. It is also have important application in the development and tolerance against several abiotic stresses in different plants. Singh et al. (2012) has studied that drought resistance varieties had rich amount of starch in the comparison of suscep-tible varieties around the vascular bundles and periphery area of roots after drought. The time when rice suffers due to drought at the flowering stage, substantial amount of carbohydrates collected in different parts of rice plants as well as other cereals also collect the carbohydrates that are of significance when plants experience water stress. Adak et al. (2011) observed in rice that under stress there is decrease in starch content accumulation however, there is accumulation of higher starch content in tolerant genotypes as compared to susceptible genotypes. Jin et al. (2013) observed that drought stress affects rice reproduction and results in severe yield loss.

The changes in the rate of starch accumulation correlated properly with the changes in the activities of sucrose synthase, invertase and sucrose phos-phate synthase within the grain filling duration (Jenga et al., 2003). Plants under submerged condition suffered due to significant loss of production by poor grain filling which was interlink to impeded carbohydrate metabolism in the plant tissues. Starch is evident that reduction of production under stress is attributed both by lower sink dimensions or sink capacity (number of panicles, in this case) as well as subdued carbohydrate metabolism in plants and its subsequent partitioning into the food grains (Adak et al., 2011).

6.7 PROTEIN AND THEIR SUB-FRACTIONS

Protein content is known to be influenced by many factors including the genotype, fertilizers, frequency of irrigation water and rainfall but is mainly influenced by the environment (Meena et al., 2012). The remobilization from leaves, stem and reproductive structure making lesser contribution for seed import, the photo assimilation was reduced in drought condition, indicating the reduction in rate of translocation of photoassimilate due to drought stress. Synthesis of protein responds to environmental stress such as heat shock, water stress by increase in some proteins with or without induction of unique stress proteins. The level of protein was found to increase in the seed obtained from the rice grains grown under high temperature stress that of compare seed obtained from irrigated and rainfed grown plants (Liao et al., 2014).

Protein is an important factor in evaluating the texture, pasting capability and sensory properties of milled rice. Several studies revealed that protein have vital role in evaluating the functional characteristics of the starch, including inhibiting the swelling of starch granules, minimizing the pasting and crystallizing ability and enhance the pasting temperature of the isolated rice starch and biochemical basis of rice quality (Ning et al., 2010). Seed-storage proteins are the crucial nitrogen source within germination and initial development of seedlings besides important roles for the configuration of the endosperm. Aleurone layer, embryo and the sub-aleurone layer of the endosperm of rice content concentrated protein in more concentrated form in comparison to the deeper starchy endosperm. According to their solubility, storage proteins can be classified into four types as water-soluble albumin, salt-soluble globulin, alkaline-soluble glutelin and alcohol soluble prolamin. Unlike other crops, up to 80% of the total stocked protein in rice is glutelin and prolamin takes up to 20–30% (Chen et al., 2012).

Commonly, prolamin has higher abundant characteristics in the upper layers, whereas glutelin higher in proportion toward the center of the endosperm. The water level of the soil has a dramatic influence on production and it also affects the qualitative characteristics of the grain. The kernels of paddy in flood-influenced samples appears soft and developed fissures which causes low head rice recoveries and the milled rice had lower kernel weight and protein content but showed higher amount of amylose and ash. Protein content is higher in non-flooded environment (Guo et al., 2007). The amount of protein among all resistant genotypes was obtained higher than susceptible ones. Water deficit condition results a significant alter in protein synthesizing apparatus of plant tissue and the capability for protein synthesis also reduces significantly as observed in response to water deficit.

Under water stress conditions globulin content, glutelin content, peak viscosity value, prolamin content and protein content get increased, whereas amylose content decreased therefore, the taste value of rice decreased as reported by Yan-dong et al. (2011). It was further supported by Weichert et al. (2010) which suggest that sucrose transporters have rate-limiting role in sucrose uptake potential into grains as sucrose transporter overexpression increases seed proteins and in rice so increasing grain protein yield.

6.8 YIELD COMPONENTS

Grain production is the conclusion of the expression and association of several plant growth constituents. Plants have several production-determining processes which react to water deficit. The scarcity of water promote to reduction in production trait of plants probably by disrupting leaf gas exchange characteristics which constraint the size of the source and sink tissues besides impairing of the phloem loading, assimilate translocation and dry matter portioning (Farooq et al., 2011). Drought stress inhibited the production of dry constituent's through its inhibitory impact on leaf expansion, and development in spite of consequent reduction in light interception. One of the significant reason of this was a decrease in assimilate flux to the developing ear beneath some threshold level essential for retaining the growth of food grains (Yadav et al., 2004). Due to water loss, substantial decrease in yield and yield components has been observed in plants (Anjum et al., 2011). Water deficit while vegetative, flowering and grain filling stages reduced mean grain yield by 21%, 50% and 21% on average in comparison to control respectively. Water scarcity at flowering stage had a higher grain yield reduction than other water deficit status. Decreasing of grain production of genotype vastly resulted from the reduction in productive panicle and filled grain percentage (Pirdashti et al., 2004).

Aerobic situations significantly inhibited structural development of root and responsible for significant change in biochemical root traits accounting for more concentration of hydrogen peroxide (24.6%) and proline (20%) besides that lesser concentration of total soluble protein (20%) that resulted in 17% reduction in production (Ghosh et al., 2012). In general, water stress results in reduced plant height and less number of productive tillers in the hybrids under aerobic condition as compared transplanted condition (Russo, 2000). It was reported that the number of panicles were significantly lower in plants, which exposed to aerobic condition at vegetative stage in comparison of the plants under flood condition. Similar results were obtained by Rajkumar and Ibrahim (2013) having high number of productive tillers per plant under transplanted conditions.

Commonly, higher numbers of panicle is associated with good number of grains panicle caused higher crop yield that is why hybrids with positive heterosis for panicle length are desirable. There was a general reduction in panicle length under aerobic condition. Under aerobic condition there was a general reduction in spikelet fertility level in most of the hybrid combinations as suggested by Singh et al. (2012). Earlier studies also suggested that the spikelet fertility is a reliable parameter for the mass screening of genotypes for yield performance under water deficit situations (Malarvizhi et al., 2009).

Weight of food grain is one of the important characteristics to decide the final grain production. It is multiplicative end product of many general constituents of production (Jenga et al., 2006). Decreasing in grain filling taking place due to decreased assimilate partitioning and activities of sucrose and starch synthesis enzymes. Impact of drought on crop production has been influenced with different factors like severity, duration and timing of stress besides of responses of plants after elimination of stress and interaction between stress and other important parameters (Jaleel et al., 2009). Water deficit led to reduction in production and yield components of plants could be ascribed to stomatal closure in response to low soil water content causes the reduction of CO_2 input results photosynthesis reduction (Anjum et al., 2011).

In an interaction between crop variety and environmental factors reveals that lowland cultivars had a robust phenological and productive traits under excess water logging areas, but the aerobic varieties had more production under aerobic aspects (Yang et al., 2007, 2008, 2012; Yang J and Zhang, 2010). It is also observed that none of the aerobic rice genotypes under aerobic conditions showed per plant more production than water logging indica rice cultivars cultivated with help of traditional method, indicating that there is essential to improve production capacity of aerobic rice. It is noticed that the rice production under aerobic conditions was reduced up to 21% in the comparison of under flooded conditions. These poor results were associated with decreased leaves growth and development of less tiller number during the vegetative stage (Okami et al., 2011). Water-saving irrigation significantly increased panicle number per meter square as aerobic conditions compared with transplanting and reduced plant height. High rice producing varieties in aerobic condition like "bunchy" appearance with profuse tillering and dwarf stature, only because of aerobic condition, not due to sowing practices (Okami et al., 2013). The 1000 grain weight is one of the important common traits which influence the yield. Based on the mean performance for 1000 grain weight was found to be less in aerobic conditions; however it was least affected trait among of all yield attributes (Patel et al., 2010).

6.9 SUMMARY AND CONCLUSION

Rice (*O. sativa* L.) serves as an important food crop for the majority of world population; nevertheless, the demand for rice production is still rising because of the continuous increase in population. To sustain present food self-sufficiency and to meet future food requirements, rice productivity has to be increased by 3% per annum. Transplanted (traditional) rice grows well in standing water with WHC of about 40–50% throughout a growing season; however, it will continue to undergo change in the coming years as a consequence of rising labor costs and increasing scarcity of water. These events are fuelling a shift from transplanting to direct seeding methods of rice establishment so as to reduce water consumption and produce rice with less water. Aerobic rice is a production system in which crop is grown from seeds directly in non-submerged, non-puddled condition in the field rather than by transplanting the seedlings in standing water. Under aerobic conditions, farmer can skip irrigation if soil moisture status is sufficient for crop even though plant undergoes stress due to less availability of water as compared to traditional methods. Rice can respond and adapt to water stress by altering its cellular metabolism and evoking various defense mechanisms. However, this results in lower plant growth rate, poor grain establishment, and hence, less yield. Antioxidant defense mechanism and metabolism of PAs along with carbohydrates play an important role in stress conditions but their relative contribution may vary genotypically along with different growing conditions (i.e., aerobic and transplanted). Measurement of such processes in response to aerobic (less water availability) stress in comparison to transplanted condition may provide valuable information on the mechanisms involved in ameliorating stress condition. Therefore, the summary of this chapter was to reviewed different cultivars of rice based on morpho-physiological and biochemical parameters under two planting conditions (aerobic *vs.* transplanted).

KEYWORDS

- **Aerobic rice and its cultivation**
- **antioxidant defense system**
- **ASA-GSH cycle**
- **biochemical evaluation**
- **biological membranes**
- **chlorophyll a/b and ratio**
- **DAO activity**
- **fertilizer use efficiency**
- **non-photosynthetic tissues**
- **oxidative polymerization and stress**
- **plant breeding programs**
- **progressive farming**
- **PUT oxidation**
- **ripening period**
- **signaling molecule**
- **spikelet sterility**
- **TBARS content**
- **water deficient period and stress**
- **water stress**
- **water-soluble antioxidants.**

REFERENCES

Adak, M. K.; Ghosh, K.; Dasgupta, N.; Gupta, S. D. K. Impeded Carbohydrate Metabolism in Rice Plants Under Submergence Stress. *Rice Sci.* **2011,** *18,* 116–126.

Ahmadi, N.; Courtois, B.; Khowaja, F.; Price, A.; Frouin, J.; Hamelin, C.; Ruiz, M. In *Meta-analysis of QTLs involved in rice root development using a QTL database.* Proceedings of the International Symposium on Root Biology and MAS Strategies for Drought Resistance Improvement in Rice, UAS-Bangalore, India, Sep 26–29. International Rice Research Institute: Los Banos Philippines, 2007, pp 29–30.

Ahmadizadeh, M. Physiological and Agro-Morphological Response to Drought Stress. *J. Sci. Res.* **2013,** *13,* 998–1009.

Alcazar, R.; Bitrian, M.; Bartels, D.; Koncz, C.; Altabella, T.; Tiburcio, F. A. Polyamine Metabolic Canalization in Response to Drought Stress in Arabidopsis and the Resurrection Plant Craterostigma Plantagineum. *Plant Signaling Behav.* **2011,** *6,* 243–250.

Al-Ghamdi, A. Evaluation of Oxidative Stres Tolerance in Two Wheat (*Triticum aestivum* L.) Cultivars in Response to Drought. *Int. J. Agric. Biol.* **2009,** *11*(1), 7–12.

Almeselmani, M.; Deshmukh, P. S.; Sairam, R. K.; Kushwaha, S. R.; Singh, T. P. Protective Role of Antioxidant Enzymes Under High Temperature Stress. *Plant Sci.* **2006,** *171,* 382–388.

Andersen, H. D.; Chunhua, W.; Lise, A.; Gunther, H. P.; Peter, W. Reconciliation of Opposing Views on Membrane–Sugar Interactions. *Proc. Natl. Acad. Sci. U. S. A.* **2011,** *108,* 1874–1878.

Anjum, S. A.; Xie, X.; Wang, L.; Saleem, M. F.; Man, C.; Lei, W. Morphological, Physiological and Biochemical Responses of Plants to Drought Stress. *Afr. J. Agric. Res.* **2011,** *6,* 2026–2032.

Apel, K.; Hirt, H. Reactive Oxygen Species: Metabolism, Oxidative Stress, and Signal Transduction. *Annu. Rev. Plant Biol.* **2004,** *55,* 373–399.

Ashraf, M.; Foolad, M. R. Roles of Glycine Betaine and Proline in Improving Plant Abiotic Stress Resistance. *Environ. Exp. Bot.* **2007,** *59,* 206–216.

Ashraf, M.; Harris, P. J. C. Photosynthesis Under Stressful Environments: an Overview. *Photosynthetica* **2013,** *51,* 163–190.

Asthir, B.; Duffus, C. M.; Smith, R. C.; Spoor, W. Diamine Oxidase is Involved in H_2O_2 Production in the Chalazal Cells During Barley Grain Filling. *J. Exp. Bot.* **2002,** *53,* 677–682.

Asthir, B.; Kaur, S.; Spoor, W.; Roitsch, T. Spatial and Temporal Dynamics of Peroxidase and Amine Oxidase Activity is Linked to Polyamines and Lignin in Wheat Grains. *Biol. Plant.* **2010,** *54,* 525–529.

Bacon, M. *Water Use Efficiency in Plant Biology;* Blackwell Publishing Ltd.: Oxford, **2004.**

Baraka, D. M. Osmotic Adjustment of Wheat Grain Germination to Hyper Osmotic Saline by Nicotine Hormone. *Res. J. Agric. Biol. Sci.* **2008,** *4,* 824–831.

Bartoli, G. C.; Facundo, G.; Martinez, D. E.; Guiamet, J. J. Mitochondria are the Main Target for Oxidative Damage in Leaves of Wheat (*Triticum aestivum* L.). *J. Exp. Bot.* **2004,** *55,* 1663–1669.

Bassard, J.; Ullmann, P.; Bernier, F.; Werck-Reichhart D Phenolamides: Bridging Polyamines to the Phenolic Metabolism. *Phytochemistry* **2010,** *71,* 1808–1824.

Basu, S.; Roychoudhury, A.; Saha, P.; Sengupta, D. N. Differential Antioxidative Responses of Indica Rice Cultivars to Drought Stress. *Plant Growth Regul.* **2010,** *60,* 51–59.

Bernier, J.; Atlin, G. N.; Serraj, R.; Kumar, A.; Spaner, D. Breeding Upland Rice for Drought Resistance. *J. Sci. Food Agric.* **2008,** *88,* 927–939.

Bhattacharjee, S. Reactive Oxygen Species and Oxidative Burst: Roles in Stress, Senescence and Signal Transduction in Plants. *Curr. Sci.* **2005,** *89,* 1113–1121.

Bhushan, L.; Ladha, J. K.; Gupta, R. K.; Singh, S.; Saharawat, Y. S.; Gathala, M.; Pathak, H. Saving of Water and Labour in a Rice–Wheat System with No Tillage and Direct Seedling Technologies. *Agron. J.* **2007,** *99,* 1288–1296.

Bijanzadeh, E.; Emam, Y. Effect of Defoliation and Drought Stress on Yield Components and Chlorophyll Content of Wheat. *Pak. J. Biol. Sci.* **2010,** *13,* 699–705.

Blair, A. W.; Smerdon, E. T. Unimodal Surface Irrigation Efficiency. *J. Irrig. Drain Eng.* **1988,** *114,* 156–168.

Bos, M. G.; Wolters, W. In *Project or Overall Irrigation Efficiency. Irrigation: Theory and Practice,* Proceedings of the International Conference held at the University of Southampton, Sept 12–15, Rydzewski, J. R., Ward, C. F., Eds.; Pentech Press: London, UK, 1989, 499–506.

Bouman, B. A. M.; Xiaoguang, Y.; Huaqui, W.; Zhiming, W.; Junfang, Z.; Changgui, W.; Bin, C. In *Aerobic Rice (Han Dao): a New Way Growing Rice in Water Short Areas,* Proceedings of the 12th International Soil Conservation Organization Conference, May 26–31, Tsinghua University: Beijing, China, 2002, 175–181.

Bouman, B. A. M.; Humphreys, E.; Tuong, T. P.; Barker, R. Rice and water. *Adv. Agron.* **2006a,** *92,* 187–237

Bouman, B. A. M.; Yang, X. G.; Wang, H.; Wang, Z.; Zhao, J.; Chen, B. Performance of Aerobic Rice Varieties Under Irrigated Conditions in North China. *Field Crop Res.* **2006b,** *97,* 53–65

Brouwer, C.; Prins, K.; Heibloem, M. Determination of the Irrigation Schedule for Paddy Rice. In: Irrigation Water Management: Irrigation Scheduling. Training Manual No.-4. Food and Agriculture Organization of the United Nations, Via delle Terme di Caracalla, 00100 Rome, Italy. URL: http://www.fao.org/docrep/t7202e/t7202e0g.jpg. Accessed on 05–05–2016. 1989.

Brown, H. *Irrigation, its Principles and Practice, as a Branch of Engineering.* 3rd ed.; London England, 1920, pp 32.

Candan, N.; Tarhan, L. The Correlation Between Antioxidant Enzyme Activities and Lipid Peroxidation Levels in Mentha Pulegium Organs Grown in Ca^{2+}, Mg^{2+}, Cu^{2+}, Zn^{2+} and Mn^{2+} Stress Conditions. *Plant Sci.* **2003,** *163,* 769–779.

Cecchini, N. M.; Monteoliva, M. I.; Alvarez, M. E. Proline Dehydrogenase Contributes to Pathogen Defense in Arabidopsis. *Plant Physiol.* **2011,** *155,* 1947–1959.

Cela, J.; Chang, C.; Munne-Bosch, S. Accumulation of Γ- Rather Than A-Tocopherol Alters Ethylene Signaling Gene Expression in the Vte4 Mutant of Arabidopsis Thaliana. *Plant Cell Physiol.* **2011,** *52,* 1389–1400.

Chandra, A.; Tyagi, A. Protein Profile of Two Rice Varieties by 2D-Gel Electrophoresis Under Moisture Stress. *Indian J. Biochem. Biophys.* **2004,** *41,* 191–194.

Chaves, M. M.; Flexas, J.; Pinheiro, C. Photosynthesis Under Drought and Salt Stress: Regulation Mechanisms from Whole Plant to Cell. *Ann. Bot.* **2009,** *103,* 551–560.

Chen, H. H.; Shen, Z. Y.; Li, P. H. Adaptability of Crop Plants to High Temperature Stress. *Crop Sci.* **1982,** *22,* 719–725.

Chen, Y.; Wang, M.; Ouwerkerk, P. B. F. Molecular and Environmental Factors Determining Grain Quality in Rice. *Food Energy Secur.* **2012,** *1,* 111–132.

Christiansen, J. E. Irrigation by sprinkling. California Agricultural Experiment Station Bulletin 670. University of California: Berkeley, CA **1942.**

Clemmens, A. J.; Solomon, K. H. Estimation of Global Irrigation Distribution Uniformity. *J. Irrig. Drain. Eng.* **1997,** *123*(6), 454–461.

Condon, A. G.; Richards, R. A.; Rebetzke, G. J.; Farquhar, G. D. Breeding for High Water-Use Efficiency. *J. Exp. Bot.* **2004,** *55,* 2447–2460.

Damanik, R. I.; Maziah, M.; Mohd, R. I.; Ahmad, S.; Zain, A. M. Responses of Antioxidative Enzymes in Malaysian Rice (*Oryza Sativa* L.) Cultivars Under Submergence Condition. *Acta. Physiol. Plant.* **2010,** *32,* 739–747.

Das, K. K.; Sarkar, R. K.; Ismail, A. M. Elongation Ability and Non-Structural Carbohydrate Levels in Relation to Submergence Tolerance in Rice. *Plant Sci.* **2005,** *168,* 131–136.

Del Rio, L. A.; Sandalio, L. M.; Corpas, F. J.; Palma, J. M.; Barroso, J. B. Reactive Oxygen Species and Reactive Nitrogen Species in Peroxisomes Production, Scavenging, and Role in Cell Signaling. *Plant Physiol.* **2006,** *141,* 330–335.

Din, J.; Khan, S. U.; Ali, I.; Gurmani, A. R. Physiological and Agronomic Response of Canola Varieties to Drought Stress. *J. Anim. Plant Sci.* **2011,** *21,* 78–82.

Dinakar, C.; Bartels, D. Desiccation Tolerance in Resurrection Plants, New Insights from Transcriptome, Proteome and Metabolome Analysis. *Front. Plant Sci.* **2013,** *4,* 1–14.

Do, P. T.; Degenkolbe, T.; Erban, A.; Heyer, A. G.; Kopka, J. Dissecting Rice Polyamine Metabolism Under Controlled Long-Term Drought Stress. *Plos One* **2013,** *8,* doi: 10.1371/journal.pone.0060325.

Dolatabadian, A.; Jouneghani, R. S. Impact of Exogenous Ascorbic Acid on Antioxidant Activity and Some Physiological Traits of Common Bean Subjected to Salinity Stress. *Not. Bot. Horti. Agrobo.* **2009,** *37,* 165–172.

Eduardo, C.; Iraida, A.; Jose, F. S.; Miguel, A. B.; Victoriano, V. Regulation of L-Ascorbic Acid Content in Strawberry Fruits. *J. Exp. Bot.* **2011,** *62,* 4191–4201.

El-Beltagi, H. S.; Mohamed, H. I. Reactive Oxygen Species, Lipid Peroxidation and Antioxidative Defense Mechanism. *Not. Bot. Horti. Agrobo.* **2013,** *41,* 44–57.

El-Enany, A. E.; AL-Anazi, A. D.; Dief, N.; Al-Taisan, W. A. Role of Antioxidant Enzymes in Amelioration of Water Deficit and Waterlogging Stresses on Vigna Sinensis Plants. *J. Biol. Earth Sci.* **2013,** *3,* 44–53.

Ellouzi, H.; Hamed, K. B.; Cela, J.; Muller, M.; Abdelly, C.; Munne-Bosch, S. Increased Sensitivity to Salt Stress in Tocopherol-Deficient Arabidopsis Mutants Growing in a Hydroponic System. *Plant Signaling Behav.* **2013,** *8.* http://dx.doi.org/10.4161/psb.23136.

Ende, W. V. E. Multifunctional Fructans and Raffinose Family Oligosaccharides. *Front. Plant Sci.* **2013,** *4,* 247.

Falk, J.; Munne-Bosch, S. Tocochromanol Functions in Plants: Antioxidation and Beyond. *J. Exp. Bot.* **2010,** *61,* 1549–1566.

Fallahi, H.; Scofield, G. N.; Badger, M. R.; Chow, W. S.; Furbank, R. T.; Ruan, Y. L. Localization of Sucrose Synthase in Developing Seed and Siliques of Arabidopsis Thaliana Reveals Diverse Roles for Sus During Development. *J. Exp. Bot.* **2008,** *59,* 3283–3295.

Farooq, M.; Wahid, A.; Kobayashi, N.; Basra, S. M. A.; Fujita, D. Plant Drought Stress: Effects, Mechanism and Management. *Agron. Sustainable Dev.* **2009,** *29,* 185–212.

Farooq, M.; Kobayashi, N.; Ito, O.; Wahid, A.; Serraj, R. Broader Leaves Result in Better Performance of Indica Rice Under Drought Stress. *J. Plant Physiol.* **2010,** *167,* 1066–1075.

Farooq, M.; Siddique, K. H. M.; Rehman, H.; Aziz, T.; Lee, D. J.; Wahid, A. Rice Direct Seeding: Experiences, Challenges and Opportunities. *Soil Till Res.* **2011,** *111,* 87–98.

Farooq, M.; Irfan, M.; Aziz, T.; Ahmad, I.; Cheema, S. A. Seed Priming with Ascorbic Acid Improves Drought Resistance of Wheat. *J. Agron. Crop Sci.* **2013,** *199,* 12–22.

Filippou, P.; Antoniou, C.; Fotopoulos, V. The Nitric Oxide Donor Sodium Nitroprusside Regulates Polyamine and Proline Metabolism in Leaves of Medicago Truncatula Plants. *Free Radical Biol. Med.* **2013,** *56,* 172–183.

Fortier, S. Irrigation Requirements of the Arid and Semi-Arid Lands of the Missouri and Arkansas River Basins. USDA Technical Bulletin No 36. Washington DC, 1928, pp.112.

Foyer, C. H.; Noctor, G. Ascorbate and Glutathione: The Heart of the Redox Hub. *Plant Physiol.* **2011,** *155,* 2–18.

Frei, M.; Tanaka, J. P.; Wissuwa, M. Genotypic Variation in Tolerance to Elevated Ozone in Rice: Dissection of Distinct Genetic Factors Linked to Tolerance Mechanisms. *J. Exp. Bot.* **2008,** *59,* 3741–3752.

Fresneau, C.; Ghashghaie, J.; Gabriel, C. Drought Effect on Nitrate Reductase and Sucrose-Phosphate Synthase Activities in Wheat (*Triticum durum* L.): Role of Leaf Internal Co_2. *J. Exp. Bot.* **2007,** *58,* 2983–2992.

Fukai, S.; Akihiko, K.; Babu, R. C.; Boopathi, N. M. Phenotypic and Genotypic Analysis of Drought-Resistance Traits for Development of Rice Cultivars Adapted to Rainfed Environments. *Field Crop Res.* **2008,** *109,* 1–23.

Garg, N.; Manchanda, G. Ros Generation in Plants: Boon Or Bane? *Plant Biosys.* **2009,** *143,* 81–96.

Geetalakshmi, V.; Ramesh, T.; Palamuthirsolai, A.; Lakshmanan, A. Productivity and Water Usage of Rice as Influenced by Different Cultivation Systems. *Madras Agric. J.* **2009,** *96,* 349–352.

Gemperlova, L.; Novakova, M.; Vankova, R.; Eder, J.; Cvikrova, M. Diurnal Changes in Polyamine Content, Arginine and Ornithine Decarboxylase and Diamine Oxidase in Tobacco Leaves. *J. Exp. Bot.* **2006,** *57,* 1413–1421.

Ghosh, A.; Dey, R.; Singh, O. N. Improved Management Alleviating Impact of Water Stress on Yield Decline of Tropical Aerobic Rice. *Agron. J.* **2012,** *104,* 584–588.

Gill, P. K.; Sharma, A. D.; Singh, P.; Bhullar, S. S. Osmotic Stress-Induced Changes in Germination, Growth and Soluble Sugar Content of *Sorghum Bicolor* (L.) Moench Seeds. *Bulg. J. Plant Physiol.* **2002,** *28,* 12–25.

Gill, S.; Tuteja, N. Reactive Oxygen Species and Antioxidant Machinery in Abiotic Stress Tolerance in Crop Plants. *Plant Physiol. Biochem.* **2010,** *48,* 909–930.

Govender, V.; Avelinga, T. A. S.; Kritzingera, Q. The Effect of Traditional Storage Methods on Germination and Vigour of Maize (*Zea mays* L.) from Northern Kwazulu-Natal and Southern Mozambique. *S. Afr. J. Bot.* **2008,** *74,* 190–196.

Gowda, V. R. P.; Henrya, A.; Yamauchic, A.; Shashidharb, H. E.; Serraj, R. Root Biology and Genetic Improvement for Drought Avoidance in Rice. *Field Crop Res.* **2011,** *122,* 1–13.

GRDC. Water Use Efficiency, Fact Sheet on Northern Region Converting Rainfall to Grain. Grain Research and Development Corporation, Australia. 2009. URL: https://grdc.com.au/. Accessed on 05–05–2016.

Grzesiak, M.; Filek, M.; Barbasz, A.; Kreczmer, B.; Hartikainen, H. Relationships Between Polyamines, Ethylene, Osmoprotectants and Antioxidant Enzymes Activities in Wheat Seedlings After Short-Term Peg- and Nacl-Induced Stresses. *Plant Growth Regul.* **2013,** *69,* 177–189.

Guan-fu, F.; Jian, S.; Yu-rong, L.; Ming-kai, Y.; Jie, X.; Long-xing, T. Alterations of Panicle Antioxidant Metabolism and Carbohydrate Content and Pistil Water Potential Involved in Spikelet Sterility in Rice Under Water-Deficit Stress. *Rice Sci.* **2010,** *17,* 303–310.

Guo, Y.; Mu, P.; Liu, J.; Lu, Y.; Li, Z. QTL Mapping and QxE Interactions of Grain Cooking and Nutrient Qualities in Rice Under Upland and Lowland Environments. *J. Genet. Genomics* **2007**, *34*, 420–428.

Gupta, K.; Dey, A.; Gupta, B. Plant Polyamines in Abiotic Stress Responses. *Acta. Physiol. Plant* **2013**, *35*, 2015–2036.

Gur, A.; Demirel, U.; Ozden, M.; Kahraman, A.; Çopur, O. Diurnal Gradual Heat Stress Affects Antioxidant Enzymes, Proline Accumulation and Some Physiological Components in Cotton (*Gossypium hirsutum* L.). *Afr. J. Biotech.* **2010**, *9*, 1008–1015.

Hall, W. A. Performance Parameters of Irrigation Systems. *Trans. ASAE.* **1960**, *3*(1), 75–76, 81.

Hansen, V. E. New Concepts in Irrigation Efficiency. Transactions of the ASAE 1960: **1960**, 55–64.

Hart, W. E.; Reynolds, W. N. Analytical Design of Sprinkler Systems. *Transactions ASAE.* **1965**, *8*, 83–85, 89.

Hollander-Czytko, H.; Grabowski, J.; Sandorf, I.; Weckermann, K.; Weiler, E. W. Tocopherol Content and Activities of Tyrosine Aminotransferase and Cystine Lyase in Arabidopsis Under Stress Conditions. *J. Plant Physiol.* **2005**, *162*, 767–770.

Hossain, M. A.; Ismail, M. R.; Uddin, M. K.; Islam, M. Z.; Ashrafuzzaman, M. Efficacy of Ascorbate-Glutathione Cycle for Scavenging H2o2 in Two Contrasting Rice Genotypes During Salinity Stress. *Aust. J. Crop Sci.* **2013**, *7*, 1801–1808.

Hou, Z. H.; Liu, G. H.; Wang, L. X.; Liu, X. Regulatory Function of Polyamine Oxidase-Generated Hydrogen Peroxide in Ethylene-Induced Stomatal Closure in *Arabidopsis thaliana*. *J. Integr. Agri.* **2013**, *12*, 251–262.

Ibrahim, A.; Quick, J. S. Heritability of Heat Tolerance in Winter and Spring Wheat. *Crop Sci.* **2001**, *41*, 1401–1405.

Israelsen, O. W. *Irrigation Principles and Practices;* Wiley and Sons: New York, 1932, pp.411.

Jain, M. Emerging Role of Metabolic Pathways in Abiotic Stress Tolerance. *J. Plant Biochem. Physiol.* **2013**, *1*, 108 http://dx.doi:10.4172/jpbp.1000108.

Jaleel, C. A.; Ksouri, R.; Raghupathi, G.; Paramasivam, P.; Jallali, I.; Hammed, J. A.; Zhao, C. X.; Shao, H. B.; Rajaram, P. Antioxidant Defense Responses: Physiological Plasticity in Higher Plants Under Abiotic Constraints. *Acta. Physiol. Plant.* **2009**, *31*, 427–436.

Jamil, M.; Rha, E. S. NaCl Stress-Induced Reduction in Grwoth, Photosynthesis and Protein in Mustard. *J. Agric. Sci.* **2013**, *5*, 102–114.

Jenga, T. L.; Tsenga, T. H.; Wanga, C. S.; Chenb, C. L.; Sung, J. M. Starch Biosynthesizing Enzymes in Developing Grains of Rice Cultivar Tainung 67 and Its Sodium Azide-Induced Rice Mutant. *Field Crop Res.* **2003**, *84*, 261–269.

Jenga, T. L.; Tsenga, T. H.; Wanga, C. S.; Chenb, C. L.; Sung, J. M. Yield and Grain Uniformity in Contrasting Rice Genotypes Suitable for Different Growth Environments. *Field Crop Res.* **2006**, *99*, 59–66.

Jin, Y.; Yang, H.; Wei, Z.; Ma, H.; Ge, X. Rice Male Development Under Drought Stress: Phenotypic Changes and Stage-Dependent Transcriptomic Reprogramming. *Mol. Plant.* **2013**, *6*, 1630–1645.

Kanwal, S.; Ashraf, M.; Shahbaz, M.; Iqbal, Y. M. Influence of Saline Stress on Growth, Gas Exchange, Mineral Nutrients and Non-Enzymatic Antioxidatns in Mungbean [(*Vigna radiata* (L.) wilczek]. *Pak. J. Bot.* **2013**, *45*, 763–771.

Kato, Y.; Katsura, K. Rice Adaptation to Aerobic Soils: Physiological Considerations and Implications for Agronomy. *Plant Prod. Sci.* **2014**, *17*, 1–12.

Kato, Y.; Okami, M. Root Morphology, Hydraulic Conductivity and Plant Water Relations of High-Yielding Rice Grown Under Aerobic Conditions. *Ann. Bot.* **2011**, *108*, 575–583.

Kaushal, N.; Awasthi, R.; Gupta, K.; Gaur, P.; Kadambot, H.; Siddique, M.; Nayyar, H. Heat-Stress-Induced Reproductive Failures in Chickpea (*Cicer arietinum*) are Associated with Impaired Sucrose Metabolism in Leaves and Anthers. *Funct. Plant Biol.* **2013**, *40*, 1334–1349.

Ke, Y.; Han, G.; He, H.; Li, J. Differential Regulation of Proteins and Phosphoproteins in Rice Under Drought Stress. *Biochem. Biophys. Res. Commun.* **2009**, *379*, 133–138.

Konwar, P. B. Characterization of some selected Ahu rice cultivars under moisture stress condition. M. Sc. Thesis Assam Agricultural University, Jorhat. 2009.

Krasensky, J; Jonak, C. Drought, Salt and Temperature Stress-Induced Metabolic Rearrangements and Regulatory Networks. *J. Exp. Bot.* **2012**, *63*, 1593–1605.

Kreye, C.; Bouman, B. A. M.; Faronilo, J. E.; Liorca, L. Causes for Soil Sickness Affected by Early Plant Growth in Aerobic Rice. *Field Crops Res.* **2009**, *114*, 182–187.

Kumar, R. R.; Goswami, S.; Sharma, S. K.; Singh, K.; Gadpayle, K. A.; Kumar, N.; Rai, G. K.; Singh, M.; Rai, R. D. Protection Against Heat Stress in Wheat Involves Change in Cell Membrane Stability, Antioxidant Enzymes, Osmolyte, H_2O_2 and Transcript of Heat Shock Protein. *Int. J. Plant Physiol. Biochem.* **2012**, *4*, 83–91.

Kumar, R. R.; Goswami, S.; Singh, K.; Rai, G. K.; Rai, R. D. Modulation of Redox Signal Transduction in Plant System Through Induction of Free Radical /ROS Scavenging Redox-Sensitive Enzymes and Metabolites. *Aust. J. Crop Sci.* **2013**, *7*, 1744–1751.

Lafitte, H. R.; Bennet, J. Requirements for aerobic rice: Physiological and molecular considerations. In: Water-Wise Rice Production" (B. A. M. Bouman, H. Hengsdijk, B. Hardy, P. S. Bindraban, T. P. Tuong, and J. K. Ladha, Eds.) 259–274. Proceedings of a Thematic Workshop on Water-Wise Rice Production, 8–11 April 2002 at IRRI Headquarters in Los Ban os, Philippines. International Rice Research Institute, Los Banos, Philippines. 2003.

Lakshmanan, M.; Zhang, Z.; Mohanty, B.; Kwon, J. Y.; Choi, H. Y.; Nam, H. J.; Kim, D.; Lee, D. Y. Elucidating the Rice Cells Metabolism Under Flooding and Drought Stresses using Flux-Based Modelling and Analysis. *Plant Physiol.* **2013**, *162*, 2140–2150. doi:http://dx.doi.org10.1104/ pp.113.220178

Lal, B.; Gautam, P.; Joshi, E. Different Rice Establishment Methods for Producing More Rice Per Drop of Water: A Review. *Int. J. Res. Biosci.* **2013**, *2*, 1–12.

Lambers, H.; Chapin, F. S.; Pons, T. L. *Plant Physiological Ecology;* Springer: New York, 2008;

Lee, M. O.; Yang, C. C.; Su, J. C.; Lee, P. D. Biochemical Characterization of Rice Sucrose Phosphate Synthase Under Illumination and Osmotic Stress. *Bot. Bull. Acad. Sin.* **2005**, *46*, 43–52.

Lehmann, S.; Funck, D.; Szabados, L.; Rentsch, D. Proline Metabolism and Transport in Plant Development. *Amino Acids* **2010**, *39*, 949–962.

Liang, Y.; Feng, H.; Yang, M.; Yu, J. Antioxidative Defenses and Water Deficit-Induced Oxidative Damage in Rice (*Oryza sativa* L.) Growing on Non-Flooded Paddy Soils with Ground Mulching. *Plant Soil* **2003**, *257*, 407–416.

Liao, J.; Zhou, H.; Zhang, H.; Zhong, P.; Huang, Y. Comparative Proteomic Analysis of Differentially Expressed Proteins in the Early Milky Stage of Rice Grains During High Temperature Stress. *J. Exp. Bot.* **2014**, *65*, 655–671.

Lotfi, N.; Vahdati, K.; Kholdebarin, B.; Hassani, D.; Amiri, R. Assessment of Droughtstress in Different Growth Stages of Persian Walnut (*Juglans regia* L.) Genotypes. Msc. Thesis, 2010, pp 193.

Lutje, S.; Bottger, M.; Doring, O. Are Plants Stacked Neutrophyles? Comparison of of Pathogen Induced Oxidative Burst in Plants and Mammals In *Progress in Botany.* Esser, K., Kadereit, J. W., Lutje, U., Runge, M., Eds.; Springer: Berlin, Heidelberg, 2000; Vol. 61, pp 187–222.

Mafakheri, A.; Siosemardeh, A.; Bahramnejad, B.; Struik, P. C.; Sohrabi, Y. Effect of Drought Stress on Yield, Proline and Chlorophyll Contents in Three Chickpea Cultivars. *Aust. J. Crop Sci.* **2010,** *4,* 580–585.

Magneschi, L.; Perata, P. Rice Germination and Seedling Growth in the Absence of Oxygen. *Ann. Bot.* **2009,** *103,* 181–196.

Malarvizhi, D.; Thiyagarajan, K.; Vijayalakshmi, C.; Manonmani, S. Exploration of Heterosis for Yield and Morpho Physiological Traits in Hybrid Rice (*Oryza sativa* L.): A Comparative Study Under Flooded and Aerobic Conditions. *Indian J. Genet. Plant Breed.* **2009,** *69,* 371–382.

Mandal, C.; Ghosh, N.; Adak, M. K.; Dey, N. Interaction of Polyamine on Oxidative Stress Induced by Exogenously Applied Hydrogen Peroxide in Salvinia Natans Linn. *Theor. Exp. Plant Physiol.* **2013,** *25,* 223–230.

Marrero-Ponce, Y.; Khan, M. T. H.; Casanola-Martin, G. M.; Ather, A.; Sultankhodzhaev, M. N.; Torrens, F.; Rotondo, R. Prediction of Tyrosinase Inhibition Spectra for Chemicals Using Novel Atom-Based Bilinear Indices. *ChemMedChem.* **2007,** *2,* 449–478.

Martin, G.; James, P. P. K.; Subramanian, E. Identification on Suitable Rice Variety Adaptability to Aerobic Irrigation. *J. Agric. Bio. Sci.* **2007,** *2,* 1–3.

Mattoo, A. K.; Minocha, S. C.; Minocha, R.; Handa, A. K. Polyamines and Cellular Metabolism in Plants: Transgenic Approaches Reveal Different Responses to Diamine Putrescine Ver. Higher Polyamines Spermidine and Spermine. *Amino Acids* **2010,** *38,* 405–413.

Meena, K. R.; Verulkar, S. B.; Chandel, G. Nutrient Characters Analysis in Rice Genotypes Under Different Environmental Conditions. *Bull. Environ. Pharmacol. Life Sci.* **2012,** *1,* 61–64.

Merriam, J. L.; Keller, J. *Farm Irrigation System Evaluation: A Guide for Management. Department of Agricultural Engineering;* Utah State University: Logan, Utah. USA, 1978.

Mikuła, A.; Niedzielski, M.; Rybczyski, J. J. The Use of TTC Reduction Assay for Assessment of Gentiana Spp. Cell Suspension Viability After Cryopreservation. *Acta. Physiol. Plant.* **2006,** *28,* 315–324.

Miro, B.; Ismail, A. M. Tolerance of Anaerobic Conditions Caused by Flooding During Germination and Early Growth in Rice (*Oryza sativa* L.). *Front. Plant Sci.* **2013,** *4,* 269.

Mittler, R. Oxidative Stress, Antioxidants and Stress Tolerance. *Trends Plant Sci.* **2002,** *7,* 405–410.

Moschou, P. N.; Paschalidis, K. A.; Roubelakis-Angelakis, K. A. The Polyamines and Their Catabolic Products are Significant Players in the Turnover of Nitrogenous Molecules in Plants. *J. Exp. Bot.* **2012,** *63,* 695–709.

Mostajeran, A.; Rahimi-Eichi, V. Effects of Drought Stress on Growth and Yield of Rice (*Oryza sativa* L.) Cultivars and Accumulation of Proline and Soluble Sugars in Sheath and Blades of Their Different Ages Leaves. *J. Agric. Environ. Sci.* **2009,** *5,* 264–272.

Munawarti, A.; Taryono; Semiarti, E.; Sismandri, H. P. Tolerance of Accessions of Glagah (Saccharum Spontaneum) to Drought Stress and Their Accumulation of Proline. *Am. J. Agric. Biol. Sci.* **2013,** *8,* 1–11.

Munne-Bosch, S.; Falk, J. New Insights Into the Function of Tocopherols in Plants. *Planta* **2004,** *218,* 323–326.

Nakagami, H.; Pitzschke, A.; Hirt, H. Emerging Map Kinase Pathways in Plant Stress Signalling. *Trends Plant Sci.* **2005,** *10,* 339–346.

Ning, H.; Qiao, J.; Liu, Z.; Lin, Z.; Li, G.; Wang, Q.; Wang, S.; Ding, Y. Distribution of Proteins and Amino Acids in Milled and Brown Rice as Affected by Nitrogen Fertilization and Genotype. *J. Cereal Sci.* **2010,** *52,* 90–95.

Nounjan, N.; Theerakulpisut, P. Effects of Exogenous Proline and Trehalose on Physiological Responses in Rice Seedlings During Salt-Stress and After Recovery. *Plant Soil Environ.* **2012,** *58,* 309–315.

Okami, M.; Kato, Y.; Yamagishi, J. Role of Early Vigor in Adaptation of Rice to Water-Saving Aerobic Culture: Effects of Nitrogen Utilization and Leaf Growth. *Field Crop Res.* **2011,** *124,* 124–131.

Okami, M.; Kato, Y.; Yamagish, J. Grain Yield and Leaf Area Growth of Direct-Seeded Rice on Flooded and Aerobic Soils in Japan. *Plant Prod. Sci.* **2013,** *16,* 276–279.

Omezzine, A.; Zaibet, L. Management of Modern Irrigation Systems in Oman: Allocative Vs. Irrigation Efficiency. *Agric. Water Manage.* **1998,** *37,* 99–107.

Ozden, M.; Demirel, U.; Kahraman, A. Effects of Proline on Antioxidant System in Leaves of Grapevine (*Vitis vinifera* L.) Exposed to Oxidative Stress by H_2O_2. *Sci. Hortic.* **2009,** *119,* 163–168.

Parvaiz, A.; Satyawati, S. Salt Stress and Phyto-Biochemical Responses of Plants—a Review. *Plant Soil Environ.* **2008,** *54,* 89–99.

Patel, D. P.; Das, A.; Munda, G. C.; Ghosh, P. K.; Bordoloi, J. S.; Kumar, M. Evaluation of Yield and Physiological Attributes of High Yielding Rice Varieties Under Aerobic and Flood Irrigated Management Practices in Mid-Hills Ecosystem. *Agric. Water Manage.* **2010,** *97,* 1269–1276.

Patel, P. K.; Hemantaranjan, A. Salicylic Acid Induced Alteration in Dry Matter Partitioning, Antioxidant Defence System and Yield in Chickpea (*Cicer arietinum* L.) Under Drought Stress. *Asian J. Crop Sci.* **2012,** *4,* 86–102.

Pattanagul, W.; Thitisakasakul, M. Effect of Salinity Stress on Growth and Carbohydrate Metabolism in Three Rice (*Oryza sativa* L) Cultivars Differing in Salinity Tolerance. *Indian J. Exp. Bio.* **2008,** *46,* 736–742.

Pirdashti, H.; Sarvestani, Z. T.; Nematzadeh, C.; Ismail, A. Study of water stress effects in different growth stages on yield and yield components of different rice (*Oryza sativa* L.) cultivars. Proc. 4th Intl. Crop Sci. Cong., 26 Sep-1 Oct, Brisbane, Australia. **2004.**

Pottosin, I.; Velarde-Buendia, A. M.; Bose, J. Cross-Talk Between Reactive Oxygen Species and Polyamines in Regulation of Ion Transport Across the Plasma Membrane: Implications for Plant Adaptive Responses. *J. Exp. Bot.* **2014,** *65,* 1271–1283. doi:10.1093/jxb/ert423.

Pourcel, L.; Routaboul, J. M.; Cheynier, V. Flavonoid Oxidation in Plants: From Biochemical Properties to Physiological Functions. *Trends Plant Sci.* **2007,** *12,* 29–36.

Pyngrope, S.; Kumari, B.; Dubey, R. S. Oxidative Stress, Protein Carbonylation, Proteolysis and Antioxidative Defense System as a Model for Depicting Water Deficit Tolerance in Indica Rice Seedlings. *Plant Growth Regul.* **2013,** *69,* 149–165.

Qin, J.; Wang, X.; Hu, F.; Li, H. Growth and Physiological Performance Responses to Drought Stress Under Non-Flooded Rice Cultivation with Straw Mulching. *Plant Soil Environ.* **2010,** *56,* 51–59.

Rahman, B.; Zhang, J. Specialty. The Forgotten Resource of Elitefeatures of Rice. *Rice* **2013,** *6,* 41.

Rajkumar, S.; Ibrahim, S. M. Aerobic Rice: Identification of Suitable Rice Hybrids Adaptability to Aerobic Condition in Water-Short Areas. *Int. J. Agric. Sci.* **2013,** *3,* 586–595.

Reguera, M.; Peleg, Z.; Yasser, M. A.; Tumimbang, E. B.; Delatorre, C. A.; Blumwald, E. Stress-Induced Cytokinin Synthesis Increases Drought Tolerance Through the Coordinated Regulation of Carbon and Nitrogen Assimilation in Rice. *Plant Physiol.* **2013,** *163,* 1609–1622.

Rolland, F.; Moore, B.; Sheen, J. Sugar Sensing and Signaling in Plants. *Plant Cell.* **2002,** *14,* 185–205.

Romero-Puertas, M. C.; Corpas, F. J.; Sandalio, L. M.; Leterrier, M.; Rodríguez-Serrano, M.; Del Río, L. A.; Palma, J. M. Glutathione Reductase from Pea Leaves: Response to Abiotic Stress and Characterization of the Peroxisomal Isozyme. *New Phytol.* **2006,** *170,* 43–52.

Rostami, A. A.; Rahemi, M. Screening Drought Tolerance in Caprifig Varieties in Accordance to Responses of Antioxidant Enzymes. *World Appl. Sci. J.* **2013,** *21,* 1213–1219.

Ruana, Y.; Jin, Y.; Yang, D. Y.; Lid, G.; Boyer, J. S. Sugar Input, Metabolism, and Signaling Mediated by Invertase: Roles in Development, Yield Potential and Response to Drought and Heat. *Mol. Plant* **2010,** *3,* 942–955.

Russo, S. Preliminary Studies on Rice Varieties Adaptability to Aerobic Irrigation. *Cah. Options Mediterr.* **2000,** *15,* 35–39.

Saeedipour, S.; Mordai, F. Comparison of the Drought Stress Responses of Tolerant and Sensitive Wheat Cultivars During Grain Filling: Impact of Invertase Activity on Carbon Metabolism During Kernel Development. *J. Agric. Sci.* **2011,** *3,* 32–44.

Sandhu, N.; Jain, S.; Battan, K. R.; Jain, R. K. Aerobic Rice Genotypes Displayed Greater Adaptation to Water-Limited Cultivation and Tolerance to Polyethyleneglycol-6000 Induced Stress. *Physiol. Mol. Biol. Plant* **2012,** *18,* 33–43.

Sarvestani, Z. T.; Pirdashti, H.; Sanavy, S. A.; Balouchi, H. Study of Water Stress Effects in Different Growth Stages on Yield and Yield Components of Different Rice (*Oryza sativa* L.) Cultivars. *Pak. J. Biol. Sci.* **2008,** *11,* 1303–1309.

Seck, P. A.; Diagne, A.; Mohanty, S.; Wopereis, M. C. S. Crops That Feed the World: Rice. *Food Secur.* **2012,** *4,* 7–24.

Shao, H B.; Chu, L. Y. Plant Molecular Biology in China: Opportunities and Challenges. *Plant Mol. Biol. Rep.* **2005,** *23,* 345–358.

Sharma, P.; Jha, A. B.; Dubey, R. S.; Pessarakli, M. Reactive Oxygen Species, Oxidative Damage, and Antioxidative Defense Mechanism in Plants Under Stressful Conditions. *J. Bot.* **2012,** http://dx.doi:10.1155/2012/217037.

Sharma, A. D.; Rakhra, G.; Mehta, S.; Mamik, S. Accumulation of Class-III Type of Boiling Stable Peroxidases in Response to Plant Growth Hormone ABA in *Triticum aestivum* Cultivars. *Plant Sci. Today* **2014,** *1,* 3–9.

Shi, H.; Ye, T.; Chan, Z. Comparative Proteomic and Physiological Analyses Reveal the Protective Effect of Exogenous Polyamines in the Bermudagrass (*Cynodon dactylon*) Response to Salt and Drought Stresses. *J. Proteome Res.* **2013,** *12,* 4951–4964.

Shi, H.; Chan, Z. Improvement of Plant Abiotic Stress Tolerance Through Modulation of the Polyamine Pathway. *J. Integr. Plant Biol.* **2014,** *56,* 114–121.

Sikuku, P. A.; Netondo, G. W.; Onyango, J. C.; Musyimi, D. M. Chlorophyll Fluorescence, Protein and Chlorophyll Content of Three Nerica Rainfed Rice Varieties Under Varying Irrigation Regimes. *J. Agric. Biol. Sci.* **2010,** *5,* 19–25.

Singh, A. K.; Dubey, R. S. Changes in Chlorophyll a and B Contents and Activities of Photosystems I and II in Rice Seedlings Induced by NaCl. *Photosynthetica* **1995,** *31,* 489–499.

Singh, G.; Sharma, N. Antioxidative Response of Various Cultivars of Sorghum (*Sorghum bicolor* L.) to Drought Stress. *J. Stress Physiol. Biochem.* **2013**, *9*, 139–151.

Singh, N.; Kaur, R.; Sharma, N.; Mahajan, G.; Bharaj, T. S. Changes in Yield and Grain Quality Characteristics of Irrigated Rice (*Oryza sativa*) Genotypes Under Aerobic Conditions. *Indian J. Agric. Sci.* **2012**, *82*, 589–595.

Sitthiwong, K.; Matsui, T.; Okuda, N.; Suzuki, H. Changes in Carbohydrate Content and the Activities of Acid Invertase, Sucrose Synthase and Sucrose Phosphate Synthase in Vegetable Soybean Fruit Development. *Asian J. Plant Sci.* **2005**, *4*, 684–690.

Smirnoff, N. Ascorbic Acid: Metabolism and Functions of a Multi-Facetted Molecule. *Curr. Opin. Plant Biol.* **2000**, *3*, 229–235.

Soleimanzadeh, H. Response of Sunflower (*Helianthus annuus* L.) to Selenium Application Under Water Stress. *World Appl. Sci. J.* **2012**, *17*, 1115–1119.

Solomon, K. H. (1983). "Irrigation unifonnity and yield theory," PhD thesis, Utah State Univ., Logan, Utah.

Solomon, K. H.; Burt, C. M. Irrigation sagacity: A Performance Parameter for Reasonable and Beneficial Use, ASAE Paper No. 97-2181, 1997.

Sridevi, V.; Parvatam, G.; Ravishankar, G. A. Endogenous Polyamine Profiles in Different Tissues of Coffea sp., and their Levels During the Ontogeny of Fruits. *Acta. Physiol. Plant.* **2009**, *31*, 757–764.

Surendar, K. K.; Devi, D. D.; Ravi, I.; Jeyakumar, P.; Velayudham, K. Water Stress Affects Plant Relative Water Content, Soluble Protein, Total Chlorophyll Content and Yield of Ratoon Banana. *Int. J. Horti.* **2013**, *3*, 96–103.

Suzuki, N.; Mittler, R. Reactive Oxygen Species and Temperature Stresses: A Delicate Balance Between Signaling and Destruction. *Physiol. Plant* **2006**, *126*, 45–51.

Szarka, A.; Tomasskovics, B.; Banhegyi, G. The Ascorbate-Glutathione-α-Tocopherol Triad in Abiotic Stress Response. *Int. J. Mol. Sci.* **2012**, *13*, 4458–4483.

Tan, G. L.; Zhang, H.; Fu, J.; Wang, Z. Q.; Liu, L. J.; Yang, J. C. Post-anthesis changes in concentration of polyamines in superior and inferior spikelets and their relation with grain filling of super rice. *Acta Agron. Sin.,* **2009**, *35*, 2225–2233.

Tewari, R. K.; Kumar, P.; Sharma, P. N.; Bisht, S. S. Modulation of Oxidative Stress Responsive Enzymes by Excess Cobalt. *Plant Sci.* **2002**, *162*, 381–388.

Tohidi-Moghadan, H. R.; Shirani, R. A. H.; Nour, M. G.; Habibi, D.; Modaries, S. S. A. M.; Mashhadi, A. B. M.; Dolatabadian, A. Response of Six Oilseed Rape Genotypes to Water Stress and Hydrogel Aspplication. *Pesqui. Agropecu. Trop.* **2009**, *39*, 243–250.

Toker, C.; Lluch, C.; Tejera, N. A.; Serraj, R.; Siddique. K. H. M. Abiotic stresses. In *Chickpea breeding and Management;* Yadav, S. S. Redden, R. J., Chen, W., Sharma, B., Eds.; CABI: Wallingford, UK, 2007; pp 474–496.

Trovato, M.; Mattioli, R.; Costantino, P. Multiple Roles of Proline in Plant Stress Tolerance and Development. *Rend. Lincei Sci. Fis.* **2008**, *19*, 325–346.

Tuberosa, R. Phenotyping for Drought Tolerance of Crops in the Genomics Era. *Front. Physiol.* **2012**, *3*, 347.

Turan, S.; Tripathy, B. C. Salt and Genotype Impact on Antioxidative Enzymes and Lipid Peroxidation in Two Rice Cultivars During De-Etiolation. *Protoplasma* **2012**, *250*, 209–222.

Turhan, E.; Gulen, H.; Eris, A. The Activity of Antioxidative Enzymes in Three Strawberry Cultivars Related to Salt-Stress Tolerance. *Acta. Physiol. Plant.* **2008**, *30*, 201–208.

Uphoff, N.; Saryanarayana, A.; Thiyagarajan, T. M. Prospects for rice sector improvement with the System of Rice Intensification, considering evidence from India. Paper presented to the 16th International Rice Conference; Bali, Indonesia, 2005, 10–14.

Vanitha, K.; Mohandass, S. Drip Fertigation Could Improve Source-Sink Relationship of Aerobic Rice (*Oryza sativa* L.). *Afr. J. Agr. Res.* **2014**, *9*, 294–301.

Veesar, N. F.; Channa, A. N.; Rind, M. J.; Larik, A. S. Influence of Water Stress Imposed At Different Stages on Growth and Yield Attributes in Bread Wheat Genotypes *Triticum aestivum* L. *Wheat Inf. Serv.* **2007**, *104*, 15–19.

Verma, S.; Dubey, R. S. Lead Toxicity Induces Lipid Peroxidation and Alters the Activities of Antioxidant Enzymes in Growing Rice Plants. *Plant Sci.* **2003**, *164*, 645–655.

Verma, D. K.; Mohan, M.; Yadav, V. K.; Asthir, B.; Soni, S. K. Inquisition of Some Physico-Chemical Characteristics of Newly Evolved Basmati Rice. *Environ. Ecol.* **2012**, *30* (1), 114–117.

Verma, D. K.; Mohan, M.; Asthir, B. Physicochemical and Cooking Characteristics of Some Promising Basmati Genotypes. *Asian J. Food Agro Ind* **2013**, *6* (2), 94–99.

Verma, D. K.; Mohan, M.; Prabhakar, P. K.; Srivastav, P. P. Physico-Chemical and Cooking Characteristics of Azad Basmati. *Int. Food Res. J.* **2015**, *22* (4), 1380–1389.

Verma, D. K.; Srivastav, P. P. Proximate Composition, Mineral Content and Fatty Acids Analyses of Aromatic and Non-Aromatic Indian Rice. *Rice Sci.* **2017**, *24* (1), 21–31.

Wang, H. X.; Liu, C. M.; Zhang, L. Water-Saving Agriculture in China: An Overview. *Adv. Agron.* **2002**, *75*, 135–171.

Weichert, N.; Saalbach, I.; Weichert, H.; Kohl, S.; Erban, A.; Kopka, J.; Hause, B.; Varshney, A.; Sreenivasulu, N.; Strickert, M.; Kumlehn, J.; Weschke, W.; Weber, H. Increasing Sucrose Uptake Capacity of Wheat Grains Stimulates Storage Protein Synthesis. *Plant Physiol.* **2010**, *152*, 698–710.

Whittlesey, N. K.; McNeal, B. L.; Obersinner, V. F. Concepts affecting irrigation management (Studies in Water Policy and Management No. 7). Energy and Water Management in Western Irrigated Agriculture. **1986**, 101–127.

Willardson, L. S. Attainable Irrigation Efficiencies. *J. Irrig. Drain. Div.* **1972**, *98*, 329–246.

Wobus, U.; Sreenivasulu, N.; Borisjuk, L.; Rolletschek, H.; Panitz, R.; Gubatz, S.; Weschke, W. Molecular Physiology Andgenomics of Developing Barley Grains. In *Recent Research Developments in Plant Molecular Biology*; Pandalai, S. G. Ed.; Research Signpost: Trivandrum, 2005; vol 2. pp 1–29.

Wu, Y. S.; Tang, K. X. MAP Kinase Cascades Responding to Environmental Stress in Plants. *Acta. Bot. Sin.* **2004**, *46*, 127–136.

Yadav, R. S.; Hash, C. T.; Bidinger, F. R.; Devos, K. M.; Howarth, C. J. Genomic Regions Associated with Grain Yield and Aspects of Post Flowering Drought Tolerance in Pearl Millet Across Environments and Tester Background. *Euphytica* **2004**, *136*, 265–277.

Yan-dong, L.; Gui-ping, Z.; Xiao-hong, G.; Da-wei, Y.; Dian-rong, M. A.; Zheng-jin, X. U.; Wen-fu, C. Effects of Lower Limit of Soil Water Potential on Grain Quality of Rice in Cold Region. *Chin J. Rice Sci.* **2011**, *5*, 515–522.

Yang, J.; Zhang, J. Crop Management Techniques to Enhance Harvets Index in Rice. *J. Exp. Bot.* **2010**, *61*, 3177–3189.

Yang, J.; Zhang, J.; Liu, K.; Wang, Z.; Liu, L. Involvement of Polyamines in the Drought Resistance of Rice. *J. Exp. Bot.* **2007**, *58*, 1545–1555.

Yang, J. C.; Liu, K.; Zhang, S. F.; Wang, X. M.; Wang, Z. Q.; Liu, L. J. Hormones in Rice Spikelets in Responses to Water Stress During Meiosis. *Acta Agron. Sin.* **2008**, *34*, 111–118.

Yang, Z. B.; Eticha, D.; Albacete, A.; Rao, I. M.; Roitsch, T.; Horst, W. J. Physiological and Molecular Analysis of the Interaction between Aluminium Toxicity and Drought Stress in Bean (Phaseolus Vulgaris). *J. Exp. Bot.* **2012**, *63*, 3109–3125.

Yativ, M.; Harary, I. Sucrose Accumulation in Watermelon Fruits: Genetic Variation and Biochemical Analysis. *J. Plant Physiol.* **2010**, *167,* 589–596.

Zaefyzadeh, M.; Quliyev, R.; Babayeva, S. M.; Abbasov, M. A. The Effect of the Interaction Between Genotypes and Drought Stress on the Superoxide Dismutase and Chlorophyll Content in Durum Wheat Landraces. *Turk. J. Biol.* **2009**, *33,* 1–7.

Zheng, J.; Fu, J.; Gou, M.; Huai, J.; Liu, Y. Genome Wide Transcriptome Analysis of Two Maize Inbred Lines Under Drought Stress. *Plant Mol. Biol.* **2010**, *72,* 407–421.

Zlatev, Z. Effects of Water Stress on Leaf Water Relations of Young Bean Plants. *J. Cent. Eur. Agric.* **2005**, *6,* 5–14.

Zlatev, Z.; Lidon, F. C. An Overview on Drought Induced Changes in Plant Growth, Water Relations and Photosynthesis. *Emirates J. Food Agric.* **2012**, *24,* 57–72.

Zulkarnain, M.; Ismail, M.; Ashrfuzzaman, M.; Saud, H.; Ismail, C. H. Rice Growth and Yield Under Rain Shelter House as Influenced by Different Water Regimes. *Int. J. Agri. Bio.* **2009**, *11,* 566–570.

Zulkarnain, W. M.; Ismail, M. R.; Saud, H. M.; Othman, R.; Habib, S. H.; Kausar, H. Growth and Yield Response to Water Availability At Different Growth Stages of Rice. *J. Food Agric. Environ.* **2013**, *11,* 540–544.

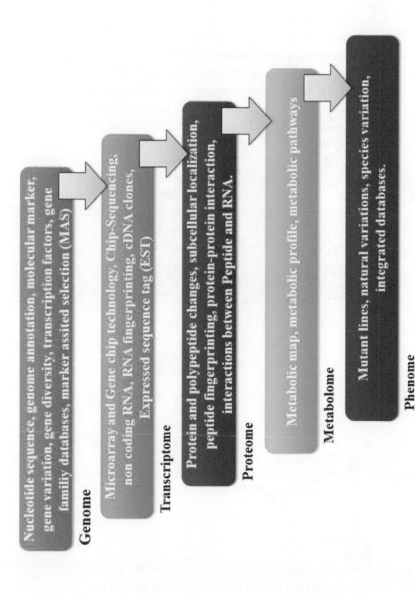

FIGURE 1.1 A conceptual model called "omic space" with layers ranging from the "genome" to "phenome."

Source: Adapted from Toyoda and Wada (2004).

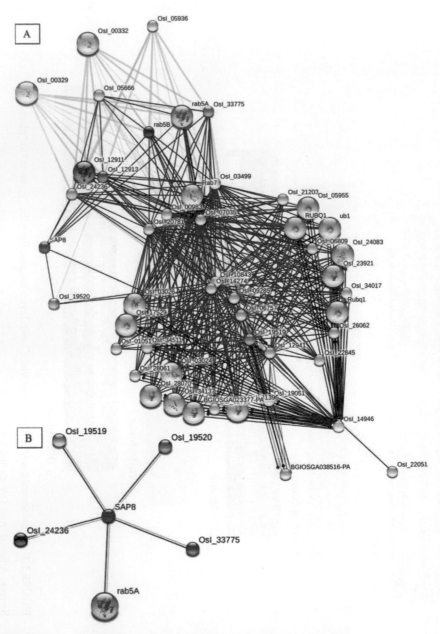

FIGURE 1.2 Multigenic nature of *OsiSAP8* predicted by string analysis.

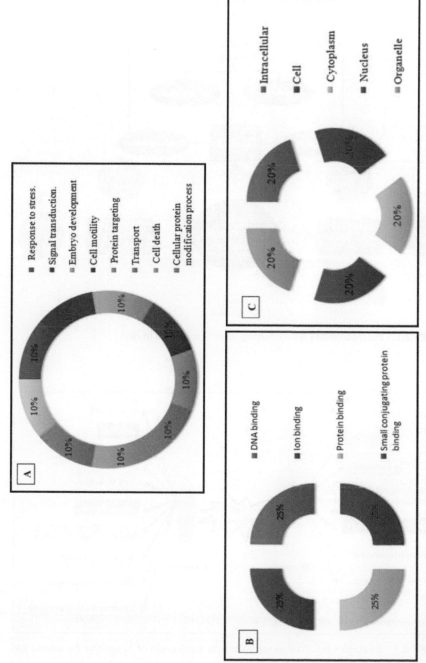

FIGURE 1.3 Gene ontology of *OsiSAP8* gene. (A) Biological process of reference SAP genes, (B) molecular function of reference SAP genes, and (C) cellular component present in reference SAP genes.

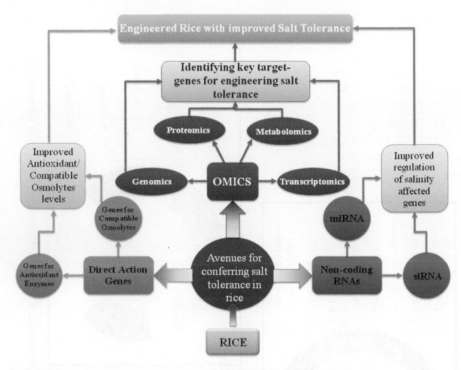

FIGURE 2.1   Potent avenues for conferring salinity tolerance in rice.

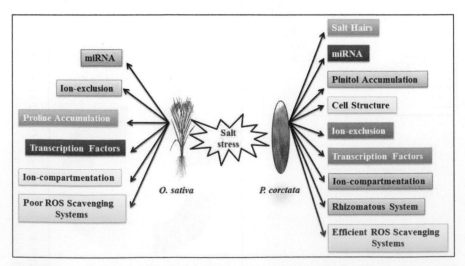

FIGURE 3.1   Strategies involved in salt tolerance mechanism of halophyte *Porteresia* and glycophyte rice.

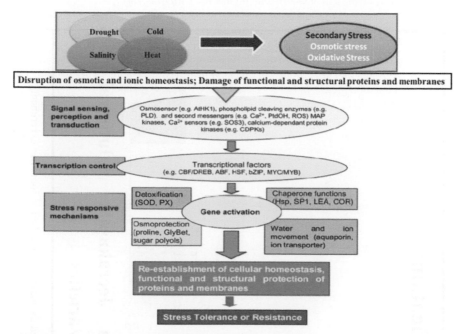

FIGURE 4.1   The complexity of plant response to abiotic stress.

Source: Adapted from Wang et al. (2003a)

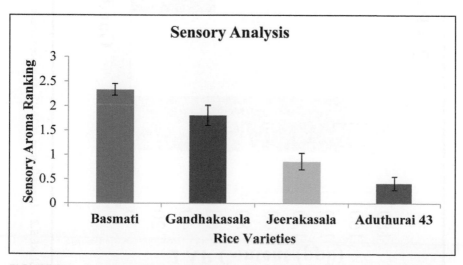

FIGURE 5.1   Sensory evaluation of aroma in the four rice varieties. Data represented the mean value ± standard error of 15 panelists as triplicates.

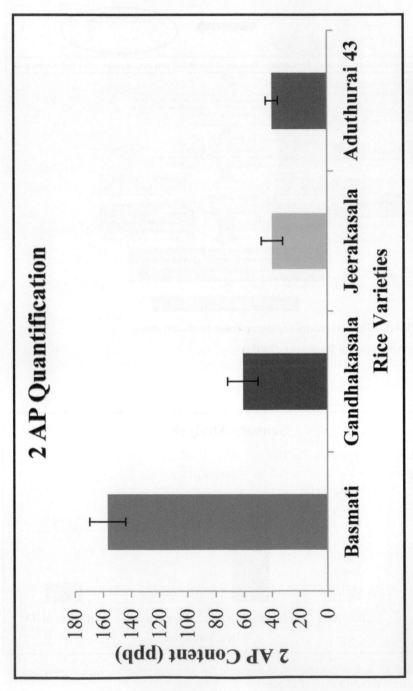

FIGURE 5.2   Quantification of 2AP in four rice varieties. Data represented the mean ± standard error from three experiments.

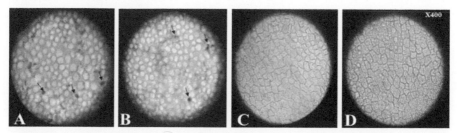

FIGURE 5.3 Histochemical staining of 2AP in rice using DNPH reagent. Arrow heads indicate the presence of 2AP. (A) Basmati 370, (B) Gandhakasala, (C) Jeerakasala, and (D) Aduthurai 43.

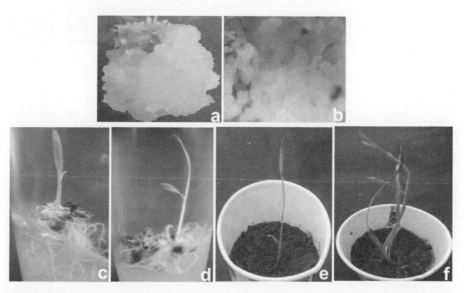

FIGURE 5.4 Stages of regeneration of rice varieties Gandhakasala and Jeerakasala from seed derived embryogenic calli. (a and b) 4 weeks old Callus of var. Gandhakasala and Jeerakasala; (c and d) Development of shooting and rooting; (e and f) Hardened plants of Gandhakasala and Jeerakasala.

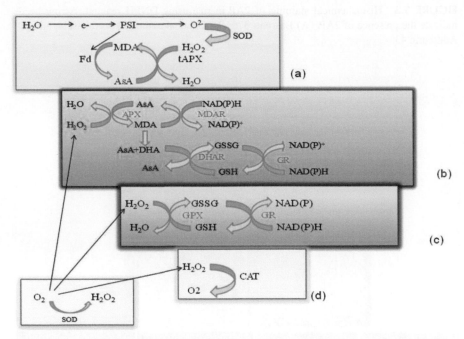

FIGURE 6.4   ROS scavenging pathways in plants. The water-water cycle (b), the ascorbate glutathione cycle (c), the glutathione-peroxidase (GPx) cycle (d) Catalase (CAT), superoxide dismutase (SODs) acts as the first line of defense converting O^{2-} into H_2O_2. In the contrast to CAT (d), APx and GPx require an ascorbate (ASA) and/or a glutathione (GSH) regenerating cycle (a-c). This cycle uses electron directly from the photosynthetic apparatus (a) or NADPH (b, c) as reducing power. [DHA: Dehydroascorbate; DHAr: DHA reductase; Fd: Ferrodoxin; GR: Glutathione reductase; GSSG: Oxidised glutathione; MDA: Monodehydro-ascorbate; MDAR: MDA reductase; PSI: Photosystem I; and tAPx: Thylakoid-bound APx]

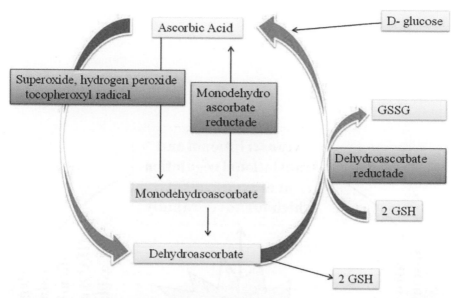

FIGURE 6.6 Synthesis and degradation pathway of L-Ascorbic acid in plant tissues. [GSH: Reduced glutathione; GSSG: Oxidized glutathione].

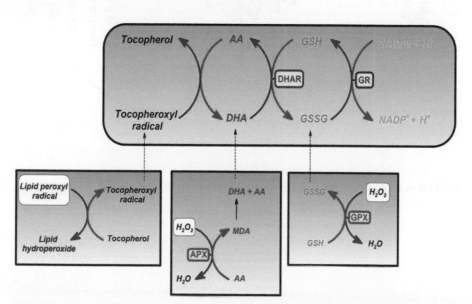

FIGURE 6.7 Ascorbic acid and tocopherol role as antioxidants. [AA: Ascorbic acid; APx: Ascorbate peroxidase, DHA: Dehydroascrobic acid; GPx: Guaiacol peroxidase; GSH: Reduced glutathione; GSSG: Oxidized glutathione; and MDA: Malondialdehyde].

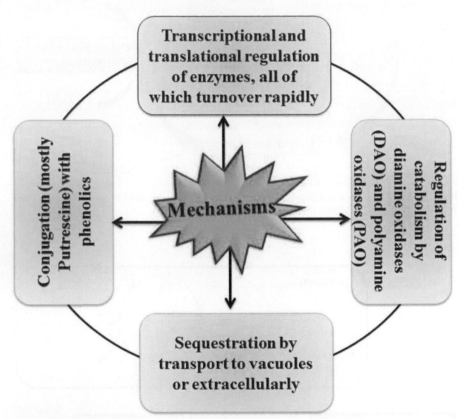

FIGURE 6.9 Mechanisms involved in homeostasis of intracellular polyamine.

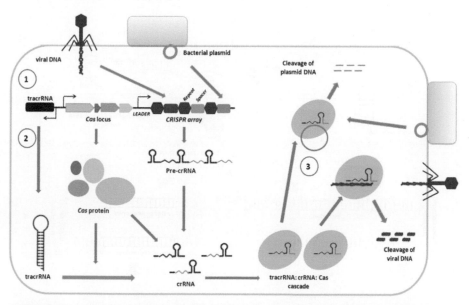

FIGURE 9.1   CRISPR/Cas mediated bacterial defense against pathogen attack. The foreign DNA is recognized and integrated as spacer at the leader end of the CRISPR locus at the acquisition step (1). Transcription of the pre-CRISPR RNA (crRNA) which is processed into crRNA by Cas proteins and *trans*-activating crRNA (tracrRNA) at the expression step (2). The crRNA: tracrRNA: Cas9 cascade target and cleave a specific genetic element from the pathogen at the interference step (3).

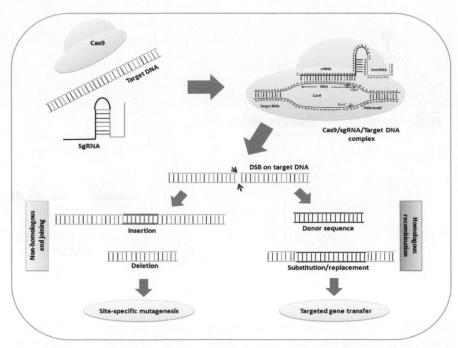

FIGURE 9.3 Cas9/sgRNA mediated cleavage and DNA repair mechanism. The HNH and RuvC nuclease domain generate a double strand break at the 5′ of a protospacer adjacent motif (PAM) sequence. The DSB at the target site is subjected to cellular DNA repair through homologous recombination (HR) or non-homologous end joining (NHEJ). NHEJ results in site specific mutagenesis through insertion or deletion (indels) while HR results in target gene replacement.

FIGURE 10.1 Biotechnology in the development of rice variety A comparative view of grain and stem of (A) Traditional type rice plant, and (B) High-yieldng type, *vis-n-vis* (C) Super Rice plant.

Source: Modified from Khush (2001).

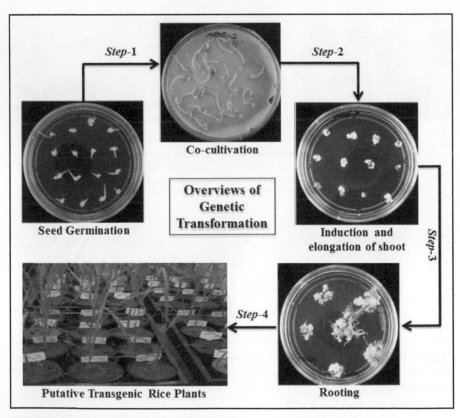

FIGURE 10.2   Genetic Transformation in the developmemt of transgenic rice plant.
Source: Adapted from Kim et al. (2012).

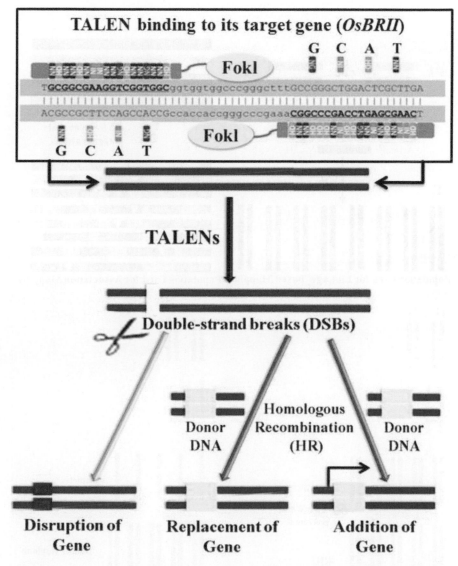

FIGURE 10.3 Overview of TALEN-mediated targeted mutagenesis (Adapted from Chen et al., 2014) (Note: For details on targeted mutagenesis mediated by TALEN, reader should refer to paper of Chen, K.; Shan, Q. and Gao, C. An efficient TALEN mutagenesis system in rice. *Methods*, 2014, 64 (1): 2–8).

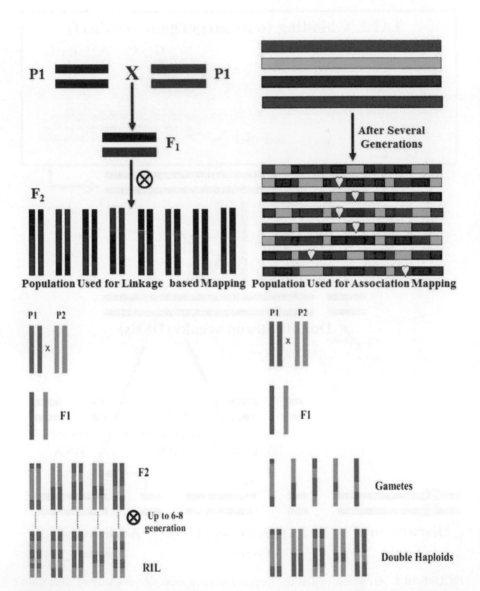

FIGURE 11.1   Difference in population structure used for linkage analysis based mapping and linkage disequilibrium based mapping.

PART III
Biotechnological Tools and Trends for Production and Rice Quality Improvement

PART III

AGROBACTERIUM-MEDIATED GENETIC TRANSFORMATION PRACTICES FOR IMPROVEMENT OF RICE QUALITY AND PRODUCTION

AZIZ AHMAD

School of Fundamental Sciences, c/o Centre for Fundamental and Liberal Education, Universiti Malaysia Terengganu, 21030, Kuala Terengganu, Terengganu, Malaysia, Tel: +609 6683176, Mob.: +6 19 3475443, Fax: +609 6683434

E-mail: mail: aaziz@umt.edu.my

7.1 INTRODUCTION

Rice is one of the main source of carbohydrate, feeds more than half of the world's population mainly in Asian region especially India, Japan, China, and Southeast part of Asia (Xue et al., 2015; Verma and Srivastav 2017). Rice is also a source for proteins, zinc, and niacin with low calcium, iron, thiamine and riboflavin (Verma and Srivastav 2017). Extensive studies have been carried out by various research institutions worldwide to increase the rice productivity and qualities to fulfill the increases of market demands (Verma et al., 2012, 2013, 2015). It was estimated that the rice production requires upto 50% increment to ensure sustainable food security by 2025 (Mohanty, 2009). Besides the inputs, diseases and post-harvest management, weather and/or climatic change is the bottleneck in rice production. Flooding and drought are two phenomenon derived from weather changes that cause an environmental stress to rice plants. Rising of sea water into agricultural land in low coastal area and accumulation of ionic fertilizer that domestically employed by farmer enlarge the environmental stress due to soil salinity (Wanjogu et al., 2001; Yadav et al., 2011). This environmental stresses poses a serious problem and limiting rice production worldwide (Kumar et al., 2003).

Germplasm improvement is the only solution to reduce environmental tension on rice production and to fulfil the high market demand. For decades, conventional breeding method has improved a lot of rice traits and benefiting farmers. Nonetheless, this traditional genetic crossing produces new rice varieties that are highly sensitive toward environmental changes. In other words, rice breeding kept pace with the growing market, despite losses caused by various biotic and abiotic factors such as salinity, drought, and temperature stress. Advances in molecular breeding emerged as a most powerful approach in the creation of desirable rice traits. Genetic transformation-mediated rice improvement gave us early maturing, high yielding, insect- and disease-resistant crops. Furthermore, accomplishment sequencing of the entire 466 megabases genomic of *Oryza sativa* L. Ssp. *indica* generates remarkable bio-information regarding genes and genome architecture and has tremendous potential for biochemical and molecular analysis during normal and unfavorable conditions (Shrawat et al., 2010). The enhanced annotation of rice genomics had remark specific genes encoding various signal transductions, transcriptional factors, metabolic pathways and ion transporters that regulate rice physiology (Kumar et al., 2003). Thus, it is an urge to explore new sources of genetic variation for the development of new rice cultivars with greater potential and stability over seasons and eco-geographical climate to maximize rice productivity. The prerequisite in this genetic modification biotechnology are the development of tissue culture protocol, the exploration of gene sources and the enhancement of gene delivery. In other words, rice manipulation by genetic engineering is a complicated process, but the basic requirements have been developed and ready to be exploited.

7.2   GENETIC TRANSFORMATION OF RICE MEDIATED BY *AGROBACTERIUM*

The basic molecular in plant genetic transformation is the introduction of foreign DNA into the host genome. Several transformation protocols have been developed and commercially implemented into rice (Hiei et al., 1997). The first report of genetic modification of rice was conducted using PEG method (Uchimiya et al., 1986). Subsequently, gene transfer through electroporation and microprojectile bombardment also employed (Christou et al., 1991; Christou, 1997). For past few decades, numerous papers published reporting many rice cultivars had been transformed with various desired gene. Nonetheles, *Agrobacterium*-mediated transformation appears as most popular protocol for the reason that inexpensive and transfer low number of gene copies minimizing chromosomal rearrangement to prevent gene silence (Hiei and Komari, 2008).

The genetic transformation mediated by *Agrobacterium* is a direct infection of *Agrobacterium tumefaciens* on rice explants (He, 2016). A comprehensive and highly efficient protocol has been established for a wide range of rice genotypes; *japonica* and *indica* (Nishimura et al., 2007; Hiei and Komari, 2008). Many factors influencing *Agrobacterium*-mediated transformation of rice including genotypes explant types, *Agrobacterium* strain, and binary vector, age of callus and medium of regeneration (He, 2010, 2006). Optimization on the transformation conditions including the number of bacterial cells, age of embryos tissues, pre-culture treatment, physical wounding, period and condition of co-cultivation, type and concentration of phenolic compounds, co-cultivation medium, ionic strength and temperature have been carried out and established (Rahman et al., 2011), shoot regeneration from transgenic callus culture, bacterial elimination and transformant selection to improve the transformation efficiency. General steps in *Agrobacterium* protocol are the infection and co-cultivation, *Agrobacterium* elimination, selection of infected tissues, conformation gene integration in putative transgenic and shoot regeneration using solid medium. Key success factors are using this system includes the rice genotypes, age and types of explants, infection methods and conditions.

7.2.1 PREPARATION OF EXPLANTS

Prior infection with *Agrobacterium*, an explant either immature embryos, calli, mature embryos, scutellum tissue, embryogenic calli, cell suspension or shoot apices must be ready. The use of shoot apices reduces the time spent to establish the transformant, and also minimize the number of somaclonal variation of transformant. Shoot apices can be obtain from rice shoot apical meristem without an intervening callus stage. Multiple shoots directly regenerated from the meristem cells cultured on MS medium containing 4 mg/L Kinetin (Fook et al., 2015) or thidiazuron (Yookongkaew et al., 2007). Trangenic plants regenerated from shoot apices explants are readily transferred to soil within 5–8 weeks after transformation (Yookongkaew et al., 2007).

Rice has their ability for callus induction, growth, and regeneration but it vary variety to variety. Manipulating of plant growth regulators (PGRs), organic components and salts are within the culture media for guarantee the production and proliferation of a great number of embryogenic calli with high regeneration capacity from mature seeds widely applicable for the tissue culture of indica varieties (Ge et al., 2006). Prior transformation an embryogenic callus is induced from mature seeds (Karthikeyan et al., 2012) or immature embryos Hoque et al. (2005) incubated on Linsmaier and Skoog (LS) medium supplemented with 2.5 mg/L 2,4-dichlorophen-oxyacetic acid (2,4-D) and 1.0 mg/L thiamine-HCl. Normally, immature embryos produces higher number of transgenics compared to the mature embryos. Callus from mature embryo with high shoot regeneration may be established on Murashige and Skoog (MS) medium supplemented with 2.0 mg/L 2,4-dichlorophenoxyacetic acid (2,4 D), 500 mg/L proline and 500 mg/L glutamine (Pawar et al., 2015). Callus from scutellum explants may be obatined on MS medium supplemented with 2.5 mg/l 2,4-D and 0.15 mg/l BAP (Sahoo and Tuteja, 2012). Embryogenic types of calli should be examined during early rice callogenesis histologically or scanning elec-tron microscopy (SEM) images. The embryogenic cells exhibited a globular, compact structure that contained tightly packed cells and thus rendered the cells suitable for transformation. The nonembryogenic callus displayed elongated morphology and a lack of cellular organization. Embryogenic calli proliferated 2-week-old cut calli are suitable for plant transformation (Bevitori et al., 2014). An excellent starting material is the callus produces from three-week old of mature embryos. On the other hand, embryogenic calli is useful for the production of somaclonal variant in *Agrobacterium*-mediated transformation (Hiei and Komari, 2008; Wei et al., 2016). The immature embryos that freshly isolated from seeds on plants grown in a greenhouse also produce high transformation rate and stable gene integra-tion of transgenic indica rice (Hiei and Komari, 2006).

7.2.2 INFECTION AND ELIMINATION OF AGROBACTERIUM

Agrobacterium tumefaciens is upadated name, previously known as *Rhizobium radiobacter* (*Agrobacterium radiobacter*) (Young et al., 2001; Aujoulat et al., 2011; NCBI, 2016), use by worldwide employed researchers for delivering a foreign gene into a plant's genome. In this purposes the nonphytogenic *Agrobacterium tumefaciens* strain is equipped with a modified T-DNA that consists of a virulen gene in a binary vector plasmid such as pTOK233 (Hoque et al., 2005). This super-binary vector is the most effective and widely used in plants genetic transformation (Slamet-Loedin et al., 2014). The ability of *Agrobacterium* in transferring DNA is only limited to "competent" cells (Hiei et al., 1994; Toki et al., 2006). Efficiency might vary depending on rice genotypes and often be enhanced by manipulating the co-cultivation protocol (Dong et al., 2012; Hiei et al., 1994; Hiei, and Komari, 2006; Toki et al., 2006).

For infection, explants are directly immersed in a bacterial broth, OD_{600} of 0.4–0.8 for 10–30 min (Ali et al., 2007; Wanichananan et al., 2010) and co-cultivated for 2–5 days in the dark at $25\pm2°C$ (Karthikeyan et al., 2012; Puhan et al., 2012; Sahoo and Tuteja, 2012; Krishnan et al., 2013; Fook et al., 2015). The high percentage of transformants can be achieved by replacing bacterium media with plant medium during cocultivation (Mohanty et al., 2002). A modified N6 medium (Jian-Jun et al., 2009) with pH5.6 may useful for callus induction, pre-cultivation and co-cultivation with *Agrobacterium* for 3 days and 7 days for a resting period of the infected calli (He et al., 2010; 2006; Zhao et al., 2011a). The infection efficiency is enhanced by inclusion of 100–200 µM acetosyringone (a phenolic compound) (Rao and Rao, 2007; Wagiran et al., 2010; Sahoo and Tuteja, 2012) and 400 mg/L of l-cysteine (Shri et al., 2013) into medium (LS or MS) during co-cultivation. Phenolic compound plays an important role in inducing *virulence* gene of *Agrobacterium* to activate the pathogenicity of bacterium during transferring T-DNA segment into plants (Joubert et al., 2002, 2004; Gelvin, 2005).

Physical treatments such as vacuum infiltration on an embryonic calli for 5 min in resulted in high transformation efficiency, stable integration of foreign genes and reduce time and effort (Safitri et al., 2016. Pretreatment by microprojectile (Uzé et al., 1997), pierced by a needle (Lin et al., 2009), heat, hydrolytic enzyme (Hiei and Komari, 2008; Clement et al., 2016) and centrifugal force enhances the efficiency of gene transfer of embryos. Short treatment by sonication prior co-cultivation also increases the transformation frequency of shoot apical meristem explants (Karthikeyan et al., 2011, 2012).

Co-cultivaing calli with *Agrobacterium* ($OD_{600}=0.04$) on filter papers moistened with enriched N6 or DKN media [5.5 mL of N6 or DKN liquid medium supplemented with 2,4-d (2 mg/L), proline (10 mM), casein hydrolysate (300 mg/L), sucrose (30 g/L), glucose (5 g/L), l-cysteine (100 mg/L) and acetosyringone (15 mg/L)] at 25°C for 3 days in the dark increases transformation efficiency 5-fold compared to solid medium (Ozawa, 2009). Using a liquid media reduces of substantial labor and time to generate transgenic plants. A one to three weeks of calli derived from mature seeds could be directly inoculated with *Agrobacterium* and the co-cultivation for 3 days, transferred onto liquid medium for *Agrobacterium* elimination and calli's selection for 3 days. Subsequently transferred to calli's growth solid media for 14 days and to shoot induction and root induction media (Yang et al., 2013a). To increase transformation efficiency with lower somaclonal variation, early infection or pre-cultured of seeds with *Agrobacterium* would helps (Toki et al., 2006; Azria and Bhalla, 2011).

Suspension culture is also one of the best sources for producing a large number of transgenic cells (Ozawa and Takaiwa, 2010) Rice suspension cultures can be established from the calli derived from mature seeds and subcultured every week. The frequency of transformation varied with the period of suspension culture and the size of the callus (Kant, 2007). Rice calli suspension-cultured less than 4 weeks is a useful targets of transformation mediated by Agrobacterium. Higher transformants can be obtained by co-cultivating 3-day old suspension-cultured cells with Agrobacterium on filter paper moistened with enriched N6 medium containing suspension-cultured cell media. Transformation efficiency using suspension-cultured cell clusters is 10-fold enhancement compared with that of calli subcultured on solid media (Ozawa and Takaiwa, 2010). A simple air-drying of the calli suspension-cultured for 10–15 min increases the transformation frequency (Urushibara et al., 2001)

Excess Agrobacterium on the putative transformant is normally elimanted by immersion in antibiotic solution and incorporated into culture medium (Khan et al., 2015). Timentin at 250 mg/L is successfully eliminated Agrobacterium growth without negative effects on the callus growth unlike other antibiotics, reduction in browning and stable transformation efficiency (Priya et al., 2012).

7.2.3   SELECTION OF TRANSFORMANTS

Various approaches have been developed for selection of transformant in the genetic transformation mediated by the *Agrobacterium*. The most powerful approach and rountinely used is based on the antibiotic resistant genes. Kanamycin and hygromycin are the commonly employed antibiotic. For selection of the shoot apical meristem transformant hygromycin is combined with carbenicillin or cefotaxime (250 mg/L). For selection of transgenic callis, the geneticin (G418: 30 mg/L) and paramomycin (70 mg/L) is also useful (Chakraborty et al., 2016b). In addition to antibiotic, herbicide resistant such as bar is also been used for selection of transformant. Average period requires for selection of transgenic rice is about three months (Toki et al., 2006; Sripriya et al., 2011).

Visual selection enables transformation in plants that are sensitive to the stress of antibiotic selection. Nonetheles, visual selection is inefficient and laborious due to the technical difficulties involved in clonal propagation of transformed calli. After antibiotic screening, the putative transformants usually stains using the β-glucuronidase (GUS) assay based reporter gene uidA. This technique causes a destructive and vanished of samples. The *bar* gene and the 5-enolpyruvylshikimate-3-phosphate synthase (epsps) (Yinxia and Te-Chato, 2015) gene are the commonly used herbicide for transgene selection. Bar activity is evaluated by the chlorophenol red test and the leaf painting test using phosphinothricin (Dedicova et al., 2015). Status for positive integration of the transgene is confirmed by polymerase chain reaction (PCR), semi-quantitative reverse-transcription PCR and Southern blot hybridization (Amin et al., 2012).

In genetic transformation systems, combination of green fluorescent protein (GFP) with antibiotic resistance genes can facilitate the selection of transformed cells (Saika and Toki, 2009). Fluorescent proteins such as GFP are useful visual markers permit direct visualization of transformed cells without the need for exogenous substrates (Saika et al., 2012a). This method is also free from antibiotic or herbicide stress. A non-destructive and highly sensitive visible marker is efficient monitoring system for stable transformation events. Selection of cells using emitting green fluorescence in *Agrobacterium* co-cultivated callus avoids sample damaging. An intrinsic pigmentation can be used as an intrinsic marker in visual selection of rice (Saika et al., 2011). The visual selection using GFP have been improved by adding enhancer trap vector and Eluc genes, the luciferase of click beethe. The Eluc luminescence and GFP fluorescene sensitivity was more than 30-fold higher than the firefyly luciferase (Saika and Toki, 2010; Saika et al., 2012b).

Other marker-free binary vectors are the *Bacillus thuringiensis* mcyr1Ac gene in the transgenic rice resistant to lepidopteran insects (Woo et al., 2015a). Transgenic conformations proceed with genomic DNA PCR and Cry1Ac ImmunoStrip assays of individual plant extracts. A spontaneous self-excision binary vector using an oxidative stress-inducible modified mFLP/FRT system with enhanced seed tocopherol content was developed by Woo et al. (2015b). This spontaneous auto-excision system can be easily adopted and used. In directly, rice grains with tocopherol are value added nutrient for human health. Wakasa and co workers (2012) have developed a system using rice-derived selection markers; the mutated acetolactate synthase gene (mALS) under the control of the callus-specific promoter (CSP) and the ferredoxin nitrite reductase gene (NiR) under the control of

NiR promoter. Previous study had shown that the NiR promotor is respsonsible for growth and regeneration of rice calli (Ozawa and Kawahigashi, 2006). The transgenic rice transformed with these genes (mALS and NiR) exhibits resistance toward bispyribac-sodium (BS), a pyrimidinyl carboxy herbicide (Endo et al., 2012). Putative transformants is easily selected on the medium containing phosphinothricin (Yang et al., 2013b).

To avoid the antibiotic markers in genetically modified (GM) rice, an alternative selection based on utilisation of mannose that catalysed by the phosphomannose isomerase (pmi) protein was utilised. The pmi gene is derived from *Escherichia coli* or *Saccharomyces cerevisiae*. During the selection of transgenic, the substrate which is a combination of mannose (10 g/L) and sucrose (5 g/L) are incorporated into the selection medium. (Ding et al., 2006; Duan et al., 2012; Wang et al., 2012a; Qiu et al., 2015). Thus, only the transgenic plants that expressing the pmi gene have an ability to metabolize mannose into fructose.

7.2.4 SHOOT REGENERATION

The ability of unorganised cells or callus to differentiate and regenerated to shoot organ is very crucial in the development of any transgenic plants, particularly rice. Varous medium have been formulated for both japonica and indica rice. Normally, medium that uses for callus induction is used for shoot regeneration, but the auxin concentration is reduced or cytokinin is added. Nonetheles, hight shoot regeneration frequency is commonly obtained on MS medium supplemented with benzylaminopurine (2.0 mg/L), 1-naphthaleneacetic acid (0.5 mg/L), proline (500 mg/L), and glutamine (500 mg/L) (Pawar et al., 2015). This medium is sutiable for callus obtained from MS medium supplemented with 2.0 mg/L 2,4-D, 500 mg/L proline and 500 mg/L glutamine. Callus arrested with absisic acid (ABA) promoted up the embryonic callus (Guruprasad et al., 2012). The percentage of shoot regeneration of transformant is also affected by the antibiotics presence in the selection medium (Tran and Sanan-Mishra, 2015). Modification of carbon sources such as maltose instead of sucrose and increases the concentration of gelling agent (8 g/L gelrite) also promotes shoot regeneration on callus (Cho et al., 2015). The shoot regeneration of indica rice such as IR64 is achieved on MS medium supplemented with 3 mg/l BAP, 1 mg/l Kinetin and 0.5 mg/l NAA (Sahoo and Tuteja, 2012). Increases the concentration of carbon sources in medium would induce the rooting process. For example, 10 g/L of glucose or 20 g/L of sucrose. Modification of auxin and/or cytokinin ratios are also benefits in initiation of shoot regeneration. In upland rice for example, the shoot regeneration is realised on callus cultured onto MS medium with combination of indolebutyric acid (0.15 mg/L), indoleacetic acid (0.15 mg/L), 6-benzyladenine (0.5 mg/L), kinetin (0.5 mg/L), Zeatin (1.0 mg/L), naphthaleneacetic acid (0.5 mg/L), and thidizazuron (TDZ: 0.5 mg/L) (Zhao et al., 2011a).

7.3 IMPROVED-RICE TRAITS VIA MOLECULAR BREEDING

As a staple food crops, rice received an extensive attention and remarkable studies have been carried out for traits improvement through either convensional or molecular breeding techniques. Rice has several advantages including a small genome (386.5 Mb), a known genome sequence (Kawakatsu et al., 2013). To date, the genetic modification with the help of *Agrobacterium*-mediated transformation is considered as the robust and widely used method to confer new rice traits. Various genes were succesfully inserted, integrated and expressed in rice genome toward abiotic and biotic stresses (Xue et al., 2015). The transformed gene is also inherited into the progeny.

7.3.1 ABIOTIC STRESS TOLERANCE

Environmental stresses such as drought, salinity and temperature have great effects on the rice productivity. For instants, drought causes the rice yield productivity loss of up to 70% (Ravikumar et al., 2014; Rachmat et al., 2014). An intensive studies have been embarked focusses on the identification of functional genes related to stress-rugulation mechanism and the transcription factors in rice. Table 7.1 shows the genes that have been succefully transformed into rice using *Agrobacterium*-mediated and the cellular activity regulates by the particular gene.

7.3.2 INSECT/PEST TOLERANCE

Rice is susceptible to many pests, particularly the coleopteran that causes an extensive destruction on rice plants and yield productivity. Rice leaf folder (RLF) (*Cnaphalocrocis medinalis* Guenee), striped stem borer (SSB) (*Chilo suppressalis* Walker) and yellow stem borer (YST) (*Tryporyza incertulas* Walker) are three most important lepidopteran pests that cause severe

TABLE 7.1　*Agrobacterium*-mediated Transformation Rice with Elite Genes for Abiotic Tolerance Traits.

Gene	Traits	References
Glyoxalase I (Gly I)	**ZnCl2, salinity, drought tolerance:** OsGly I upregulated the reduction of malondialdehyde, plants performed higher seed setting rate and yield	Zeng et al. (2016)
DEAD-box helicase gene	**Salinity tolerance:** Increases ion homeostasis, chlorophyll content, leaf K^+/Na^+ ratio, fertility and yield. Decreases the root length at seedling and reproductive stages.	Amin et al. (2012)
AtbZIP60,	**Salinity tolerance:** Increases the activities of antioxidant enzymes; ascorbate peroxidase and superoxide dismutase. Decreases lipid peroxidation	Tang et al. (2012)
MnSOD	**Salinity and drought tolerance:** Increases NaMnSOD and catalase activity, higher efficiency of ROS elimination under salt stress and photosynthesis. Decreases the superoxide anion radicals (O_2^-) and hydrogen peroxide (H_2O_2), lower relative ion leakage and MDA content compared to wild-type plants.	Wang et al. (2005); Chen et al. (2013)
AtEm6	**Salt tolerance:** Regulating expression of Ca^{2+}–dependent protein kinase genes; OsCPK6, OsCPK9, OsCPK10, OsCPK19, OsCPK25, and OsCPK26 under salinity conditions. Decreases lipid peroxidation, increments antioxidant enzymes ascorbate peroxidase, glutathione reductase and superoxide dismutase activities.	Tang and Page (2013a,b)
TPS gene	**Salinity tolerance:** osmolytes	Zhao et al. (2013)
SOS1	**Salinity tolerance:** Regulates ion homeostasis, the plasma membrane Na^+/H^+ antiporter, plays role in germination and growth in	Razzaque et al. (2013)
PsCBL and PsCIPK	**Salinity tolerance:** Improved rice survive exposure to NaCl	Sikdar et al. (2015)
SmAPX	**Flood tolerance:** increases the APX enzyme activities and total chlorophyll contents and less oxidative injury.	Chiang et al. (2015)
Br-UGE1 gene	**Drought tolerance:** Increases the biomass production, filled spikelets, number of productive tillers, number of spikelets per panicle, panicle length, higher photosynthetic rate, higher CO_2 assimilatio accumulation and soluble sugars, delayed leaf rolling and drying.	Abdula et al. (2016); Sailila et al. (2016)

TABLE 7.1 *(Continued)*

Gene	Traits	References
AtDREB1A/ OsDREB1G/ OsDREB2B	**Salinity and drought tolerance:** Increases the accumulation of higher grain yield and spikelet fertility; improved tolerance to water deficit stress; increased relative water content and decreased ion leakage; maintenance of chlorophyll; and osmotic substance proline.	Chen et al. (2008); Datta et al. (2012); Hussain et al. (2014); Ravikumar et al. (2014)
OsNAC6	**Salinity and drought tolerance:** Upregulated the stress-associated regulatory genes (AP2, Zincfinger protein and MYB).	Rachmat et al. (2014)
PpEXP1	**Heat tolerance:** Reduces structural damage to cells, lower the electrolyte leakage, hydrogen peroxide content and membrane lipid peroxidation. Increases the activity of antioxidant enzyme, chlorophyll content, net photosynthetic rate, relative water content and seed germination rates.	Xu et al. (2014)
OsGL1-6	**Drought resistance:** Droopy leaves at the booting stage, decreased leaf cuticular wax deposition, thinner cuticle membrane, increased chlorophyll leaching and water loss rates.	Zhou et al. (2013)
naat	**Low iron (Alkaline soil) tolerance:** Expressing higher nicotianamine aminotransferase activity and secreted larger amounts of phytosiderophores, greater grain yields.	Takahashi et al. (2001)
ApGSMT+ ApDMT	**Salinity and cold tolerance:** Increases the glycine betaine biosynthesis (GB); an osmoprotectant	Niu et al. (2014)
OsMYB14	**Salinity, drough and UV tolerance;** ABA-dependent, growth regulatory and stress-responsive MYB transcription factor.	Miao et al. (2013)
PgNHX1	**Salinity tolerance;** Increase salinity tolerance at seedling and reproductive stages.	Touhidul Islam et al. (2009)
SAMDC	**Salinity tolerance;** ABA-induced in transgenic plants under NaCl stress, and increased seedling growth.	Roy and Wu (2002)
GST+CAT1	**Cold tolerance;** Enhanced low temperature resistance, higher level of photosynthetic capacity, lower H_2O_2 and MDA (malondialdehyde) content and lower relative electrolyte leakage through the plasma membrane. Higher GST and glutathione peroxidase activities, germination and growth rates at low temperature.	Takesawa et al. (2002); Zhao et al. (2006)

TABLE 7.1 *(Continued)*

Gene	Traits	References
codA gene	***Salinity and drought tolerance:*** increase accumulation of choline oxidase product, glycine betaine, an osmoprotectant and increased root growth,	Mohanty et al. (2002); Sawahel (2004); Shirasawa et al. (2006)
SfIAP	***Salinity tolerance:*** Preventing cell death at the onset of salt stress and; ion homeostasis; maintaining the cell membrane's integrity; photosynthetic efficiency and growth; and retain plant water status.	Hoang et al. (2014)
HAL2	***Salinity tolerance:*** Increases salt tolerance during the seedling and booting stage, less damaged in the cytomembrane and stronger in leaf tissue viability during booting period.	Li et al. (2002)
OsCBL8	***Salinity tolerance:*** Ca^{2+} sensors in plant-specific calcium signaling.	Gu et al. (2008)
ICE1,	***Cold tolerance:*** Reduces mortality rate and increases proline content.	Xiang et al. (2008)
AtHsp101/OsHsp101	***Thermotolerance:*** Better growth performance in the recovery phase following the stress.	Katiyar-Agarwal et al. (2003); Chang et al. (2007)
sHSP17.7	***Thermotolerance:*** Increases the survival rate, heat-stable chaperone production, greater resistance to UV-B stress	Murakami et al. (2004)
mtlD	***Salinity and drought tolerance:*** Increases mannitol accumulation, better seed germination and seedlings growth.	Pujni et al. (2007)
MTH1745	***Hg tolerance:*** Reduces the levels of hydrogen peroxide and malondialdehyde in leaves or roots. Increases the superoxide dismutase and peroxidase activity, non-protein thiols, reduced-and oxidized glutathione ratio in seedlings of transgenic in various concentrations of mercuric chloride.	Chen et al. (2012)

damage to rice in many areas of the world (Tang et al., 2006; Ignacimuthu and Raveendar, 2011). The application of molecular breeding techniques is by introducing a specific gene that encods for metabolites that able to kill the pest or inhibites its growth become an interest. To date, an extensive transgenic rice plants transformed with endotoxin genes, cry1 from *Bacillus thuringiensis* (*Bt*) have been established (Table 7.2). Other gene that have been suscesfully transferred into rice including pta gene (*Pinellia tenata* agglutinin)—inhibiting the livability and viability of brown planthopper (Zhang et al., 2003), pin2 gene (potato chymotrypsin inhibitor)—resistant against yellow stem borer (Bhutani et al., 2006; Rao et al., 2009), PINII-2x gene (proteinase inhibitor II *Solanum phurejia* L.)—ressistant againt stripe stem borer (*Chilo suppressalis* Walker) (Bu et al., 2006). Sitophilus oryzae causes severe damage to rice seeds during storage. Transformation of α-amylase inhibitor gene of *Phaseolus vulgaris* into Basmati rice (PB1) reduces the survival rate of rice weevil, *S. oryzae* reared on transgenic rice seeds (Ignacimuthu and Arockiasamy, 2006).

7.3.3 RICE SHEATH BLIGHT AND BLAST RESISTANCE

Rhizoctonia solani and *Magnaporthe oryzae* are among the most important fungal pathogens of rice that cause serious rice diseases severely limiting its productivity. Rice sheath blight and blast are two most destructive fungal diseases in rice, respectively. Bacterial leaf streak (BLS) is caused by *Xanthomonas oryzae* pv. oryzicola (Xooc). It is a destructive bacterial disease in China. *X. oryzae* pv. oryzae PXO99 is the causal agent of rice bacterial blight disease (Khan et al., 2007). For decade, few anti-microbial genes have been successfully isolated and transformed into rice genome. The transgenic rice that entertaining the particular gene (Table 7.3) demonstrates higer levels of resistanc toward the respective pathogen. Interestingly, few transgenic rice plants exhibites ad broad-spectrum disease resistance such as those transformed with the Np gene.

7.3.4 HERBICIDES TOLERANCE

Weeds are the main problem during the first month of growth. An expenditure of labor and herbicides is normally heavy. Hand picking is usually practiced to remove the remaining weeds after the application of herbicides. Thus, there is weed competition in plot from early tillering to harvest.

TABLE 7.2 The Type of Endotoxin Gene Transformed into Rice and the Insect-Resistant.

Endotoxin Gene	Insect-resistant transgenic	References
Cry1Ab	Yellow stem borer, striped stem borer, and rice leaf folder	Qi et al. (2009)
Cry1Ab and Cry1Ac	European corn borer	Ahmad et al. (2002)
Cry1Ab/Ac	Rice leaffolder	Qiu et al. (2010)
Cry1Ab/cry1Ac	Yellow Stem borer	Ramesh et al. (2004a,b)
Cry1Ac + Cry1I	Striped stem borer and rice leaf roller, rice leaf folder	Zhao et al. (2015)
Cry1C*	Leaf folders, stem borers and yellow stemborer	Tang et al. (2006); Ye et al. (2009)
Cry1Ca1	Striped stem borer and *Spodoptera litura* larvae	Zaidi et al. (2009)
Cry2A*	Striped stem borer and rice leaf roller, rice leaf folder	Chen et al. (2005)
Cry2AX1	Yellow stem borer, rice leaf folder, and oriental army worm (OAW) *Mythimna separata*	Chakraborty et al. (2016a)
Cry2AX1 (cry2A + cry2Ac)	Striped stem borer and rice leaf roller, rice leaf folder	Manikandan et al. (2014)
CryIA (c)	Striped stem borer and rice leaf folder	Yu et al. (2009)
CryIIIA	Rice water weiver larvae and pupae *Lissorhoptrus oryzophilus* Kuschel	Lee et al. (2013)
Synthetic cry2A*	Rice striped stem borer	Jin et al. (2015)
Synthetic truncated cry1Ac	Rice leaf folder, rice green caterpillar, rice skipper,	Kim et al. (2009)

TABLE 7.3 Gene Transformed into Rice for Sheath Blight and Blast Resistance.

Gene Source	Pathogen/Disease	References
AFP -	*Magnaporthe grisea*; rice blast-resistance	Coca et al. (2004)
Pikh gene	*Magnaporthe oryza*; rice blast-resistance	Azizi et al. (2016)
Dm-AMP1; Dm-AMP1 + Rs-AFP2	*Magnaporthe oryza* and *Rhizoctonia solani*; broad-spectrum disease resistance	Jha et al. (2009); Jha and Chattoo (2009)
cht42 endochitinase	*Rhizoctonia solani*; sheath blight resistance	Shah et al. (2009)
chi11 - rice chitinase	*Rhizoctonia solani*; sheath blight resisitance	Kumar et al. (2003); Sridevi et al. (2008, 2003); Nandakumar et al. (2007); Rao et al. (2011)
Pi-d2	Rice leaf blast and neck blast resistance	Chen et al. (2010)
BjNPR1.	Rice blast, sheath blight and bacterial leaf blight resistance	Sadumpati et al. (2013)
RCH10 + AGLU1	Sheath blight and blast resistance	Mao et al. (2014)
MK1	Rice blast resistance	Lee et al. (2004)
Asthi1	Seed-transmitted phytopathogenic bacteria.	Mitsuhara et al. (2006)
OsBAK1	Blight resistance	Liao et al. (2016)
Xa 21 gene	*X. oryzae* pv. oryzae (Xoo); Bacterial blight disease resistance.	Xia et al. (2006); Rafique et al. (2010); Gao et al. (2016)
Np3 and Np5	*X. oryzae* pv. oryzae broad spectrum resistance to rice bacterial blight	Wang et al. (2011)

Development of herbicide-tolerant cultivars is an effective stratergy to improve rice production. Various methods have been successfully applied to produce rice cultivars with herbicide resistance charateristics (Table 7.4). For example, the transgenic rice plants that transformed with Epsps is survive when spraying with glyohosate herbicide, while bar gene transgenic rice is resilient toward glufosinate herbicide.

TABLE 7.4   The Herbicides Resistance Rice Harbouring the Respective Gene Transferred Using *Abrobacterium* Mediated Protocol.

Herbicide	Gene	References
Glyphosate-resistant at 8 g/L dose	EPSPS; VHb + tzs + EPSPS; CP4-ESPS	Cao et al. (2004); Zhao et al. (2011b); Chhapekar et al. (2015)
Glufosinate-resistant (0.375 g/m²) and Glyphosate-resistant (0.833 g/m².	Epsps + Bar gene	Deng et al. (2014)
Carfentrazone-ethyl and oxyfluorfen- resistant	Protoporphyrinogen oxidase (Protox)	Kuk et al. (2005)
Metolachlor-resistant (5 μM), Chloroacetamides and mefenacet, pyributicarb, amiprofos-methyl, trifluralin, pendimethalin, norflurazon, and chlorotoluron.	CYP2B6 of rice	Hirose et al. (2005)
BASTA-resistant	PPT-resistant gene (Bar); *SsGDH*	Fu et al. (2001)

7.4   ADVANCES OF TRANSGENIC RICE

Advances of *Agrobacterium*-mediated genetic transformation not only alters the agronomic traits but also utilised for gene function analysis. few modification and utilization of the tools in advancement of rice cultivars are listed in subsequent text.

7.4.1   RICE STRIPE VIRUS RESISTANCE

Rice is also prone toward virus occurrences. There are 25 types of viruses that affect rice production (Abo and Sy, 1997). Among these, one of the transgenic rice that is resistant to the rice stripe virus (RSV) has been developed (Ma et al., 2011). The RSV causes the rice stripe disease, one of the most widespread and severe virus disease. It is transmitted by the *Laodelphax striatellus*, a small brown planthopper. The trangenic rice plant that resilient toward RSV was established by RNA interference (RNAi) based on chimeric CP/SP gene sequence, coat protein (CP), and special-disease protein (SP) (Huang, 2013). The gene was inserted in a binary vector and transformed into rice using the Agrobacterium-mediated protocols (Ma et al. 2011).

7.4.2 PHYTOREMEDIATION

Heavy metal, such as cadmium had an adverse effect on growth and development of rice. To resolve this problem a transgenic rice plants with cadmium tolerance gene (YCF1) was produced. The transgenic was obtained by co-cultivating of an embryogenic calli with *Agrobacterium* harbouring the binary vector pCAMBIA 1303-YCF1 [the hygromycin phosphotransferase (*hpt*) gene and yeast cadmium factor 1 (YCF1) gene] in 100 mM acetosyringone. The rooted plantlets was transferred to pot-soil, hardened and grown in a greenhouse (Islam and Khalekuzzaman, 2015). When planted in cadmium presence soil, the transgenic exhibites a phytoremediation capability toward cadmium, where cadmium taken up from soil, store into cell vacuoles and protect rice grain from cadmium. Thus, the transgenic rice assists the phytoremediation process and the planted soil is free from cadmium (Islam and Khalekuzzaman, 2015).

7.4.3 FASTER FLOWERING

AtFPF1 is a gene that promotes flowering in *Arabidopsis*. Transgenic rice transformed with this gene exhibites reducing in flowering time, enhanced adventitious root formation but inhibited root growth in rice during the seedling stage (Xu et al., 2005).

7.4.4 CAROTENOID AND β-CAROTENE

A indica rice grains with higher contents of carotenoid and and β-carotene have been establised by insertation of phytoene synthase (psy) and phytoene desaturase (crtI) gene using the *Agrobacterium* system. Datta and co-workers (2006) had introduces these gene into IR64 and BR29 and using pmi markers during the selection process. The total carotenoids and β-carotene levels in the progenies of trangenic BR29 had increased up to 9.34 and 3.92 µg/g in polished grains, respectively. The total carotenoid in transgenic progenies of IR64 was 2.32 µg/g in polished grains (Datta et al., 2006, 2014).

7.4.5   SEROTONIN PRODUCTION

Rice is also used for the production of serotonin, a monoamine neurotransmitter derived from tryptophane. Using the *Agrobacterium*-mediated techniques, the rice embryos were transformed with the serotonin N-hydroxycinnam-oyltransferase (SHT) (Lee et al., 2008) and serotonin N-acetyletransferase (SNA) genes (Kang et al., 2010).

7.4.6   IMPROVE FATTY ACID CONTENTS

Rice bran oil (RBO), being naturally rich in antioxidants, is currently regarded as one of the health-beneficial edible oils. The RBO has essential linoleic acid (ω-6, C18:2) and α-linolenic acid (ω-3, C18:3) in nutritionally disproportionate level (~25:1), WHO/FAO's recommendation of ~5:1. The BjFad3 independent transgenic lines of rice plants were developed by *Agrobacterium*-mediated transformation. The BjFad3 transgene produces higher C18:3 fatty acid content in RBO, improving the nutritionally desirable ω-6:ω-3 ratio (~2:1) in the transgenic rice lines (Bhattacharya et al., 2014).

7.4.7   PURIFY HYBRID SEED

The purity of hybrid seed is a crucial limiting factor when developing hybrid japonica rice (*Oryza sativa* L.). To chemically control hybrid seed purity, a transgenic was developed by transferring the atrazine chlorohydro-lase gene (atzA) from *Pseudomonas* ADP using *Agrobacterium*-mediated transformation. In the presence of the investigated atrazine concentrations (150 µM), the transgenic lines produced larger seedlings, with higher germination percentages than the controls. Rice grown in soil containing 2–5 mg/kg atrazine increases plant height and chlorophyll contents. Wild type is killed by spraying with 0.15% atrazine at the vegetative growth stage (Zhang et al., 2014).

7.4.8 XYLANASES PRODUCTION

Transgenic rice plant able to produce xylanase has also been produced using the protocols mediated by *Agrobacterium*. To establish the transgenic rice plants, a vector constructs pCAMBIA1301-ATX-Ru3ep consists of thermostable xylanse ATX gene was introduced into rice variety Zhinghua 11 (Weng et al., 2013). As a result, the endogenous xylanase has increases in the transgenic and it has no effects on the growth and development of transgenic (Jiang et al., 2013a,b; Weng et al., 2013).

7.4.9 AMINO ACID (TRYPTOPHAN)

A specific amino acid such as tryptophan that requires for normal growth in infants and nitogen balancing in adults has been succesfully prorduced in transgenic rice. The key enzyme in tryptophan biosynthesis, anthranilate synthase from rice (OsASA2) was transferred into Dongjinbyeo cultivar using *Agrobacterium*. The selection of transgenic was performed based on kanamycin selection. An increases of tryptophan content was detected in leaves of transgenic rice plants together with two other metabolites the indole-3-acetonitrile and indole-3-acetic acid (Jung et al., 2015).

7.4.10 AMYLOSE AND AMYLOPECTIN COMPOSITION

Plant starch consists of amylose and amylopectin. The amylose:amylopectin ratio determines the physicochemical properties of the starch. While, the formation of amylopectin is catalysed by the starch branching enzymes, reducing the amylose content will reduce the glutinous and viscidity characteristics of rice. A rice variety with higher composition of amylopectin was established by introducing the starch branching *Sbe1* gene into embryogenic calli of Nakdong cv rice using *Agrobacterium* (Kim et al., 2005a,b). The selection of transgenic was based on the herbicide resistance, the phosphinothricin. The transgenic lines produce higher amylopectin and frequency of branching (Kim et al., 2005a,b; Jiang et al., 2013a,b). Attempts have also been made to alter the amylose level in rice. The antisense fragment of waxy gene was introduced into rice by Shen and co-workers (2004). The amylose content of seeds derived from transgenic plants seeds reduced up to 28.81% over the wild type (Li et al., 2005). The anti-waxy gene also increase protein content in the rice grains, and develop good soft rice varieties with low amylose content (Li et al., 2009; Wang et al., 2010).

7.4.11 BIOREACTOR FOR PROTEIN PRODUCTION

Transgenic rice has also been used for the production of heterologous protein, which is suitable for development of edible vaccines. For this purpose, cell suspension culture that exhibites fast cell proliferation and growing rate is the best system. The transgenic cell lines harbouring the gene of interest established using *Agrobacterium* transformation protocol is up-scale and cost-affective as compared to the conventional one (Liu et al., 2012). Table 7.5 shows the protein produced by rice cell cultures. Rice is also able to produce ginsenosides by inserting the key rate-limiting enzyme; β-amyrin synthase (βAS). The transgenic rice managed to convert the 2,3-oxidosqualene to ginsenoside. This "ginseng rice" the cv "Taijing 9" produces the oleanane-type sapogenin, oleanolic acid at 8.3–11.5 mg/100 g/dw (Huang et al., 2015).

TABLE 7.5   Foreign Protein that have been Introduced into Rice Cell Using *Agrobacterium*-Mediated Technique.

Protein	References
Mouse granulocyte-macrophage colony-stimulating factor (mGM-CSF) to proliferate murine myeloblastic leukemia cell line, NSF-60	Liu et al. (2012).
JEV subunit vaccine	Wang et al. (2009).
Dexamethasone-inducible Norwalk virus capsid protein (NVCP)	Tang and Page (2013a,b).
Foot-and-mouth disease virus (FMDV), subunit vaccines	Wang et al. (2012b)
Vacine against *Neisseria gonorrhoeae*	Yiu et al. (2008).
Bevacizumab, a humanized monoclonal antibody (mAb) targeting to the vascular endothelial growth factor (VEGF)	Chen et al. (2016)
Human T-cell epitopes the protein body I (PB-I)	Takaiwa et al. (2007)
Fusion protein gene of Newcastle disease virus (NDV F) subunit vaccine	Yang et al. (2007, 2005)
Iron-binding glycoprotein, human lactoferrin (hLF)	Rachmawati et al. (2005)

7.5 SUMMARY AND CONCLUSION

The topic highlights the current development of molecular breeding approaches for improvement of rice quality toward various biotic and abiotic factors. The topic covers key parameters required in such technique for successful development and production of new/novel germplasms, the application of such techniques in insertion of foreign genes and/or overexpression of genes related to the issues, and disadvantages of the system and products. The system allows transformation of high copy number of gene, produces stable transformation, and multiple copies gene.

KEYWORDS

- 2,4-dichlorophenoxy acetic acid
- antibiotics
- Bacillus thuringiensis
- BASTA
- Calcineurin B-like (CBL) proteins
- cauliflower mosaic virus
- Chilo suppressalis Walker
- Cnaphalocrocis medinalis Guenec
- European corn borer
- ferredoxin nitrite reductase gene (NiR)

- geneticin
- green fluorescent protein (GFP)
- kinetin
- Murashigae and Skoog
- Proteinase inhibitor II
- serotonin N-hydroxycinnamoyltransferase
- super-binary vector
- Southern blot hybridization
- DEAD-box helicase gene

REFERENCES

Abdula, S. E.; Lee, H. J.; Kim, J.; Niño, M. C.; Jung, Y. J.; Cho, Y. C.; Nou, I.; Kang, K. K.; Cho, Y. G. Bruge1 Transgenic Rice Showed Improved Growth Performance with Enhanced Drought Tolerance. *Breed. Sci.* **2016**, *66*(2), 226–233.

Abo, M. E.; Sy, A. A. Rice Virus Diseases: Epidemiology and Management Stratergies. *J. Sustainable Agric.* **1997**, *11*(2–3), 113–134.

Ahmad, A.; Maqbool, S. B.; Riazuddin, S.; Sticklen, M. B. Expression of Synthetic Cry1Ab and Cry1Ac Genes in Basmati Rice (*Oryza Sativa* L.) Variety 370 Via Agrobacterium-Mediated Transformation for the Control of the European Corn Borer (Ostrinia nubilalis). *In Vitro Cell. Dev. Biol. Plant* **2002**, *38*(2), 213–220.

Ali, S.; Xianyin, Z.; Xue, Q.; Hassan, M. J.; Qian, H. Investigations for Improved Genetic Transformation Mediated by Agrobacterium Tumefaciens in Two Rice Cultivars. *Biotechnology* **2007**, *6*(1), 138–147.

Amin, M.; Elias, S. M.; Hossain, A.; Ferdousi, A.; Rahman, M. S.; Tuteja, N.; Seraj, Z. I. Over-Expression of a Dead-Box Helicase, Pdh45, Confers Both Seedling and Reproductive Stage Salinity Tolerance to Rice (Oryza Sativa L.). *Mol. Breed.* **2012**, *30*(1), 345–354.

Aujoulat, F.; Jumas-Bilak, E.; Masnou, A.; Sallé, F.; Faure, D.; Segonds, C.; Marchandin, H.; Teyssier, C. Multilocus Sequence-Based Analysis Delineates a Clonal Population of Agrobacterium (Rhizobium) Radiobacter (Agrobacterium Tumefaciens) of Human Origin. *J. Bacteriol* **2011**, *193*(10), 2608–2618.

Azizi, P.; Rafii, M. Y.; Abdullah, S. N. A.; Hanafi, M. M.; Maziah, M.; Sahebi, M.; Ashkani, S.; Taheri, S.; Jahromi, M. F. Over-Expression of the Pikh Gene with a CaMV 35S Promoter Leads to Improved Blast Disease (*Magnaporthe oryzae*) Tolerance in Rice. *Front. Plant Sci.* **2016**, *7*, 773 (14p).

Azria, D.; Bhalla, P. L. Agrobacterium-Mediated Transformation of Australian Rice Varieties and Promoter Analysis of Major Pollen Allergen Gene, Ory S1. *Plant Cell. Rep.* **2011**, *30*(9), 1673–1681.

Bevitori, R.; Popielarska-Konieczna, M.; Dos Santos, EM.; Grossi-De-sá, M. F.; Petrofeza, S. Morpho-Anatomical Characterization of Mature Embryo-Derived Callus of Rice (Oryza Sativa L.) Suitable for Transformation. *Protoplasma* **2014**, *251*(3), 545–554.

Bhattacharya, S.; Chattopadhyaya, B.; Koduru, L.; Das, N.; Maiti, M. K. Heterologous Expression of Brassica Juncea Microsomal Ω-3 Desaturase Gene (Bjfad3) Improves the Nutritionally Desirable Ω-6: Ω-3 Fatty Acid Ratio in Rice Bran Oil. *Plant Cell Tissue Organ Cult.* **2014**, *119*(1), 117–129.

Bhutani, S.; Kumar, R.; Chauhan, R.; Singh, R.; Chowdhury, V. K.; Chowdhury, J. B.; Jain, R. K. Development of Transgenic Indica Rice Plants Containing Potato Proteinase Inhibitor 2 (Pin2) Gene with Improved Defense Against Yellow Stem Borer. *Physiol. Mol. Biol. Plants* **2006**, *12*(1), 43–52.

Bu, Q. Y.; Wu, L.; Yang, S. H.; Wan, J. M. Cloning of a Potato Proteinase Inhibitor Gene Pinii-2X from Diploid Potato (Solanum Phurejia L.) and Transgenic Investigation of Its Potential to Confer Insect Resistance in Rice. *J. Integr. Plant Biol.* **2006**, *48*(6), 732–739.

Cao, M. X.; Huang, J. Q.; Wei, Z. M.; Yao, Q. H.; Wan, C. Z.; Lu, J. A. Engineering Higher Yield and Herbicide Resistance in Rice by Agrobacterium-Mediated Multiple Gene Transformation. *Crop Sci.* **2004**, *44*(6), 2206–2213.

Chakraborty, M.; Reddy, P. S.; Mustafa, G.; Rajesh, G.; Narasu, V. M. L.; Udayasuriyan, V.; Rana, D. Transgenic Rice Expressing the cry2AX1 Gene Confers Resistance to Multiple Lepidopteran Pests. *Transgenic Res.* **2016a**, *25*(5), 665–678.

Chakraborty, M.; Sairam Reddy, P.; Laxmi Narasu, M.; Krishna, G.; Rana, D. Agrobacterium-Mediated Genetic Transformation of Commercially Elite Rice Restorer Line Using nptII Gene as a Plant Selection Marker. *Physiol. Mol. Biol. Plants* **2016b**, *22*(1), 51–60.

Chang, C. C.; Huang, P. S.; Lin, H. R.; Lu, C. H. Transactivation of Protein Expression by Rice Hsp101 in Planta and Using Hsp101 as a Selection Marker for Transformation. *Plant Cell Physiol.* **2007**, *48*(8), 1098–1107.

Chen, D. X.; Chen, X. W.; Ma, B. T.; Wang, Y. P.; Zhu, L. H.; Li, S. G. Genetic Transformation of Rice with Pi-D2 Gene Enhances Resistance to Rice Blast Fungus Magnaporthe Oryzae. *Rice Sci.* **2010**, *17*(1), 19–27.

Chen, H.; Tang, W.; Xu, C.; Li, X.; Lin, Y.; Zhang, Q. Transgenic Indica Rice Plants Harboring a Synthetic Cry2a* Gene of Bacillus Thuringiensis Exhibit Enhanced Resistance Against Lepidopteran Rice Pests. *Theor. Appl. Genet.* **2005**, *111*(7), 1330–1337.

Chen, J. Q.; Meng, X. P.; Zhang, Y.; Xia, M.; Wang, X. P. Over-Expression of Osdreb Genes Lead to Enhanced Drought Tolerance in Rice. *Biotechnol. Lett.* **2008**, *30*(12), 2191–2198.

Chen, L.; Yang, X.; Luo, D.; Yu, W. Efficient Production of a Bioactive Bevacizumab Monoclonal Antibody Using the 2A Self-Cleavage Peptide in Transgenic Rice Callus. *Front Plant Sci.* **2016**, *7*, 1156.

Chen, Z.; Pan, Y.; Wang, S.; Ding, Y.; Yang, W.; Zhu, C Overexpression of a Protein Disulfide Isomerase-Like Protein from Methanothermobacter Thermoautotrophicum Enhances Mercury Tolerance in Transgenic Rice. *Plant Sci.* **2012**, *197*, 10–20.

Chen, Z.; Pan, Y. H.; An, L. Y.; Yang, W. J.; Xu, L. G.; Zhu, C. Heterologous Expression of a Halophilic Archaeon Manganese Superoxide Dismutase Enhances Salt Tolerance in Transgenic Rice. *Russ J. Plant Physiol.* **2013**, *60*(3), 359–366.

Chhapekar, S.; Raghavendrarao, S.; Pavan, G.; Ramakrishna, C.; Singh, V. K.; Phanindra, M. L. V.; Dhandapani, G.; Sreevathsa, R.; Ananda Kumar, P. Transgenic Rice Expressing a Codon-Modified Synthetic Cp4-Epsps Confers Tolerance to Broad-Spectrum Herbicide, Glyphosate. *Plant Cell Rep.* **2015**, *34*(5), 721–731.

Christou, P. Rice transformation: bombardment. *Plant Mol. Biol.* **1997**, *35*, 195–203.

Christou, P.; Ford, T. L.; Kofron, M. Rice genetic engineering: a review. *Trends Biotech.* **1992**, *10*, 239–246.

Chiang, C. M.; Chen, L. F. O.; Shih, S. W.; Lin, K. H. Expression of Eggplant Ascorbate Peroxidase Increases the Tolerance of Transgenic Rice Plants to Flooding Stress. *J. Plant Biochem. Biotechnol.* **2015**, *24*(3), 257–267.

Cho, A. R.; Lee, D. K.; Kim, K. M. High-Frequency Plant Regeneration from Transgenic Rice Expressing Arabidopsis Thaliana Bax Inhibitor (Atbi-1) Tissue Cultures. *J. Plant Biotechnol.* **2015**, *42*(2), 83–87.

Clement, W..; Lai, K. S.; Wong, M. Y.; Maziah, M. Heat and Hydrolytic Enzymes Treatment Improved the Agrobacterium-Mediated Transformation of Recalcitrant Indica Rice (*Oryza Sativa* L.). *Plant Cell Tissue Organ Cult.* **2016**, *125*, 183. DOI: 10.1007/s11240-015-0926-9.

Coca, M.; Bortolotti, C.; Rufat, M.; Peñas, G.; Eritja, R.; Tharreau, D.; Martinez Del Pozo, A.; Messeguer, J.; San Segundo, B. Transgenic Rice Plants Expressing the Antifungal Afp Protein from Aspergillus Giganteus Show Enhanced Resistance to the Rice Blast Fungus Magnaporthe Grisea. *Plant Mol. Biol.* **2004**, *54*(2), 245–259.

Datta, K.; Rai, M.; Parkhi, V.; Oliva, N.; Tan, J.; Datta, S. K. Improved 'Golden' Indica Rice and Post-Transgeneration Enhancement of Metabolic Target Products of Carotenoids (B-Carotene) in Transgenic Elite Cultivars (Ir64 and Br29). *Curr. Sci.* **2006**, *91*(7), 935–939.

Datta, K.; Sahoo, G.; Krishnan, S.; Ganguly, M.; Datta, S. K. Genetic Stability Developed for B-Carotene Synthesis in Br29 Rice Line Using Dihaploid Homozygosity. *PLoS One* **2014**, *9*(6), e100212.

Datta, K.; Baisakh, N.; Ganguly, M.; Krishnan, S.; Shinozaki, K. Y.; Datta, S. K. Overexpression of Arabidopsis and Rice Stress Genes' Inducible Transcription Factor Confers Drought and Salinity Tolerance to Rice. *Plant Biotechnol. J.* **2012**, *10*(5), 579–586.

Dedicova, B.; Bermudez, C.; Prias, M.; Zuniga, E.; Brondani, C. High-Throughput Transformation Pipeline for a Brazilian Japonica Rice with Bar Gene Selection. *Protoplasma* **2015**, *252*(4), 1071–1083.

Deng, L. H.; Weng, L. S.; Xiao, G. Y. Optimization of Epsps Gene and Development of Double Herbicide Tolerant Transgenic Pgms Rice. *J. Agric. Sci. Technol.* **2014**, *16*(1), 217–228.

Ding, Z. S.; Ming, Z.; Jing, Y. X.; Li, L. B.; Kuang, T. Y. Efficient Agrobacterium-Mediated Transformation of Rice by Phosphomannose Isomerase/Mannose Selection. *Plant Mol. Biol. Rep.* **2006**, *24*(3–4), 295–303.

Dong, R. X.; Chen, J.; Wang, X. M.; Li, J. S.; Zhou, J.; Yang, Y.; Yu, C. L.; Cheng, Y.; Yan, C. Q.; Chen, J. P. Agrobacterium-Mediated Transformation Efficiency is Altered in a Novel Rice Bacterial Blight Resistance Cultivar and is Influenced by Environmental Temperature. *Physiol. Mol. Plant Pathol.* **2012**, *77*(1), 33–40.

Duan, Y.; Zhai, C.; Li, H.; Li, J.; Mei, W.; Gui, H.; Ni, D.; Song, F.; Li, L.; Zhang, W.; Yang, J. An Efficient and High-Throughput Protocol for Agrobacterium-Mediated Transformation Based on Phosphomannose Isomerase Positive Selection in Japonica Rice (Oryza Sativa L.). *Plant Cell Rep.* **2012**, *31*(9), 1611–1624.

Endo, M.; Shimizu, T.; Toki, S. Selection of Transgenic Rice Plants Using a Herbicide Tolerant Form of the Acetolactate Synthase Gene. *Methods Mol. Biol.* **2012**, *847*, 59–66.

Fook, C. W. K.; Song, L. K.; Yun, W. M.; Mahmood, M. Efficient Regeneration and Agrobacterium-Mediated Transformation Protocol for Recalcitrant Indica Rice (Oryza sativa L.). *Emirates J. Food Agric.* **2015**, *27*(11), 837–848.

Fu, Y.; Si, H.; Zhu, Z.; Hu, G.; Sun, Z. Homozygous Transgenic Plants Regenerated from Rice Pollen Derived Callus Via Agrobacterium Tumefaciens. *J. Zhejiang Univ. Agric. Life Sci.* **2001**, *27*(4), 407–410.

Gao, L.; Liu, P.; Xia, Z.; Zhao, J.; Shi, J.; Jiang, G.; Liu, G.; Zhai, W. Integration and Expression of Xa21 in Transgenic Rice Cx8621. *Shengwu Gongcheng Xuebao/Chinese J. Biotechnol.* **2016**, *32*(9), 1255–1263.

Ge, X.; Chu, Z.; Lin, Y.; Wang, S. A Tissue Culture System for Different Germplasms of Indica Rice. *Plant Cell Rep.* **2006**, *25*(5), 392–402.

Gelvin, S. B. Agricultural biotechnology: gene exchange by design. *Nature*, **2005**, *433*, 583–584.

Gu, Z.; Ma, B.; Jiang, Y.; Chen, Z.; Su, X.; Zhang, H. Expression Analysis of the Calcineurin B-Like Gene Family in Rice (Oryza Sativa L.) Under Environmental Stresses. *Gene* **2008**, *415*(1–2), 1–12.

Guruprasad, M.; Sandeep Raja, D.; Jaffar, S. K.; Shanthisri, K. V.; Srinu Naik, M. An Efficient and Regeneration Protocol for Agrobacterium Mediated Transformation of Indica Rice. *Int. J. Pharm. Technol.* **2012**, *4* (2), 4280–4286.

Ha, S. B.; Lee, S. B.; Lee, D. E.; Guh, J. O.; Back, K. Transgenic Rice Plants Expressing Bacillus Subtilis Protoporphyrinogen Oxidase Gene Show Low Herbicide Oxyfluorfen Resistance. *Biol. Plant.* **2004a**, *47*(2), 277–280.

Ha, S. B.; Lee, S. B.; Lee, Y.; Yang, K.; Lee, N.; Jang, S. M.; Chung, J. S.; Jung, S.; Kim, Y. S.; Wi, S. G.; Back, K. The Plastidic Arabidopsis Protoporphyrinogen IX Oxidase Gene, with or without the Transit Sequence, Confers Resistance to the Diphenyl Ether Herbicide in Rice. *Plant Cell Environ.* **2004b**, *27*(1), 79–88.

He, R. Multigene eEngineering in Rice Using High-Capacity Agrobacterium Tumefaciens BIBAC Vectors. *Protoc. Recomb. Proteins Plants Methods Mol. Biol.* **2016**, 29–37.

He, R. F.; Wang, Y. Y.; Du, B.; Tang, M.; You, A. Q.; Zhu, L. L.; He, G. C. Development of Transformation System of Rice Based on Binary Bacterial Artificial Chromosome (Bibac) Vector. *Acta Genet. Sin.* **2006**, *33*(3), 269–276.

He, R.; Pan, J.; Zhu, L.; He, G. Agrobacterium-Mediated Transformation of Large DNA Fragments Using a Bibac Vector System in Rice. *Plant Mol. Biol. Rep.* **2010**, *28*(4), 613–619.

Hiei, Y.; Komari, T. Improved Protocols for Transformation of Indica Rice Mediated by Agrobacterium Tumefaciens. *Plant Cell Tissue Organ Cult.* **2006**, *85*(3), 271–283.

Hiei, Y.; Komari, T. Agrobacterium-Mediated Transformation of Rice Using Immature Embryos or Calli Induced from Mature Seed. *Nat. Protoc.* **2008**, *3*(5), 824–834.

Hiei, Y.; Ishida, Y.; Kasaoka, K.; Komari, T. Improved Frequency of Transformation in Rice and Maize by Treatment of Immature Embryos with Centrifugation and Heat Prior to Infection with Agrobacterium Tumefaciens. *Plant Cell Tissue Organ Cult.* **2006**, *87*(3), 233–243.

Hiei Y.; Ohta S.; Komari, T.; Kumashiro, T Efficient Transformation of Rice (Oryza Sativa L.) Mediated by Agrobacterium and Sequence Analysis of the Boundaries of the T-Dna. *Plant J.* **1994**, *6*(2), 271–282.

Hiei, Y.; Komari, T.; Kubo, T. Transformation of rice mediated by *Agrobacterium tumefaciens*. *Plant Mol Biol.,* **1997,** *35*(1–2), 205–218.

Hirose, S.; Kawahigashi, H.; Ozawa, K.; Shiota, N.; Inui, H.; Ohkawa, H.; Ohkawa, Y. Transgenic Rice Containing Human Cyp2b6 Detoxifies Various Classes of Herbicides. *J. Agric. Food Chem.* **2005**, *53*(9), 3461–3467.

Hoang, T. M. L.; Williams, B.; Khanna, H.; Dale, J.; Mundree, S. G. Physiological Basis of Salt Stress Tolerance in Rice Expressing the Antiapoptotic Gene Sfiap. *Funct. Plant Biol.* **2014**, *41*(11), 1168–1177.

Hoque, M. E.; Mansfield, J. W.; Bennett, M. H. Agrobacterium-Mediated Transformation of Indica Rice Genotypes: an Assessment of Factors Affecting the Transformation Efficiency. *Plant Cell Tissue Organ Cult.* **2005**, *82*(1), 45–55.

Huang, L.; Niu, X.; Xiong, F.; Zou, X.; Liu, Y. Construction and Genetic Transformation of the RNA Interference Vector and Functional Analysis of Rice Expansin Gene Osexpb7. *Chin. J. Appl. Environ. Biol.* **2013**, *19*(2), 198–204.

Huang, Z.; Lin, J.; Cheng, Z.; Xu, M.; Guo, M.; Huang, X.; Yang, Z.; Zheng, J. Production of Oleanane-Type Sapogenin in Transgenic Rice Via Expression of B-Amyrin Synthase Gene from Panax Japonicus C. A. Mey. *BMC Biotechnol.* **2015**, *15*, 45. DOI: 10.1186/s12896-015-0166-4.

Hussain, Z.; Ali, S.; Hayat, Z.; Zia, M. A.; Iqbal, A.; Ali, G. M. Agrobacterium Mediated Transformation of Drebla Gene for Improved Drought Tolerance in Rice Cultivars (Oryza Sativa L.). *Australian. J. Crop Sci.* **2014**, *8*(7), 1114–1123.

Ignacimuthu, S.; Raveendar, S. Agrobacterium Mediated Transformation of Indica Rice (Oryza sativa L.) for Insect Resistance. *Euphytica* **2011**, *179*(2), 277–286.

Ignacimuthu, S.; Arockiasamy, S. Agrobacterium-Mediated Transformation of an Elite Indica Rice for Insect Resistance. *Curr. Sci.* **2006**, *90*(6), 829–835.

Islam, M. M.; Khalekuzzaman, M. Development of Transgenic Rice (Oryza Sativa L.) Plant Using Cadmium Tolerance Gene (Ycfi) Through Agrobacterium Mediated Transformation for Phytoremediation. *Asian J. Agric. Res.* **2015**, *9*(4), 139–154.

Islam, S. M. T.; Tammi, R. S.; Singla-Pareek, S. L.; Seraj, Z. I. Enhanced Salinity Tolerance and Improved Yield Properties in Bangladeshi Rice Binnatoa Through Agrobacterium-Mediated Transformation of PgNHX1 from Pennisetum Glaucum. *Acta Physiol. Plant.* **2010**, *32*(4), 657–663.

Jha, S.; Chattoo, B. B. Transgene Stacking and Coordinated Expression of Plant Defensins Confer Fungal Resistance in Rice. *Rice* **2009**, *2*(4), 143–154.

Jha, S.; Tank, H. G.; Prasad, B. D.; Chattoo, B. B. Expression of Dm-Amp1 in Rice Confers Resistance to Magnaporthe Oryzae and Rhizoctonia Solani. *Transgenic Res.* **2009**, *18*(1), 59–69.

Jiang, H. Y.; Zhang, J.; Wang, J. M.; Xia, M.; Zhu, S. W.; Cheng, B. J. RNA Interference-Mediated Silencing of the Starch Branching Enzyme Gene Improves Amylose Content in Rice. *Genet Mol. Res.* **2013a**, *12*(3), 2800–2808.

Jiang, Y.; Sun, L.; Jiang, M.; Li, K.; Song, Y.; Zhu, C. Production of Marker-Free and RSV-Resistant Transgenic Rice Using a Twin T-DNA System and RNAi. *J. Biosci.* **2013b**, *38*(3), 573–581.

Jian-Jun, M.; Xiu-Fen, Y.; Hong-Mei, Z.; Jing-Jing, Y.; De-Wen, Q. Medium Selection in Tissue Culture of Japonica Rice Cultivar Nipponbare and Preparation of Transgenic Plants with an Elicitor-Coding Gene from Magnaporthe Grisea. *Chin. J. Agric. Biotechnol.* **2009**, *6*(2), 103–109.

Fluorescence in Situ Hybridization Analysis of Alien Genes in Agrobacterium-Mediated Cry1a (B)-Transformed Rice. *Ann Bot (Lond)* **2002**, *90*(1), 31–36.

Jin, Y. M.; Ma, R.; Yu, Z. J.; Wang, L.; Jiang, W. Z.; Lin, X. F. Development of Lepidopteran Pest-Resistant Transgenic Japonica Rice Harboring a Synthetic Cry2a* Gene. *J. Integr. Agric.* **2015**, *14*(3), 423–429.

Joubert, P.; Beaupere, D.; Lelievre, P.; Wadouachi, A.; Sangwan, R. S.; Sangwan- Norreel, B. S. Effects of phenolic compounds on *Agrobacterium* vir genes and gene transfer induction a plausible molecular mechanism of phenol binding protein activation. *Plant Sci.,* **2002**, *162*, 733–743.

Joubert, P.; Beaupère, D.; Wadouachi, A.; Chateau, S.; Sangwan, R. S.; Sangwan-Norreel, B. S. Effect of phenolic glycosides on *Agrobacterium tumefaciens virH* gene induction and plant transformation. *J. Natu. Prod.,* **2004**, *67*, 348–351.

Jung, Y. J.; Nogoy, F. M.; Cho, Y. G.; Kang, K. K Development of High Tryptophan Gm Rice and Its Transcriptome Analysis. *J. Plant Biotechnol.* **2015**, *42*(3), 186–195.

Kang, K.; Lee, K.; Park, S.; Kim, Y. S.; Back, K. Enhanced Production of Melatonin by Ectopic Overexpression of Human Serotonin N-Acetyltransferase Plays a Role in Cold Resistance in Transgenic Rice Seedlings. *J. Pineal Res.* **2010**, *49*(2), 176–182.

Kant, P.; Kant, S.; Jain, R. K.; Chaudhury, V. K. Agrobacterium-Mediated High Frequency Transformation in Dwarf Recalcitrant Rice Cultivars. *Biol. Plant* **2007**, *51*(1), 61–68.

Karthikeyan, A.; Pandian, S. K.; Ramesh, M. Agrobacterium-Mediated Transformation of Leaf Base Derived Callus Tissues of Popular Indica Rice (Oryza Sativa L. Sub Sp. Indica Cv. Adt 43). *Plant Sci.* **2011**, *181*(3), 258–268.

Karthikeyan, A.; Shilpha, J.; Pandian, S. K.; Ramesh, M. Agrobacterium-Mediated Transformation of Indica Rice Cv. Adt 43. *Plant Cell Tissue Organ Cult.* **2012**, *109*(1), 153–165.

Katiyar-Agarwal, S.; Agarwal, M.; Grover, A. Heat-Tolerant Basmati Rice Engineered by Over-Expression of Hsp101. *Plant Mol. Biol.* **2003**, *51*(5), 677–686.

Kawakatsu, T.; Kawahara, Y.; Itoh, T.; Takaiwa, F. A Whole-Genome Analysis of a Transgenic Rice Seed-Based Edible Vaccine Against Cedar Pollen Allergy. *DNA Res.* **2013**, *20*(6), 623–631.

Khan, M.; Jan, A.; Hashim, S.; Marwat, K. B. Cloning and Agrobacterium Mediated Transformation of the Putative Promoter Region of Oryza Sativa C3h52 Gene Into Local Rice Variety. *Pak. J. Bot.* **2015**, *47*(SI), 211–218.

Khan, M. H.; Rashid, H.; Swati, Z. A.; Chaudhry, Z. Agrobacterium Mediated Transformation to Build Resistance Against Bacterial Blight in Rice. *Pak. J. Bot.* **2007**, *39*(4), 1285–1292.

Kim, E. H.; Suh, S. C.; Park, B. S.; Shin, K. S.; Kweon, S. J.; Han, E. J. Park, S. H;. Kim, Y. S.; Kim, J. K. Chloroplast-Targeted Expression of Synthetic Cry1ac in Transgenic Rice as an Alternative Strategy for Increased Pest Protection. *Planta* **2009**, *230*(2), 397–405.

Kim, K. M.; Park, Y. H.; Kim, C. K.; Hirschi, K.; Sohn, J. K. Development of Transgenic Rice Plants Overexpressing the Arabidopsis H +/Ca2+ Antiporter CAX1 Gene. *Plant Cell Rep.* **2005b**, *23*(10–11), 678–682.

Kim, W. S.; Kim, J.; Krishnan, H. B.; Nahm, B. H. Expression of Escherichia Coli Branching Enzyme in Caryopses of Transgenic Rice Results in Amylopectin with an Increased Degree of Branching. *Planta* **2005a**, *220*(5), 689–695.

Krishnan, S. R.; Priya, A. M.; Ramesh, M. Rapid Regeneration and Ploidy Stability of 'Cv Ir36' Indica Rice (Oryza Sativa. L) Confers Efficient Protocol for in Vitro Callus Organogenesis and Agrobacterium Tumefaciens Mediated Transformation. *Bot. Stud.* **2013**, *54*(1), 47.

Kuk, Y. I.; Lee, H. J.; Chung, J. S.; Kim, K. M.; Lee, S. B.; Ha, S. B.; Back, K.; Guh, J. O. Expression of a Bacillus Subtilis Protoporphyrinogen Oxidase Gene in Rice Plants Reduces Sensitivity to Peroxidizing Herbicides. *Biol. Plant* **2005**, *49*(4), 577–583.

Kumar, K. K.; Poovannan, K.; Nandakumar, R.; Thamilarasi, K.; Geetha, C.; Jayashree, N.; Kokiladevi, E.; Raja, J. A. J.; Samiyappan, R.; Sudhakar, D.; Balasubramanian, P. A High Throughput Functional Expression Assay System for a Defence Gene Conferring Transgenic Resistance on Rice Against the Sheath Blight Pathogen, Rhizoctonia Solani. *Plant Sci.* **2003**, *165*(5), 969–976.

Lee, D. E.; Lee, I. J.; Han, O.; Baik, M. G.; Han, S. S.; Back, K. Pathogen Resistance of Transgenic Rice Plants Expressing Mitogen-Activated Protein Kinase 1, Mk1, from Capsicum Annuum. *Mol. Cells* **2004**, *17*(1), 81–85.

Lee, J. H.; Shin, K. S.; Suh, S. C.; Rhim, S. L.; Lee, Y. H.; Lim, M. H.; Woo, H. J.; Qin, Y.; Cho, H. S. Cryiiia Toxin Gene Expression in Transgenic Rice Confers Resistance to Rice Water Weevil. *Plant Cell Tissue Organ Cult.* **2013**, *115*(2), 243–252.

Lee, K.; Kang, K.; Park, M.; Woo, Y. M.; Back, K. Endosperm-Specific Expression of Serotonin N-Hydroxycinnamoyltransferase in Rice. Plant Foods Human Nutrition **2008**, *63*(2), 53–57.

Li, J. Y.; Xu, S. Z.; Yang, L. J.; Zhou, Y. G.; Fan, S. J.; Zhang, W. Breeding Elite Japonica-Type Soft Rice with High Protein Content Through the Introduction of the Anti- Waxy Gene. *Afr. J. Biotechnol.* **2009**, *8*(2), 161–166.

Li, J; Mao, W; Yang, L; Zhou, G; Liu, J; Yan, Q; Mi D. Introducing Antisense Waxy Gene Into Rice Seeds Reduces Grain Amylose Contents Using a Safe Transgenic Technique. *Chin. Sci. Bull.* **2005**, *50*(1), 39–44.

Li, R.; Zhang, Z.; Zhang, Q Transformation of Japonica Rice with Rhl Gene and Salt Tolerance of the Transgenic Rice Plant. *Chin. Sci. Bull.* **2002**, *47*(12), 998–1002.

Liao, H.; Xiao, X.; Li, X.; Chen, Y.; Fu, X.; Lin, D.; Niu, X.; Chen, Y.; He, C. Osbak1 is Involved in Rice Resistance to Xanthomonas Oryzae Pv. Oryzae Pxo99. *Plant Biotechnol. Rep.* **2016**, *10*(2), 75–82.

Lin, J.; Zhou, B.; Yang, Y.; Mei, J.; Zhao, X.; Guo, X.; Huang, X.; Tang, D.; Liu, X. Piercing and Vacuum Infiltration of the Mature Embryo: a Simplified Method for Agrobacterium-Mediated Transformation of Indica Rice. *Plant Cell Rep.* **2009**, *28*(7), 1065–1074.

Liu, Y. K.; Huang, L. F.; Ho, S. L.; Liao, C. Y.; Liu, H. Y.; Lai, Y. H.; Yu, S. M.; Lu, C. A. Production of Mouse Granulocyte-Macrophage Colony-Stimulating Factor by Gateway Technology and Transgenic Rice Cell Culture. *Biotechnol. Bioeng.* **2012**, *109*(5), 1239–1247.

Ma, J.; Song, Y.; Wu, B.; Jiang, M.; Li, K.; Zhu, C.; Wen, F. Production of Transgenic Rice New Germplasm with Strong Resistance Against Two Isolations of Rice Stripe Virus by Rna Interference. *Transgenic Res.* **2011**, *20*(6), 1367–1377.

Manikandan, R.; Sathish, S.; Balakrishnan, N.; Balasubramani, V.; Sudhakar, D.; Udayasuriyan, V. Agrobacterium Mediated Transformation of Indica Rice with Synthetic Cry2ax1 Gene for Resistance Against Rice Leaf Folder. *J. Pure Appl. Microbiol.* **2014**, *8*(4), 3135–3142.

Mao, B.; Liu, X.; Hu, D.; Li, D. Co-Expression of Rch10 and Aglu1 Confers Rice Resistance to Fungal Sheath Blight Rhizoctonia Solani and Blast Magnorpathe Oryzae and Reveals Impact on Seed Germination. *World J. Microbiol. Biotechnol.* **2014**, *30*(4), 1229–1238.

Miao, Y.; Li, R.; Xiong, F.; Zeng, Z.; Niu, X.; Liu, Y. Rna Interference Vector Construction, Genetic Transformation and Function Analysis of Rice Transcription Factor Osmyb14. *Chin. J. Appl. Environ. Biol.* **2013**, *19*(6), 960–968.

Mitsuhara, I.; Yatou, O.; Iwai, T.; Naito, Y.; Nawa, Y.; Ohashi, Y. Genetic Studies of Transgenic Rice Plants Overproducing an Antibacterial Peptide Show That a High Level of Transgene Expression Did Not Cause Inferior Effects on Host Plants. *Plant Biotechnol.* **2006**, *23*(1), 63–69.

Priya, A. M.; Pandian, S. K.; Manikandan, R. The Effect of Different Antibiotics on the Elimination of Agrobacterium and High Frequency Agrobacterium-Mediated Transformation of Indica Rice (Oryza Sativa L.). *Czech J. Genet. Plant Breed.* **2012**, *48*(3), 120–130.

Mohanty S. Rice and the global financial crisis. *Rice Today* **2009**, *8*(1), 40.

Mohanty, A; Kathuria, H; Ferjani, A; Sakamoto, A; Mohanty, P; Murata, N; Tyagi, A. K. Transgenics of an Elite Indica Rice Variety Pusa Basmati 1 Harbouring the Coda Gene are Highly Tolerant to Salt Stress. *Theor. Appl. Genet.* **2002**, *106*(1), 51–57.

Murakami, T.; Matsuba, S.; Funatsuki, H.; Kawaguchi, K.; Saruyama, H.; Tanida, M.; Sato, Y. Over-Expression of a Small Heat Shock Protein, sHSP17.7, Confers Both Heat Tolerance and UV-B Resistance to Rice Plants. *Mol. Breed.* **2004**, *13*(2), 165–175.

Nandakumar, R.; Babu, S.; Kalpana, K.; Raguchander, T.; Balasubramanian, P.; Samiyappan, R. Agrobacterium-Mediated Transformation of Indica Rice with Chitinase Gene for Enhanced Sheath Blight Resistance. *Biol. Plant.* **2007**, *51*(1), 142–148.

NCBI (National Center for Biotechnology Information). Taxonomy browser (Agrobacterium radiobacter K84). **2016**. https://www.ncbi.nlm.nih.gov/Taxonomy/Browser/wwwtax.cgi?id=311403. (accessed Nov 17, 2016).

Nishimura, A.; Aichi, I.; Matsuoka, M. A Protocol for Agrobacterium-Mediated Transformation in Rice. *Nat. Protoc.* **2007**, *1*(6), 2796–2802.

Niu, X.; Xiong, F.; Liu, J.; Sui, Y.; Zeng, Z.; Lu, B. R.; Liu, Y Co-Expression of Apgsmt and Apdmt Promotes Biosynthesis of Glycine Betaine in Rice (Oryza Sativa L.) and Enhances Salt and Cold Tolerance. *Environ. Exp. Bot.* **2014**, *104*, 16–25.

Ozawa, K. Establishment of a High Efficiency Agrobacterium-Mediated Transformation System of Rice (*Oryza Sativa* L.). *Plant Sci.* **2009**, *176*(4), 522–527.

Ozawa, K.; Takaiwa, F. Highly Efficient Agrobacterium-Mediated Transformation of Suspension-Cultured Cell Clusters of Rice (Oryza Sativa L.). *Plant Sci.* **2010**, *179*(4), 333–337.

Ozawa, K.; Kawahigashi, H. Positional Cloning of the Nitrite Reductase Gene Associated with Good Growth and Regeneration Ability of Calli and Establishment of a New Selection System for Agrobacterium-Mediated Transformation in Rice (Oryza Sativa L.). *Plant Sci.* **2006**, *170*(2), 384–393.

Pawar, B.; Kale, P.; Bahurupe, J.; Jadhav, A.; Kale, A.; Pawar, S. Proline and Glutamine Improve in Vitro Callus Induction and Subsequent Shooting in Rice. *Rice Sci.* **2015**, *22*(6), 283–289.

Pujni, D.; Chaudhary, A.; Rajam, M. V. Increased Tolerance to Salinity and Drought in Transgenic Indica Rice by Mannitol Accumulation. *J. Plant Biochem. Biotechnol.* **2007**, *16*(1), 1–7.

Qi, Y. B.; Ye, S. H.; Lu, Y. T.; Jin, Q. S.; Zhang, X. M. Development of Marker-Free Transgenic Cry1ab Rice with Lepidopteran Pest Eesistance by Agrobacterium Mixture-Mediated Co-Transformation. *Rice Sci.* **2009**, *16*(3), 181–186.

Qiu, C.; Li, H.; Song, F.; Duan, Y.; Sun, Y.; Yang, Y.; Qin, R.; Li, L.; Wei, P.; Yang, J. A Novel Plant Code Optimization Phosphomannose Isomerase (Ppmi) and Its Application in Rice (Oryza Sativa L.) Transformation as Selective Marker. *Plant OMICS* **2015**, *8*(1), 30–36.

Qiu, C.; Sangha, J. S.; Song, F.; Zhou, Z.; Yin, A.; Gu, K.; Tian, D.; Yang, J.; Yin, Z. Production of Marker-Free Transgenic Rice Expressing Tissue-Specific Bt Gene. *Plant Cell Rep.* **2010**, *29*(10), 1097–1107.

Rachmat, A; Nugroho, S; Sukma, D; Aswidinnoor, H Sudarsono. Overexpression of Osnac6 Transcription Factor from Indonesia Rice Cultivar Enhances Drought and Salt Tolerance. *Emirates J. Food Agric.* **2014**, *26*(6), 519–527.

Rachmawati, D.; Mori, T.; Hosaka, T.; Takaiwa, F.; Inoue, E.; Anzai, H. Production and Characterization of Recombinant Human Lactoferrin in Transgenic Javanica Rice. *Breed. Sci.* **2005**, *55*(2), 213–222.

Rafique, M. Z.; Zia, M.; Rashid, H.; Chaudhary, M. F.; Chaudhry, Z. Comparison of Transgenic Plant Production for Bacterial Blight Resistance in Pakistani Local Rice (Oryza Sativa L.) Cultivars. *Afr. J. Biotechnol.* **2010**, *9*(13), 1892–1904.

Rahman, Z. A.; Seman, Z. A.; Basirun, N.; Julkifle, A. L.; Zainal, Z.; Subramaniam, S. Preliminary Investigations of Agrobacterium-Mediated Transformation in Indica Rice Mr219 Embryogenic Callus Using Gusa Gene. *Afr. J. Biotechnol.* **2011**, *10*(40), 7805–7813.

Ramesh, S.; Nagadhara, D.; Pasalu, I. C.; Kumari, A. P.; Sarma, N. P.; Reddy, V. D.; Rao, K. V. Development of Stem Borer Resistant Transgenic Parental Lines Involved in the Production of Hybrid Rice. *J. Biotechnol.* **2004b**, *111*(2), 131–141.

Ramesh, S.; Nagadhara, D.; Reddy, V. D.; Rao, K. V. Production of Transgenic Indica Rice Resistant to Yellow Stem Borer and Sap-Sucking Insects, Using Super-Binary Vectors of Agrobacterium Tumefaciens. *Plant Sci.* **2004a**, *166*(4), 1077–1085.

Rao, M. V. R; Rao, G. J. N. Agrobacterium-Mediated Transformation of Indica Rice Under Acetosyringone-Free Conditions. *Plant Biotechnol.* **2007**, *24*(5), 507–511.

Rao, M. V. R.; Behera, K. S.; Baisakh, N.; Datta, S. K.; Rao, G. J. N. Transgenic Indica Rice Cultivar 'Swarna' Expressing a Potato Chymotrypsin Inhibitor Pin2 Gene Show Enhanced Levels of Resistance to Yellow Stem Borer. *Plant Cell Tissue Organ Cult.* **2009**, *99*(3), 277–285.

Rao, M. V. R.; Parameswari, C.; Sripriya, R.; Veluthambi, K. Transgene Stacking and Marker Elimination in Transgenic Rice by Sequential Agrobacterium-Mediated Co-Transformation with the Same Selectable Marker Gene. *Plant Cell Rep.* **2011**, *30*(7), 1241–1252.

Ravikumar, G.; Manimaran, P.; Voleti, S. R.; Subrahmanyam, D.; Sundaram, R. M.; Bansal, K. C.; Viraktamath, B. C.; Balachandran, S. M. Stress-Inducible Expression of Atdreb1a Transcription Factor Greatly Improves Drought Stress Tolerance in Transgenic Indica Rice. *Transgenic Res.* **2014**, *23*(3), 421–439.

Razzaque, S; Elias, S. M; Biswas, S; Haque, T; Seraj, Z. I. Cloning of the Plasma Membrane Sodium/Hydrogen Antiporter Sos1 for It'S Over Expression in Rice. *Plant Tissue Cult. Biotechnol.* **2013**, *23*(2), 263–273.

Roy, M.; Wu, R. Overexpression of S-Adenosylmethionine Decarboxylase Gene in Rice Increases Polyamine Level and Enhances Sodium Chloride-Stress Tolerance. *Plant Sci.* **2002**, *163*(5), 987–992.

Sadumpati, V.; Kalambur, M.; Vudem, D. R.; Kirti, P. B.; Khareedu, V. R. Transgenic Indica Rice Lines, Expressing Brassica Juncea Nonexpressor of Pathogenesis-Related Genes 1 (Bjnpr1), Exhibit Enhanced Resistance to Major Pathogens. *J. Biotechnol.* **2013**, *166*(3), 114–121.

Safitri, F. A.; Mohammad, U.; Kyung-Min, K. Efficiency of Transformation Mediated by Agrobacterium Tumefaciens Using Vacuum Infiltration in Rice (Oryza Sativa L.). *J. Plant Biotechnol.* **2016**, *43*, 66–75. http://dx.doi.org/10.5010/JPB.2016.43.1.66.

Sahoo, R. K.; Tuteja, N. Development of Agrobacterium-Mediated Transformation Technology for Mature Seed-Derived Callus Tissues of Indica Rice Cultivar IR64. *GM Crops Food* **2012** *3*(2), 123–128.

Saika H.; Toki S. Mature Seed-Derived Callus of the Model Indica Rice Variety Kasalath is Highly Competent in Agrobacterium-Mediated Transformation. *Plant Cell Rep.* **2010**, *29*(12), 1351–1364.

Saika, H.; Toki, S. Visual Selection Allows Immediate Identification of Transgenic Rice Calli Efficiently Accumulating Transgene Products. *Plant Cell Rep.* **2009**, *28*(4), 619–626.

Saika, H.; Onodera, H.; Toki, S. Visual Selection in Rice: A Strategy for the Efficient Identification of Transgenic Calli Accumulating Transgene Products. *Transgenic Plants: Methods Mol. Biolo.* **2012b**, 67–74.

Saika, H.; Nonaka, S.; Osakabe, K.; Toki, S. Sequential Monitoring of Transgene Expression Following Agrobacterium-Mediated Transformation of Rice. *Plant Cell Physiology* **2012a**, *53*(11), 1974–1983.

Saika, H.; Sakamoto, W.; Maekawa, M.; Toki, S. Highly Efficient Visual Selection of Transgenic Rice Plants Using Green Fluorescent Protein Or Anthocyanin Synthetic Genes. *Plant Biotechnol.* **2011**, *28*(1), 107–110.

Sailila, E. A.; Hye, J. L.; Joonki, K.; Marjohn, C. N.; Yu-Jin, J.; Young-Chan, C.; Illsup, N.; Kwon-Kyoo, K.; Yong-Gu, C. Bruge1 Transgenic Rice Showed Improved Growth Performance with Enhanced Drought Tolerance. *Breed. Sci.* **2016**, *66*(2), 226–233.

Sawahel, W. Improved Performance of Transgenic Glycinebetaine-Accumulating Rice Plants Under Drought Stress. *Biol. Plant* **2004**, *47*(1), 39–44.

Shah, J. M.; Raghupathy, V.; Veluthambi, K. Enhanced Sheath Blight Resistance in Transgenic Rice Expressing an Endochitinase Gene from Trichoderma Virens. *Biotechnol. Lett.* **2009**, *31*(2), 239–244.

Shen, G. Z.; Wang, X. Q.; Yin, L. Q.; Cai, X. L.; Wang, Z. Y. Rapidly Obtaining the Markerless Transgenic Rice with Reduced Amylose Content by Co-Transformation and Anther Culture. *Zhi wu sheng li yu fen zi sheng wu xue xue bao J. Plant Physiol. Mol. Biol.* **2004**, *30*(6), 637–643.

Shirasawa, K.; Takabe, T.; Takabe, T.; Kishitani, S. Accumulation of Glycinebetaine in Rice Plants That Overexpress Choline Monooxygenase from Spinach and Evaluation of Their Tolerance to Abiotic Stress. *Ann Bot (Lond)* **2006**, *98*(3), 565–571.

Shrawat, A. K.; Good, A. G. A High-Throughput Agrobacterium Tumefaciens-Mediated Transformation System for Molecular Breeding and Functional Genomics of Rice (Oryza Sativa L.,). *Plant Biotechnol.* **2010**, *27*(1), 47–58.

Shri, M.; Rai, A.; Verma, P. K.; Misra, P.; Dubey, S.; Kumar, S.; Verma, S.; Gautam, N.; Tripathi, R. D.; Trivedi, P. K.; Chakrabarty, D. An Improved Agrobacterium-Mediated

Transformation of Recalcitrant Indica Rice (Oryza sativa L.) Cultivars. *Protoplasma* **2013**, *250*(2), 631–636.

Sikdar, S. U.; Zobayer, N.; Nasrin, S.; Prodhan, S. H. Agrobacterium-Mediated PsCBL and PsCIPK Gene Transformation to Enhance Salt Tolerance in Indica Rice (Oryza sativa). *In Vitro Cell. Dev. Biol. Plant* **2015**, *51*(2), 143–151.

Slamet-Loedin, I. H.; Chadha-Mohanty, P.; Torrizo, L. Agrobacterium-Mediated Transformation. *Rice Transform. Methods Mol. Biol.* **2014**, *1099*, 261–271.

Sridevi, G.; Parameswari, C.; Sabapathi, N.; Raghupathy, V.; Veluthambi, K. Combined Expression of Chitinase and B-1,3-Glucanase Genes in Indica Rice (Oryza Sativa L.) Enhances Resistance Against Rhizoctonia Solani. *Plant Sci.* **2008**, *175*(3), 283–290.

Sridevi, G.; Sabapathi, N.; Meena, P.; Nandakumar, R.; Samiyappan, R.; Muthukrishnan, S.; Veluthambi, K Transgenic Indica Rice Variety Pusa Basmati 1 Constitutively Expressing a Rice Chitinase Gene Exhibits Enhanced Resistance to Rhizoctonia Solani. *J. Plant Biochem. Biotechnol.* **2003**, *12*(2), 93–101.

Sripriya, R.; Sangeetha, M.; Parameswari, C.; Veluthambi, B.; Veluthambi, K. Improved Agrobacterium-Mediated Co-Transformation and Selectable Marker Elimination in Transgenic Rice by Using a High Copy Number Pbin19-Derived Binary Vector. *Plant Sci.* **2011**, *180*(6), 766–774.

Takahashi, M.; Nakanishi, H.; Kawasaki, S.; Nishizawa, N. K.; Mori, S. Enhanced Tolerance of Rice to Low Iron Availability in Alkaline Soils Using Barley Nicotianamine Aminotransferase Genes. *Nat. Biotechnol.* **2001**, *19*(5), 466–469.

Takaiwa, F.; Takagi, H.; Hirose, S.; Wakasa, Y. Endosperm Tissue is Good Production Platform for Artificial Recombinant Proteins in Transgenic Rice. *Plant Biotechnol. J.* **2007**, *5*(1), 84–92.

Takesawa, T.; Ito, M.; Kanzaki, H.; Kameya, N.; Nakamura, I. Over-Expression of Z Glutathione S-Transferase in Transgenic Rice Enhances Germination and Growth At Low Temperature. *Mol. Breed.* **2002**, *9*(2), 93–101.

Tang, W.; Page, M. Inducible Expression of Norwalk Virus Capsid Protein Gene in Plant Cell Suspension Cultures. *In Vitro Cell. Dev. Biol. Plant* **2013a**, *49*(2), 129–136.

Tang, W.; Page, M. Overexpression of the Arabidopsis AtEm6 Gene Enhances Salt Tolerance in Transgenic Rice Cell Lines. *Plant Cell. Tissue Organ Cult.* **2013b**, *114*(3), 339–350.

Tang, W.; Page, M.; Fei, Y.; Liu, L.; Xu, F.; Cai, X.; Yuan, L.; Wu, Q. S.; Zhou, M.Overexpression of Atbzip60deltac Gene Alleviates Salt-Induced Oxidative Damage in Transgenic Cell Cultures. *Plant Mol. Biol. Rep.* **2012**, *30*(5), 1183–1195.

Tang, W.; Chen, H.; Xu, C.; Li, X.; Lin, Y.; Zhang, Q. Development of Insect-Resistant Transgenic Indica Rice with a Synthetic Cry1C* Gene. *Mol. Breed.* **2006**, *18*(1), 1–10.

Toki, S.; Hara, N.; Ono, K.; Onodera, H.; Tagiri, A.; Oka, S.; Tanaka, H. Early Infection of Scutellum Tissue with Agrobacterium Allows High-Speed Transformation of Rice. *Plant J.* **2006**, *47*(6), 969–976.

Touhidul Islam, S. M.; Tammi, R. S.; Singla-Pareek, S. L.; Seraj, Z. I. Agrobacterium-Mediated Transformation and Constitutive Expression of PgNHX1 from Pennisetum glaucum L. in Oryza sativa L. Cv. Binnatoa. *Plant Tissue Cult. Biotechnol.* **2009**, *19*(1), 25–33.

Uchimiya, H.; Fushimi, T.; Hashimoto, H.; Harada, H.; Syono, K.; Sugawara, Y. Expression of a foreign gene in callus derived from DNA-treated protoplasts of rice (Oryza sativa L.). MOI. Gen. Genet. **1986**, *204*, 204–207.

Urushibara, S.; Tozawa, Y.; Kawagishi-Kobayashi, M.; Wakasa, K. Efficient Transformation of Suspension-Cultured Rice Cells Mediated by Agrobacterium Tumefaciens. *Breed. Sci.* **2001**, *51*(1), 33–38.

Uzé, M.; Wünn, J.; Puonti-Kaerlas, J.; Potrykus, I.; Sautter, C. Plasmolysis of Precultured Immature Embryos Improves Agrobacterium Mediated Gene Transfer to Rice (Oryza Sativa L.). *Plant Sci.* **1997**, *130*(1), 87–95.

Verma, D. K.; Srivastav, P. P. Proximate Composition, Mineral Content and Fatty Acids Analyses of Aromatic and Non-Aromatic Indian Rice. *Rice Sci.* **2017**, *24*(1), 21–31.

Verma, D. K.; Mohan, M.; Asthir, B Physicochemical and Cooking Characteristics of Some Promising Basmati Genotypes. *Asian J. Food Agro-Indus* **2013**, *6*(2), 94–99.

Verma, D. K.; Mohan, M.; Prabhakar, P. K.; Srivastav, P. P. Physico-Chemical and Cooking Characteristics of Azad Basmati. *Int. Food Res. J.* **2015**, *22*(4), 1380–1389.

Verma, D. K.; Mohan, M.; Yadav, V. K.; Asthir, B.; Soni, S. K. Inquisition of Some Physico-Chemical Characteristics of Newly Evolved Basmati Rice. *Environ. Ecol.* **2012**, *30*(1), 114–117.

Wagiran, A.; Ismail, I.; Zain, C. R. C. M.; Abdullah, R. Agrobacterium Tumefaciens-Mediated Transformation of the Isopentenyltransferase Gene in Japonica Rice Suspension Cell Culture. *Aust. J. Crop Sci.* **2010**, *4*(6), 421–429.

Wakasa, Y.; Ozawa, K.; Takaiwa, F Agrobacterium-Mediated Co-Transformation of Rice Using Two Selectable Marker Genes Derived from Rice Genome Components. *Plant Cell Rep.* **2012**, *31*(11), 2075–2084.

Wang, F. Z.; Wang, Q. B.; Kwon, S. Y.; Kwak, S. S.; Su, W. A. Enhanced Drought Tolerance of Transgenic Rice Plants Expressing a Pea Manganese Superoxide Dismutase. *J. Plant Physiol.* **2005**, *162*(4), 465–472.

Wang, T.; Liu, L. Y.; Tang, Y. Y.; Zhang, X. B.; Zhang, M. D.; Zheng, Y. L.; Zhang, F. D. Using the Phosphomannose Isomerase (Pmi) Gene from Saccharomyces Cerevisiae for Selection in Rice Transformation. J. Integr. Agric. **2012b**, *11*(9), 1391–1398.

Wang, W.; Wu, C.; Liu, M.; Liu, X. R.; Hu, G. C.; Si, H. M.; Sun, Z. X.; Liu, W. Z.; Fu, Y. P. Resistance of Antimicrobial Peptide Gene Transgenic Rice to Bacterial Blight. *Rice Sci.* **2011**, *18*(1), 10–16.

Wang, X. Q.; Yin, L. Q.; Shen, G. Z.; Xu, L.; Liu, Q. Q. Determination of Amylose Content and its Relationship with RVA Profile Within Genetically Similar Cultivars of Rice (Oryza sativa L. ssp. japonica). *Agric. Sci. China* **2010**, *9*(8), 1101–1107.

Wang, Y.; Deng, H.; Zhang, X.; Xiao, H.; Jiang, Y.; Song, Y.; Fang, L.; Xiao, S.; Zhen, Y.; Chen, H. Generation and Immunogenicity of Japanese Encephalitis Virus Envelope Protein Expressed in Transgenic Rice. *Biochem. Biophys. Res. Commun.* **2009**, *380*(2), 292–297.

Wang, Y.; Shen, Q.; Jiang, Y.; Song, Y.; Fang, L.; Xiao, S.; Chen, H. Immunogenicity of Foot-And-Mouth Disease Virus Structural Polyprotein P1 Expressed in Transgenic Rice. *J. Virol. Methods* **2012b**, *181*(1), 12–17.

Wanichananan, P.; Teerakathiti, T.; Roytrakul, S.; Kirdmanee, C.; Peyachoknagul, S. A Highly Efficient Method for Agrobacterium Mediated Transformation in Elite Rice Varieties (Oryza Sativa L. Spp. Indica). *Afr. J. Biotechnol.* **2010**, *9*(34), 5488–5495.

Wanjogu, S. N.; Muya, E. M.; Gicheru, P. T.; Waruru, B. K. In *Soil degradation: Management and rehabilitation in Kenya*, Proceedings of the FAO/ISCW expert Consultation on Management of Degraded Soil in Southern and Eastern Africa (MADS-SEA) 2nd Networking Meeting Sept 18–20, 2000; Pretoria, South Africa. PR, 2001, pp 102–113.

Wei, F. J.; Kuang, L. Y.; Oung, H. M.; Cheng, S. Y.; Wu, H. P.; Huang, L. T.; Tseng, Y. T.; Chiou, W. Y.; Hsieh-Feng, V.; Chung, C. H.; Yu, S. M.; Lee, L.; Gelvin, S. B.; Hsing, Y. I. C. Somaclonal Variation Does not Preclude the use of Rice Transformants for Genetic Screening. *Plant J.* **2016**, *85*(5), 648–659.

Weng, X.; Huang, Y.; Hou, C.; Jiang, D Effects of an Exogenous Xylanase Gene Expression on the Growth of Transgenic Rice and the Expression Level of Endogenous Xylanase Inhibitor Gene Rixi. *J. Sci. Food Agric.* **2013**, *93*(1), 173–179.

Woo, H. J.; Lee, S. B.; Qin, Y.; Lim, M. H.; Lee, J. H.; Shin, K. S.; Cho, H. S.; Park, S. K. Generation and Molecular Characterization of Marker-Free Bt Transgenic Rice Plants by Selectable Marker-Less Transformation. *Plant Biotechnol. Rep.* **2015a**, *9*(6), 351–360.

Woo, H. J.; Qin, Y.; Park, S. Y.; Park, S. K.; Cho, Y. G.; Shin, K. S.; Lim, M. H.; Cho, H. S. Development of Selectable Marker-Free Transgenic Rice Plants with Enhanced Seed Tocopherol Content Through FLP/FRT-Mediated Spontaneous Auto-Excision. *PLoS One* **2015b**, *10*(7), e0132667.

Xia, Z. H.; Li, X. B.; Chen, C. Y.; Fan, H. K.; Jiang, G. H.; Zhu, L. H.; Zhai, W. X. Generation of Selectable Marker-Free and Vector Backbone Sequence-Free Xa21 Transgenic Rice. *Sheng Wu Gong Cheng Xue Bao Chin. J. Biotechnol.* **2006**, *22*(2), 204–210.

Xiang, D. J.; Hu, X. Y.; Zhang, Y.; Yin, K. D. Over-Expression of Ice1 Gene in Transgenic Rice Improves Cold Tolerance. *Rice Sci.* **2008**, *15*(3), 173–178.

Xu, M. L.; Jiang, J. F.; Ge, L.; Xu, Y. Y.; Chen, H.; Zhao, Y.; Bi, Y. R.; Wen, J. Q.; Chong, K. Fpf1 Transgene Leads to Altered Flowering Time and Root Development in Rice. *Plant Cell Rep.* **2005**, *24*(2), 79–85.

Xu, Q.; Xu, X.; Shi, Y.; Xu, J.; Huang, B. Transgenic Tobacco Plants Overexpressing a Grass Ppexp1 Gene Exhibit Enhanced Tolerance to Heat Stress. *PLoS One* **2014**, *9*(7), e100792.

Xue, D.; Jiang, H.; Qian, Q. Rice Genomics and Biotechnology. *Appl. Plant Genomics Biotechnol.* **2015**, 167–178.

Yadav, S.; Irfan, M.; Ahmad, A.; Hayat, S. Causes of salinity and plant manifestations to salt stress: A review. *J. Environ. Biol.* **2011**, *32*, 667–685.

Yang, D. H.; Chang, A. C.; Ahn, I. P.; Kim, H. J.; Kim, D. H.; Lee, H. Y.; Suh, S. C. Rapid Agrobacterium-Mediated Genetic Rice Transformation Method Using Liquid Media. *J. Plant Biotechnol.* **2013a**, *40*(1), 37–42.

Yang, R.; Zhou, Y.; Cao, Y.; Yin, Z.; Yang, L.; Li, J. The Transformation of the Photo-Thermo Sensitive Genic Male-Sterile Line 261S of Rice Via an Expression Vector Containing the Anti-Waxy Gene. *Breed. Sci.* **2013b**, *63*(2), 147–153.

Yang, Z. Q.; Liu, Q. Q.; Pan, Z. M.; Yu, H. X.; Jiao, X. A. Expression of the Fusion Glycoprotein of Newcasstle Disease Virus in Transgenic Rice and Its Immunogenicity in Mice. *Vaccine* **2007**, *25*(4), 591–598.

Yang, Z. Q.; Liu, Q. Q.; Yu, H. X.; Pan, Z. M.; Jiao, X. A. Expression and Immunization Testing of Fusion Protein of Newcastle Disease Virus in Leaf Tissue of Transgenic Rice. *Acta Genet. Sin.* **2005**, *32*(12), 1305–1311.

Ye, R.; Huang, H.; Yang, Z.; Chen, T.; Liu, L.; Li, X.; Chen, H.; Lin, Y. Development of Insect-Resistant Transgenic Rice with Cry1c*-Free Endosperm. *Pest Manage. Sci.* **2009**, *65*(9), 1015–1020.

Yinxia, Z.; Te-Chato, S. Optimization of Certain Parameters for Transformation of Indica Rice Hom Kra Dang Ngah Variety Via Agrobacterium-Mediated Transformation. *Kasetsart J. Nat. Sci.* **2015**, *49*(5), 676–686.

Yiu, J. C.; Tseng, M. J.; Yang, C. Y.; Liu, C. W.; Lai, Y. S.; Kuo, C. T.; Liu, H. H. Transgenic Rice Expresses an Antigenic Lipoprotein of Neisseria Gonorrhoeae. *J. Sci. Food Agric.* **2008**, *88*(9), 1603–1613.

Yookongkaew, N.; Srivatanakul, M.; Narangajavana, J. Development of Genotype-Independent Regeneration System for Transformation of Rice (Oryza Sativa Ssp. Indica). *J. Plant Res.* **2007**, *120*(2), 237–245.

Young, J. M.; Kuykendall, L. D.; Martínez-Romero, E.; Kerr, A.; Sawada, H. A Revision of Rhizobium Frank 1889, with an Emended Description of the Genus, and the Inclusion of All Species of Agrobacterium Conn 1942 and Allorhizobium Undicola De Lajudie Et Al., 1998 as New Combinations: Rhizobium Radiobacter, R. Rhizogenes, R. Rubi, R. Undicola and R. Vitis. *Int. J. Syst. Evol. Microbiol.* **2001**, *51*(Pt (1)), 89–103.

Yu, H.; Yao, Q.; Wang, L.; Zhao, Z.; Gong, Z.; Tang, S.; Liu, Q.; Gu, M. Generation of Selectable Marker-Free Transgenic Rice Resistant to Chewing Insects Using Two Co-Transformation Systems. *Prog. Nat. Sci.* **2009**, *19*(11), 1485–1492.

Zaidi, M. A.; Ye, G.; Yao, H.; You, T. H.; Loit, E.; Dean, D. H.; Riazuddin, S.; Altosaar, I. Transgenic Rice Plants Expressing a Modified Cry1ca1 Gene are Resistant to Spodoptera Litura and Chilo Suppressalis. *Mol. Biotechnol.* **2009**, *43*(3), 232–242.

Zeng, Z.; Xiong, F.; Yu, X.; Gong, X.; Luo, J.; Jiang, Y.; Kuang, H.; Gao, B.; Niu, X.; Liu, Y Overexpression of a Glyoxalase Gene, Osgly I, Improves Abiotic Stress Tolerance and Grain Yield in Rice (Oryza Sativa L.). *Plant Physiol. Biochem.* **2016**, *109*, 62–71.

Zhang, H. Y.; Wu, X. J.; Tang, K. X.; Wang, X. D.; Sun, X. F.; Zhou, K. D. A Primary Study of Transferring the Pinellia Tenata Agglutinin (Pta) Gene Into Rice and Expression. *Acta Genet. Sin.* **2003**, *30*(11), 1013–1019.

Zhang, L.; Chen, H.; Li, Y.; Li, Y.; Wang, S.; Su, J.; Liu, X.; Chen, D.; Chen, X. Evaluation of the Agronomic Performance of Atrazine-Tolerant Transgenic Japonica Rice Parental Lines for Utilization in Hybrid Seed Production. *PLoS One* **2014**, *9*(9), e108569.

Zhao, F. Y.; Wang, X. Y.; Zhao, Y. X.; Zhang, H. Transferring the Suaeda Salsa Glutathione S-Transferase and Catalase Genes Enhances Low Temperature Stress Resistance in Transgenic Rice Seedlings. *J. Plant Physiol. Mol. Biol.* **2006**, *32*(2), 231–238.

Zhao, F.; Li, Q.; Weng, M.; Wang, X.; Guo, B.; Wang, L.; Wang, W.; Duan, D.; Wang, B. Cloning of TPS Gene from Eelgrass Species Zostera Marina and its Functional Identification by Genetic Transformation in Rice. *Gene* **2013**, *531*(2), 205–211.

Zhao, Q. C.; Liu, M. H.; Zhang, X. W.; Lin, C. Y.; Zhang, Q.; Shen, Z. C. Generation of Insect-Resistant and Glyphosate-Tolerant Rice by Introduction of a T-Dna Containing Two Bt Insecticidal Genes and an Epsps Gene. *J. Zhejiang Uni. Sci. B* **2015**, *16*(10), 824–831.

Zhao, T.; Lin, C. Y.; Shen, Z. C. Development of Transgenic Glyphosate-Resistant Rice with G6 Gene Encoding 5-Enolpyruvylshikimate-3-Phosphate Synthase. *Agric. Sci. China* **2011b**, *10*(9), 1307–1312.

Zhao, W.; Zheng, S.; Ling, H. Q. An Efficient Regeneration System and Agrobacterium-Mediated Transformation of Chinese Upland Rice Cultivar Handao297. *Plant Cell Tissue Organ Cult.* **2011a**, *106*(3), 475–483.

Zhou, Y.; Cai, H.; Xiao, J.; Li, X.; Zhang, Q.; Lian, X. Over-Expression of Aspartate Aminotransferase Genes in Rice Resulted in Altered Nitrogen Metabolism and Increased Amino Acid Content in Seeds. *Theor. Appl. Genet.* **2009**, *118*(7), 1381–1390.

CHAPTER 8

MOLECULAR MARKERS AND MARKER-ASSISTED SELECTION TOWARD YIELD AND QUALITY IMPROVEMENT IN RICE

RUKMINI MISHRA[1], JATINDRA NATH MOHANTY[1,2],
DEEPAK KUMAR VERMA[3,*], RAJ KUMAR JOSHI[1,4,*] and
PREM PRAKASH SRIVASTAV[3,5]

[1]*Centre of Biotechnology, Siksha O Anusandhan University,
Bhubaneswar, Odisha 751030, India, Mob.: +91-9556458058,
Email: rukmini.mishra@gmail.com*

[2]*jatindranathmohanty@gmail.com, Mob.:+91-7751831025*

[3]*Department of Agricultural and Food Engineering, Indian Institute of
Technology, Kharagpur, West Bengal 721302, India,
Tel: +91 3222281673, Mob.: +91 7407170259.
Fax: +91 3222282224*

[4]*Mob.: +91-9437684176*

[5]*pps@agfe.iitkgp.ernet.in, Mob.: +91 9434043426,
Tel.: +91 3222-283134, Fax: +91 3222-282224*

*Corresponding author. E-mail: rajjoshi@soauniversity.ac.in,
deepak.verma@agfe.iitkgp.ernet.in; rajadkv@rediffmail.com*

8.1 INTRODUCTION

Rice (*Oryza sativa* L.) is the most important food crop and a carbohydrate source for more than half of the world's population (Feng et al., 2013). Global rice consumption is expected to rise from 763 million t in 2020 to approximately 850 million t 2035 (Khush, 2013). Ever-increasing population, environmental pollution, emergence of new races of pathogens and pests, and climate changes are the major issues that have threatened the food security and human health worldwide. Therefore, developing rice varieties with high yield and quality, durable resistance to pathogens and pests, and resilience to environmental fluctuations is the necessity of the hour to meet these problems. With gradual reduction in cropland area due to urbanization and industrialization, it is required to increase the productivity levels per unit area. Hence, increasing productivity on existing land can be an option to satisfy the growing demand for food worldwide.

Grain production is mostly influenced by the yield potential of rice varieties. Therefore, improving the yield potential of rice varieties can be a suitable alternative to increase world rice production. Traditional hybridization and selection, ideotype breeding, hybrid breeding, exploitation of wild species, enhanced rate of photosynthesis, genomic and physiological approaches are a few methods for enhancing the yield potential (Khush, 2013). Conventional breeding usually involves visual selection of plants which is time consuming, labor intensive and heavily influenced by the environment. The advent of molecular marker technology has played an important role in overcoming the problems in conventional methods and offers great promise for plant breeding. Molecular markers are specific regions of DNA that can be easily identified in a population if linked to a particular gene or trait of interest (Jiang et al., 2013). Molecular marker-assisted selection (MAS) is one of the most promising approach that uses DNA based genetic markers linked to the gene of interest in breeding programs (Choudhary et al., 2008). MAS not only help in minimizing the number and size of populations but also make the selection method comparatively easier (Collard and Mackill, 2008). It is a technique that enhances the efficiency of traditional breeding and further provides tools for targeted selection of the existing plant material rather than transferring the isolated gene sequences as in case of genetic engineering (Wijerathna, 2015).

Yield and quality are considered to be important traits in rice breeding. Quality traits are usually complex in nature and get affected by the environment easily; hence, selection based on field observations is quite difficult and inefficient. However, finding quantitative trait loci (QTLs) have been much easier with the help of markers and linkage maps of rice. Several QTLs related to agronomic traits have been identified by MAS and subsequently used for rice improvement (Bernardo, 2008; Ashraf and Foolad, 2013). In this chapter, an overview of molecular markers and MAS, its applications in rice breeding and role towards yield and quality improvement of rice has been discussed.

8.2 MOLECULAR MARKERS

A marker is a piece of DNA fragment that is linked to a particular location within the genome and transmitted from one generation to the next. During past few decades, the molecular markers have played an important role in plant biotechnology. Morphological markers are usually based on phenotypic characteristics and biochemical markers are based on enzymatic assays, molecular markers depend on a DNA assay (Choudhary et al., 2008). They are especially advantageous for agronomic traits related to pathogenic resistance, insects and nematodes, abiotic stress tolerance, quality parameters and quantitative traits. Other important uses of molecular markers includes gene pyramiding through backcross method, characterization of huge germplasm, genetic diagnostics, characterization of transformants, genome organization and phylogenetic studies (Jain et al., 2002).

Molecular markers are usually categorized into two types: polymerase chain reaction (PCR)-based and non-PCR based markers. PCR-based markers use small amount of template DNA and are more efficient in handling large size populations, hence more effective and efficient for MAS technique. An ideal DNA marker should be highly polymorphic, co-dominant, highly reproducible, frequently occurring in the genome and easily accessible (Kumar et al., 2014). There are several molecular markers with different principles, methodologies and applications and it is difficult to choose a single marker which can meet all the above requirements needed by researchers. Among all the markers, simple sequence repeat (SSR) markers have been recommended as marker of choice for breeding purposes (Gupta and Varshney, 2000). SSR markers are mostly useful for marker-based selection and genome mapping since they are informative in many types of genetic crosses and effective in high-throughput analysis. The desirable properties such as co-dominance, highly reproducible, multi-allelic in nature, and wide genomic distribution makes them preferable for molecular breeding purposes (Parida et al., 2006; Miah et al., 2013). SSRs are efficient tools for inking phenotypic and genotypic variations and have been widely used for integrating the genetic, physical, and sequence-based maps of rice (Temnykh et al., 2001). A total of 18,828 Class one di, tri and tetra-nucleotide SSRs, were identified on the rice genome (IRGSP, 2005).

Amplified fragment length polymorphisms (AFLPs), sequence characterized amplified regions (SCARs), sequence tagged sites (STS), inter-simple sequence repeats (ISSRs), and single nucleotide polymorphism (SNP) based markers are frequently used in molecular tagging and mapping of genes in rice. Especially, SNP genotyping is an efficient and effective tool for mapping of genes, map-based cloning, and MAS in a wide range of crops (Hayashi et al., 2004; Varshney et al., 2006). SNPs are usually converted into cleaved amplified polymorphism (CAPs) markers. Novel PCR-based insertion/deletion (InDel) markers have been routinely used in MAS and mapping studies due to their simplicity in design and sharp discrimination in genotyping (Shen et al., 2004). InDel markers show polymorphisms mainly between *Indica* cultivars and *Japonica* cultivars.

8.3 MARKER ASSISTED SELECTION (MAS)

MAS is an indirect method of selecting a gene of interest tightly linked to a marker. It combines both traditional selection methods and molecular biology to enhance the effectiveness of the crop breeding programs. Plants can be identified at the molecular level rather than the morphological traits by the help of DNA markers. A molecular marker can be part of the gene of interest or linked to a gene determining a trait of interest. Marker-assisted selection is mostly used to detect the presence of the gene in a breeding population (Jena and Mackill, 2008). Marker-assisted backcross breeding (MABC) has been used for introgression of major genes or QTLs into widely grown varieties. Furthermore, MAS has also helped in the establishment of major techniques in plant breeding such as gene tagging and gene pyramiding (Khan et al., 2015).

Pyramiding of genes through a conventional approach may not be suitable for crop improvement due to dominance, epistasis and other genetic effects (Rajpurohit et al., 2011). In contrast, DNA-based genetic markers could be more convenient in the introgression of specific traits by using marker-associated breeding (MAB) method (Nogoy et al., 2016). MABC and marker-assisted gene pyramiding (MAGP) are the two basic MAS strategies used for crop improvement. MABC make use of molecular markers to develop resistant rice varieties or varieties with high yield and quality by introgressing a gene of interest into cultivars (Collard and Mackill, 2008; Hasan et al., 2015). This has helped in the development of many high yielding, stress tolerant, quality and fragrance related rice varieties in major rice growing countries within a short period of time. Since the introduction of a bacterial leaf blight (BB) resistance gene Xa21 in a Chinese rice variety (Chen et al., 2000), MABC have been effectively used in rice breeding through introgression of many important trait related genes including *Sub* 1 QTL, *Saltol* QTL and *waxy* gene into elite cultivars (Neeraja et al., 2007; Thomson et al., 2010). Gene pyramiding is the process of incorporating more than one number of genes together into a single genotype and further selecting them using DNA markers (Collard and Mackill, 2008). It is an effective approach to develop broad-spectrum resistance against pathogens and pests (Das and Rao, 2015). MAB and MAS-QTL pyramiding have enabled quick and accurate selection of lines with the target trait (Chen et al, 2000; Ashikari and Matsuoka, 2006).

8.4 QTL MAPPING

QTL mapping is an efficient technique in which both the genotypic and phenotypic data are collected from a segregating mapping population, followed by a statistical analysis to reveal all possible marker loci where the allelic state correlates with the phenotype (Salvi et al., 2003). It is a well-established method in quantitative genetics and mostly used to identify genetic regions that affects complex quantitative traits (Hackett, 2002; Collard et al., 2005). QTL mapping is used to understand the genetic basis of rice yield traits like grain number per panicle, grain weight and tillers per plant. In recent years, several QTLs for yield-related components of rice have been reported (Gutiérrez et al., 2010; Saleem et al., 2011). Fu et al. (2010) detected 26 QTLs for yield traits and the alleles of 10 QTLs originating from *O. rufipogon* showed a beneficial effect for yield traits in the *indica* 93–11 genetic background. Tong et al. (2011) identified almost 57 main-effect QTLs, and 33 digenic interactions for yield components under three N levels. Till date, more than 200 QTLs for 1000-grain weight have been located across all the rice chromosomes (Ghafoor et al., 2005).

Yield and quality traits in rice are highly correlated and influenced by shape and size of the grain. Many rice QTLs have been previously mapped for grain size and weight particularly on chromosomes 2, 3, and 5 (Tan et al., 2000; Wacera et al., 2015). Identification of QTLs has led to the development of DNA markers and linkage maps which contribute towards genetic improvement of grain quality (Wacera et al., 2015). Panicle length is an important trait responsible for improving panicle architecture and grain yield in rice. It is controlled by both major and minor QTLs and inherited in a quantitative manner (Liu et al., 2011a). Almost 200 QTLs for panicle length have been detected and distributed on all the 12 chromosomes (Yao et al., 2015; Zhang et al., 2015). Agronomic traits such as crop yield and stress tolerance are usually affected by the environment and controlled simultaneously by QTLs (Song and Ashikari, 2008). Isolation and characterization of each QTL is essential in order to have an understanding of the agronomic traits

(Miura et al., 2011). Several reports suggest that QTLs with major effect on agriculturally important traits such as grain number (Gn1a), grain size (GS3), grain weight (qSW5/GW5), and heading date (Hd1, Hd3a), and a major QTL (i.e., Ghd7) with large pleiotropic effects on heading date, plant height, and number of spikelets per panicle have been cloned (Fan et al., 2006; Weng, 2008; Shomura et al., 2008; Xue et al., 2008).

8.5 MARKER-ASSISTED SELECTION IN RICE IMPROVEMENT

In the past two decades, the use of molecular markers and marker-assisted selection approaches have played a vital role in developing new and advanced varieties of rice with much greater precision than conventional breeding. The tightly linked molecular markers have been identified and successfully introgressed into rice cultivars by marker assisted backcross breeding method. This has helped in attaining strong and broad range resistance against foremost diseases including bacterial leaf blight, rice blast and gall midge in addition to abiotic stresses such as drought, submergence, and salinity (Das and Rao, 2015). Several studies have been reported where multiple resistance genes have been successfully incorporated into rice cultivars by MAS and MABC method (Hittalmani et al., 2000; Neeraja et al., 2007; Sundaram et al., 2010; Singh et al., 2014; Divya et al., 2014; Pradhan et al., 2015). In a study, scientists have successfully pyramided five different genes/QTLs (*Gm1, Gm4, Pi2, Pi9, Saltol*and*Sub1*) conferring resistance to blast, gall midge, submergence and salinity respectively into a single rice variety (Das and Rao, 2015). The released variety is named as CRMAS2621-7-1 (Improved Lalat) and the study shows the importance of MAS as a technique for pyramiding more than one gene into a single line. BB is the most serious rice disease and affects the yield due to poor grain filling (Pradhan et al., 2015). So far, 38 BB resistance genes (designated in a series from *Xa1* to *Xa38*) have been identified (Guo et al., 2010; Miao et al., 2010; Bhasin et al., 2012; Kumar et al., 2014). By pyramiding several of these genes into one line, plants have expressed broad spectrum resistance to BB (Kottapalli et al., 2010; Shanti et al., 2010). A perfect marker, *pTA248*, has been developed to discriminate BB resistance and susceptibility based on the functional polymorphism of *Xa21* (Kottearachchi, 2013). Pradhan and his group have successfully incorporated three bacterial blight resistance genes into the blight susceptible elite deep water cultivar, Jalmagna by using an MABC breeding strategy (Pradhan et al., 2015).

Rice blast caused by *Magnaporthe gresea* is a major constraint to rice production causing yield losses as high as 70–80% in severe cases (Babujee and Gnanamanickham, 2000). Existing literature reports the mapping and cloning of more than 100 genes conferring blast resistance in rice (Hua et al., 2015). Sundaram and his group have recently introgressed a major gene for blast resistance (*Pi54*) in addition to two bacterial blight resistance genes (*Xa21* and *xa13*) into a single Indian rice variety MTU1010 through MAB breeding (Sundaram et al., 2008; Arunakumari et al., 2016). More recently, Zhongchao Yin and his group have developed a new line WH6725 resistant to blast disease through backcrossing and marker-assisted selection (Luo et al., 2016). The introgressed line reported enhanced resistance to different biotic and abiotic stresses, enhanced fragrance and has also been recommended to replace Mianhui 725, a restorer lineused to produce hybrid rice in China. Previous studies indicates that the rice blast *R* gene *Pi9* which shows resistance to almost 43 *Magnaportheoryzae* isolates has been successfully used in rice breeding programs (Luo and Yin, 2013; Khanna et al., 2015; Ni et al., 2015). As many as 28Brown Plant Hopper (BPH) resistance genes have been identified till date (Cheng et al., 2013; Wu et al. 2014) and have been subsequently used for introgression and pyramiding through MAS (Qiu et al., 2010). Recently, Liu et al. (2016b) used marker-assisted selection tointrogress Bph27 (t), a dominant BPH resistance gene into a susceptible variety and further pyramided Bph27 (t) and a durable BPH resistance gene *Bph3* through MAS for enhanced disease resistance in rice crop (Eathington et al. 2007).

Salinity and submergence are the major environmental constraints in rice production leading to more than 50% yield loss (Molla et al., 2015; Iftekharuddaula et al., 2015). Salinity may lead to variations in multiple yield components such as spikelet number per panicle, panicle length beside delaying the emergence of panicles as well as flowering (Zcng and Shannon, 2000). Currently, *Saltol*, a major salt tolerant QTL, identified in rice chromosome 1, is being incorporated into popular salt stress-sensitive varieties through MABC (Vu et al., 2012). In a recent study, saltol has been successfully incorporated into a basmati rice cultivar "Pusa Basmati 1121" through marker assisted breeding for improved salt tolerance of seedling stage (Babu et al., 2017). Furthermore, several other QTLs were also identified on different chromosomes for salinity tolerance at the seedling stage (Thomson et al., 2010; Hasan et al., 2015). Similarly, the rice gcnotype FR13A is the most widely used source of submergence tolerance with the discovery of major QTLs including *Sub1* (Xu and Mackill, 1996; Nandi et al., 1997; Toojinda et al., 2003). Subsequently, three putative genes encoding ethylene response factors (ERFs)- *Sub1A*, *Sub1B*, and *Sub1C* were fine-mapped and cloned. All these three genes from the *Sub1* locus have since been successfully incorporated through MABC into known high-yielding rice varieties to confer tolerance to excessive water condition (Neeraja et al., 2007; Septiningsih et al., 2009; Das and Rao, 2015).

Drought is a major hinderance to rice production mostly in rainfed areas and is directly related to yield instability and quality deterioration. MAB can be used for identifying yield-related QTLs effective against stress and could be subsequently used to develop high-yielding rice cultivars suitable for drought affected areas. (Bernier et al., 2007) identified a major QTL (*qtl12.1*) against drought by using a (Vandana × Way Rarem) mapping population. This QTL explained for more than 50% of the genetic variance between the two genotypes and confers as much as 47% enhancement in rice yield under severe drought conditions. (Swamy and Kumar, 2013) pyramided two drought yield related QTLs (qDTYs) in the IR64 background resulting in an yield advantage of 1200–2000 kg ha^{-1} under stress conditions. Several yield related QTLs including *qtl12.1* and *qDTYs* have been successfully introgressed for the development of rice genotypes with substantial yield under natural drought conditions (Sandhu et al., 2014; Prince et al., 2015). A group of scientists from the International Rice Research Institute (IRRI), The Philippines have recently developed a drought-tolerant pyramided line in the background of a Malaysian rice variety MR219that demonstrated an yield advantage of more than 1500 kg ha^{-1} (Shamsudin et al., 2016). MR219 introgressed with three yield related QTLs namely qDTY2.2, qDTY3.1, and qDTY12.1 showed consistency in yield under reproductive stage drought stress.

8.6 YIELD IMPROVEMENT IN RICE THROUGH MARKER-ASSISTED SELECTION

Yield improvement of cultivars is a major challenge for rice breeders. Conventional breeding methods are used in improvisation of agronomical traits that contribute to yield but the task seems difficult due to the epistatic interactions between several genes contributing to yield (Mei et al., 2006). Use of DNA markers and crop simulation model which defines the best combination of yield components for a range of agro ecosystems can be a solution to the problem. There is an urgent need to select the genes responsible for adaptation to the natural environment and enhancing grain yield. Grain yield is mainly governed by QTLs and several mapping populations have been developed to determine the QTLs like *Gn1a*, *GS3*, *qSW5/GW5*, *Ghd7* and *DEP1*controlling yield related traits in rice (Bai et al., 2012). A large-effect QTL named qDTY6.1, is mainly used in molecular breeding of rice under aerobic environment (Venuprasad et al., 2011). Grain weight is an integrated index for grain length, width and thickness and is identified as a complex trait controlled by multiple genes (Yuan et al., 2014). At least 89 QTLs related to grain weight and associated traits have been detected across the 12 chromosomes of rice genome (Brondani et al., 2002; Thomson et al., 2003; Xu et al., 2010). A series of genes including *GS3* (Fan et al., 2006), *GW2* (Song et al., 2007), *qSW5* (Shomura et al., 2008), qGW8 (Wang et al., 2012b), *Ghd7* (Xue et al., 2008), and *qGL3* (Zhang et al., 2012) encoding the traits related to grain weight or grain shape have been isolated via map-based cloning strategies. MAS in the background of an Indica parent 9311 and a japonica variety Zhonghua 11 (ZH11) has resulted in the mapping of many QTLs related to rice yield and some of them have been cloned and functionally characterized. In addition, a few QTLs, such as *GS3*, *GW5* and *GW8* have been exclusively used in the breeding programs of China rice varieties.

1000-Grain weight and spikelet number per panicle are two important yield components of rice. Eight QTLs related to spikelet number per panicle and 1000-grain weight were mapped by using recombinant inbred lines (RILs) as mapping population through sequencing-based genotyping (Zong et al., 2012). The scientists further proposed a QTL pyramid breeding scheme with marker assisted and phenotype selections (MAPS), which improved the efficiency of conventional phenotype selection. Under the scheme as many as 24 QTLs were pyramided with a single hybridization without extensive cross work. This study indicates about the molecular basis of rice grain yield for high-yielding rice breeding. A thousand-grain weight (TGW) QTL has been identified from back crossed inbred lines derived from a *japonica-indica* cross (Ishimaru, 2003). Other than grain weight, grain size and shape are important factors that contribute to grain yield and quality. Wang et al. (2012a) reported a yield related QTLGW8, which is highly synonymous with *OsSPL16* that encodes a protein conferring positive regulation of cell proliferation (Wang et al., 2012a). The study showed that higher expression of the protein promotes cell division and grain filling, which enhances the grain width and yield in rice. Further, it also indicated that a marker-assisted strategy can be effectively used to improve the grain yield and quality by targeting the alleles of *GS3* and *OsSPL16* underlying grain size and shape.

Molecular design breeding and QTLs/genes pyramiding using linked or functional markers has become a straightforward approach for improving plant type and yield. Guo et al. (2009) has fine mapped a major QTL, *GW6* for grain length and weight. In this study, 9311-*GW6* (SSL-1) and ZH11-*GW6* (R1, R2 and R3) NILs were developed using MAB, which laid a foundation for gene cloning and function analysis. *GW6* is controlled by a single dominant gene which was confirmed in BC_4F_2 population using ZH11 as the recurrent parent. Thus, *GW6* has a high potential to increase the yield of hybrid rice. In another study, Yuan et al. 2014, successfully transferred *GW6* gene into an *indica* recurrent parent 9311 and a *japonica* variety Zhonghua 11 (ZH11) from Baodali using MAB. Furthermore, several genes linked to yield related QTLs such as*qGL3* for grain length, *qGS3* for grain length and weight, *qgw3* for grain weight, *qGW2* for grain width and weight, *qGnl* for grain number and *PhJ* for plant height have been identified and fine mapped (Li et al., 2004; Fan et al., 2006; Wan et al., 2006; Song et al., 2007).

8.7 QUALITY IMPROVEMENT IN RICE THROUGH MARKER-ASSISTED SELECTION

Beside yield, quality improvement is also one of the important aspects for enhancing the market value of rice genotypes. Grain quality traits that influence the market value includes various physico-chemical properties such as hulling and milling, cooking-eating, nutritional quality, elongation after cooking, palatability, appearance in addition to other socio-cultural factors (Lou et al., 2009; Lau et al. 2015; Verma et al., 2012, 2013, 2015; Verma and Srivastav, 2017). Amylose content (AC), gel consistency (GC), and gelatinization temperature (GT) are the three major physico-chemical characteristics that helps in evaluating the eating–cooking quality of rice (Pang et al., 2016). The waxy (*Wx*) gene located on chromosome 6 is an important determinant of eating-cooking quality and mainly controls the amylose content of rice (Wang et al., 2007; Verma et al., 2012, 2013, 2015). Besides amylose content, *Wx* is also responsible for affecting the gel consistency and gelatinization temperature of rice starch (Su et al., 2011). By using marker assisted selection technique, waxy (*Wx*) gene marker has been used for quality improvement of rice (Jantaboon et al., 2011). Since then, several marker resources linked to this gene have been developed for molecular breeding to improve the cooking and eating quality of rice (Tian et al., 2010; Jin et al., 2010). Marker assisted backcross method was used to improve the fragrance and intermediate amylose content of a non-fragrant rice cultivar-Manawthukha, by transferring the positive alles of *badh2* and *Wx* from Basmati 370.

Low amylose content results in improved cooking and eating quality of hybrid rice and also helps in maintaining good agronomical traits of the parent lines (Sattari et al., 2015). Jianbo Yang and his group used MAS to lower the amylose content in the original hybrid XieqingzaoA (GG) X 57 (GG) from 26 to 19% by modifying the *Wx* genes in both parents of Xieyou 57, an elite hybrid rice with high grain yield and broad eco-adaptability (Ni et al., 2011). Marker assisted selection was also used to produce TS4, an improved rice line, designed to contain several genes like*sd1, Wxb, Xa4* and *Xa21*conferring resistance to dwarf ness, amylose content and blight disease (Luo et al., 2014). The introgression of these resistant genes into TS4 have not only enhanced yield and quality of local rice varieties but also provided wide resistance to almost all *Xoo* strains. By using marker assisted selection, PCR marker *AccI* was effectively used for grain quality improvement of Gang 46 B, a maintainer line with good agronomic traits and high combining ability but with high amylose content and poor quality (Aiqiu et al., 2006).

Grain quality is a complex quantitative trait which is controlled by various genes with low heritability and expression influenced significantly by the environment (Lou et al., 2009). Large number of quality trait related QTLs have been identified in a wide range of mapping population in rice (Xing and Zhang, 2010), and subsequently used to improve breeding efficiency through marker assisted selection. Several reports have indicated the identification of QTLs for appearance traits and thousand-grain weights (TGW) using different rice populations (Rabiei et al., 2004; Fan et al., 2006; Yoon et al., 2006; Song et al., 2007; Shomura et al., 2008; Weng, 2008; Bai et al., 2010; Jiao, 2010; Li et al., 2011).

Chalkiness is an important trait for appearance quality of rice (Verma et al., 2012, 2013, 2015). A QTL named qPGWC-7, has been finely mapped for grain chalkiness and located to a 44-kb DNA fragment containing 13 genes (Zhou et al., 2009). 79 QTLs associated with chalkiness traits were identified using five rice populations across two environments (Peng et al., 2014). Recently, Zhenyu Gao and his group, have identified a major QTL, *qACE9*, for the area of chalky endosperm (ACE) (Gao et al., 2016). The candidate gene not only determines the area of chalky endosperm in rice but also plays an important role during starch synthesis in endosperm. Similarly, fragrance has been considered as one of the most highly valued grain quality traits in rice (Michael et al., 2009; Verma and Srivastav, 2016). Fragrance in rice has resulted due to the loss-of-function of *BADH2* enzyme which in turn accumulates, 2-acetyl-1-pyrroline (2-AP), the compound responsible for fragrance in rice (Bradbury et al., 2005; Kovach et al., 2009; Verma and Srivastav, 2016). Tanksley and his group were the first to tag a RFLP marker, *RG28*, to the aroma gene (fgr) (Tanksley et al., 1989; Ahn et al., 1992). Molecular markers are reliable and not influenced by the environmental conditions, hence; facilitate the selection of complex traits during the breeding process. The fragrant gene markers, ESP, IFAP, INSP, EAP developed from the cloned mutated allele of the *badh2* gene in rice, is one of the most frequently used perfect markers in discriminating Basmati type aromatic rice (Bradbury et al., 2005). Functional markers for *fgr* have been developed and effectively used to transfer this gene from aromatic rices to the target non-aromatic rices in *indica* cultivars (Salgotra et al., 2012; Jantaboon et al., 2011).

8.8 SUMMARY AND CONCLUSION

The advent of gene sequencing, functional and comparative genomics has led to the development of a plethora of molecular marker systems that has revolution-ized the research related to rice yield and quality. Both random and functional markers are being widely used in the recent times for unambiguous detection of yield related traits. QTL mapping has led to the identification of 23 yield related genes and their subsequent validations have highlighted the genetic mechanism involved in yield improvement. Various MAB strategies including recurrent selection (MARS), MABC, and MAGP are commonly being used for selecting rice types carrying the preferred yield related allele and allowing early selection of improved genotypes. Backcrossing schemes have successfully improved the amylase content, grain yield, 1000 grain weight, fragrance through successful introgression of multiple traits using functional markers and other yield related properties. As the functional markers are linked with the target genes, their usage in MAS significantly decrease the risk of linkage drag and recombina-tion. However, these markers may not be found suitable for identifying QTLs or genes with smaller effects and environmental pressures. The variability of the allele specific amplification, pleotrophic effect of one or more traits and epistasis are some of the major drawbacks with the involvement of functional markers in yield related breeding of rice. Besides, the current methods for the development and establishment of functional markers linked to specific yield traits incur Hugh cost. However, with the advancement of sequencing tech-nologies, the cost for discovery of genes and functional marker has been greatly reduced during the past few years. Recently, an alternative MAS strategy called as genomic selection (GS) has been introduced, which takes into account all the markers and phenotypic data to predict for the best individuals to be used as parents in a mapping population (Meuwissen et al. 2001). Such a predic-tion model could be developed for quality and yield related traits to select for specific yield improved genotypes without any phenotypic records. Quantita-tive trait improvement through genome wide selection of functional markers has been recently introduced in rice and positively simulated in other crops including wheat and barley (Hayes et al. 2012; Spindel et al. 2015). With the availability of these tools, it is now possible to discover new yield related genes and QTLs and their subsequent incorporation into specific rice genotypes to meet the demand for food securities of the future.

ACKNOWLEDGMENT

Jatindra Nath Mohanty is thankful for the award of Junior Research Fellowship (JRF) from the Department of Biotechnology, Government of India. Rukmini Mishra, Jatindra Nath Mohanty, Raj Kumar Joshi are also thankful to Prof. Manoj Ranjan Nayak, President, Siksha O Anusandhan University, Bhubaneswar, India for his guidance and support. We also thank DST-FIST, Govt. of India, for the research infrastructure facilities provided to Centre of Biotechnology, Siksha O Anusandhan University. Deepak Kumar Verma and Prem Prakash Srivastav are indebted to Department of Science and Technology, Ministry of Science and Technology, Govt. of India for an individual research fellowship (INSPIRE Fellowship Code No.: IF120725; Sanction Order No. DST/INSPIRE Fellowship/2012/686 & Date: 25/02/2013).

KEYWORDS

- BADH2 enzyme
- Chalkiness
- Cleaved amplified polymorphism
- Crop breeding
- Elite cultivars
- Ethylene response factors
- Gene sequences
- Genome mapping
- Ideotype breeding
- InDel markers
- ISSRs
- Map-based cloning

- Marker assisted backcross breeding
- MAS technique
- MAS-QTL pyramiding
- Morphological markers
- PCR-based markers
- Pleiotropic effects
- QTL mapping
- Recombinant inbred lines
- Single Nucleotide Polymorphism
- SNP genotyping
- Yield potential

REFERENCES

Ahn, S. N.; Bollich, C. N.; Tanksley, S. D. Rflp Tagging of a Gene for Aroma in Rice. *Theor. Appl. Genet.* **1992**, *84*, 825–828.

Aiqiu, L.; Fengjun, L.; Pingreng, W.; Xiaojian, D. Relationship Between Pcr-Acci Marker of Wx Gene and Amylose Content of Rice and Its Application in Improving Rice Quality. *Chin. J. Appl. Environ. Biol.* **2006**, *12*, 318–321.

Arunakumarim, K.; Durgaranim, C. V.; Satturu, V.; Sarikonda, K. R.; Chittoor, P. D. R.; Vutukuri, B.; Laha, G. S.; Nelli, A. P. K.; Gattu, S.; Jamal, M.; Prasadbabu, A.; Hajira, S.; Sundaram, M. R. Marker-Assisted Pyramiding of Genes Conferring Resistance Against Bacterial Blight and Blast Diseases Into Indian Rice Variety Mtu1010. *Rice Science* **2016**, *23*, 306–316.

Ashikari, M.; Matsuoka, M. Identification, Isolation and Pyramiding of Quantitative Trait Loci for Rice Breeding. *Trends Plant Sci.* **2006**, *11*, 344–350.

Ashraf, M.; Foolad, M. R. Crop Breeding for Salt Tolerance in the Era of Molecular Markers and Marker-Assisted Selection. *Plant Breed.* **2013**, *132*, 10–20.

Babu, N. N.; Krishnan, S. G.; Vinod, K. K.; Krishnamurthy, S. L.; Singh, V. K.; Singh, M. P.; Singh, R.; Ellur, R. K.; Rai, V.; Bollinedi, H.; Bhowmick, P. K.; Yadav, A. K.; Nagarajan, M.; Singh, N. K.; Prabhu, K. V.; Singh, A. K. Marker Aided Incorporation of Saltol, a Major Qtl Associated with Seedling Stage Salt Tolerance, Into Oryza Sativa 'Pusa Basmati 1121'. *Front Plant Sci.* **2017**, *8*(41), 1–14. DOI: 10.3389/fpls.2017.00041.

Babujee, L.; Gnanamanickham, S. S. Molecular Tools for Characterization of Rice Blast Pathogen (Magnaporthe Grisea) Population and Molecular Marker-Assisted Breeding for Disease Resistance. *Curr. Sci.* **2000**, *78*, 248–257.

Bai, X. F.; Luo, L. J.; Yan, W. H.; Rao, K. M.; Zhan, W.; Xing, Y. Z. Genetic Dissection of Rice Grain Shape Using a Recombinant Inbred Line Population Derived from Two Contrasting Parents and Fine Mapping a Pleiotropic Quantitative Trait Locus Qgl7. *BMC Genet.* **2010**, *11*, 16–26.

Bai, X. F.; Wu, B.; Xing, Y. Z. Yield-Related Qtls and Their Applications in Rice Genetic Improvement. *J. Integr. Plant Biol.* **2012**, *54*, 300–311.

Bernardo, R. Molecular Markers and Selection for Complex Traits in Plant: Learning from the Last 20 Years. *Crop Sci.* **2008**, *48*, 1649–1664.

Bernier, J.; Kumar, A.; Venuprasad, R.; Spaner, D.; Atlin, G. N. A Large-Effect Qtl for Grain Yield Under Reproductive-Stage Drought Stress in Upland Rice. *Crop Sci.* **2007**, *47*, 507–516.

Bhasin, H.; Bhatia, D.; Raghuvanshi, S.; Singh, K. New Pcr-Based Sequence Tagged Site Marker for Bacterial Blight Resistance Gene Xa38 of Rice. *Mol. Breed.* **2012**, *30*, 607–611.

Bradbury, L. M. T.; Henry, R. J.; Jin, Q.; Reinke, R. F.; Waters, D. L. E. A Perfect Marker for Fragrance Genotyping in Rice. *Mol. Breed.* **2005**, *16*, 279–283.

Brondani, C.; Rangel, N.; Brondani, V.; Ferreira, E. Qtl Mapping and Introgression of Yield-Related Traits from Oryza Glumaepatula to Cultivated Rice (Oryza Sativa) Using Microsatellite Markers. *Theor. Appl. Genet.* **2002**, *104*, 1192–1203.

Chen, S.; Lin, X. H.; Xu, C. G.; Zhang, Q. F. Improvement of Bacterial Blight Resistance of 'Minghui 63', an Elite Restorer Line of Hybrid Rice, by Molecular Marker-Assisted Selection. *Crop Sci.* **2000**, *40*, 239–244.

Cheng, X. Y.; Zhu, L. L.; He, G. C. Towards Understanding of Molecular Interactions Between Rice and the Brown Plant Hopper. *Mol. Plant* **2013**, *6*, 621–634.

Choudhary, K.; Choudhary, O. P.; Shekhawat, N. S. Marker Assisted Selection: a Novel Approach for Crop Improvement. *Am.-Eurasian J. Agron.* **2008**, *1*, 26–30.

Collard, B. C. Y.; Mackill, D. J. P. Marker-Assisted Selection: an Approach for Precision Plant Breeding in the Twenty-First Century. *Trans. R. Soc.* **2008**, *363*, 557–572.

Collard, B. C. Y.; Jahufer, M. Z. Z.; Brouwer, J. B.; Pang, E. C. K. An Introduction to Markers, Quantitative Trait Loci (Qtl) Mapping and Marker Assisted Selection for Crop Improvement: the Basic Concepts. *Euphytica* **2005**, *142*, 169–196.

Das, G.; Rao, G. J. N. Molecular Marker Assisted Gene Stacking for Biotic and Abiotic Stress Resistance Genes in an Elite Rice Cultivar. *Front Plant Sci.* **2015**, *6*(698), 1–18.

Divya, B.; Robin, S.; Rabindran, R.; Senthil, S.; Raveendran, M.; Joel, A. J. Marker Assisted Back Cross Breeding Approach to Improve Blast Resistance in Indian Rice (Oryza Sativa) Variety Adt43. *Euphytica* **2014**, *200*, 61–77.

Eathington, S. R.; Crosbie, T. M.; Edwards, M. D.; Reiter, R. S.; Bull, J. K. Molecular Markers in a Commercial Breeding Program. *Crop Sci.* 47, **2007**, S–S154. DOI: 10.2135/cropsci2007.04.0015IPBS.

Fan, C.; Xing, Y.; Mao, H.; Lu, T.; Han, B.; Xu, C.; Li, X.; Zhang, Q. GS3, a major QTL for grain length and weight and minor QTL for grain width and thickness in rice, encodes a putative transmembrane protein. *Theor. Appl. Genet.* **2006**, *112*, 1164–1171.

Feng, Y.; Zhai, R. R.; Lin, Z. C.; Cao, L. Y.; Wei, X. H.; Cheng, S. H. Qtl Analysis for Yield Traits in Rice Under Two Nitrogen Levels. *Chin. J. Rice Sci.* **2013**, *27*(6), 577–584.

Fu, Q.; Zhang, P.; Tan, L.; Zhu, Z.; Ma, D.; Fu, Y.; Zhan, X.; Cai, H.; Sun, C. Analysis of Qtls for Yield-Related Traits in Yuanjiang Common Wild Rice (Oryza Rufipogon Griff.). *J. Genet. Genomics* **2010**, *37*, 147–157.

Gao, Y.; Liu, C.; Li, Y.; Zhang, A.; Dong, G.; Xie, L.; Zhang, B.; Ruan, B.; Hong, K.; Xue, D.; Zeng, D.; Guo, L.; Qian, Q.; Gao, Z. Qtl Analysis for Chalkiness of Rice and Fine Mapping of a Candidate Gene for Qace9. *Rice* **2016**, *9*, 41.

Ghafoor, A.; Ahmad, Z.; Afzal, M. Use of Sds-Page Markers for Determining Quantitative Traits Loci in Blackgram [Vigna Mungo (L.) Hepper] Germplasm. *Pak. J. Bot.* **2005**, *37*, 263–269.

Guo, L.; Ma, L.; Jiang, H.; Zeng, D.; Hu, J.; Wu, L.; Gao, Z.; Zhang, G.; Qian, Q. Genetic Analysis and Fine Mapping of Two Genes for Grain Shape and Weight in Rice. *J. Integr. Plant Biol.* **2009**, *51*, 45–51.

Guo, S. B.; Zhang, D. P.; Lin, X. H. Identification and Mapping of a Novel Bacterial Blight Resistance Gene Xa35(T) Originated from Oryza Minuta. *Sci. Agric. Sin.* **2010**, *43*, 2611–2618.

Gupta, P. K.; Varshney, R. K. The Development and Use of Microsatellite Markers for Genetic Analysis and Plant Breeding with Emphasis on Bread Wheat. *Euphytica* **2000**, *113*, 163–185.

Gutiérrez, A. G.; Carabali, S. J.; Giraldo, O. X.; Martinez, C. P.; Correa, F.; Prado, G.; Tohme, J.; Lorieux, M. Identification of a Rice Stripe Necrosis Virus Resistance Locus and Yield Component Qtls Using Oryza Sativa X O. Glaberrima Introgression Lines. *BMC Plant Biol.* **2010**, *8*, 10–16.

Hackett, C. A. Statistical Methods of Qtl Mapping in Cereals. *Plant Mol. Biol.* **2002**, *48*, 585–599.

Hasan, M. M.; Rafii, M. Y.; Ismail, M. R.; Mahmood, M..; Rahim, H. A.; Alam, M. A.; Ashkani, S.; Malek, M. A.; Abdul, M. Marker-Assisted Backcrossing: A Useful Method for Rice Improvement. *Biotechnol. Biotechnol. Equip.* **2015**, *29*, 237–254.

Hayashi, K.; Hashimoto, N.; Daigen, M.; Ashikawa, I Development of Pcr-Based Snp Markers for Rice Blast Resistance Genes At the Piz Locus. *Theor. Appl. Genet* **2004**, *108*, 1212–1220.

Hayes, B. J.; Lewin, H. A.; Goddard, M. E. The Future of Livestock Breeding: Genomic Selection for Efficiency, Reduced Emissions Intensity, and Adaptation. *Trends Genet.* **2012**, *29*(4), 206–214.

Hittalmani, S.; Parco, A.; Mew, T. V.; Zeigler, R. S.; Huang, N. Fine Mapping and DNA Marker-Assisted Pyramiding of the Three Major Genes for Blast Resistance in Rice. *Theor. Appl. Genet.* **2000**, *100*, 1121–1128.

Hua, L. X.; Liang, L. Q.; He, X. Y.; Wang, L.; Zhang, W. S.; Liu, W.; Liu, X. Q.; Lin, F. Development of a Marker Specific for the Rice Blast Resistance Gene Pi39 in the Chinese Cultivar Q 15 and Its Use in Genetic Improvement. *Biotechnol. Biotechnol. Equip.* **2015**, *29*, 448–456.

Iftekharuddaula, K. M.; Ahmed, H. U.; Ghosal, S.; Moni, Z. R.; Amin, A.; Ali, M. S. Development of New Submergence Tolerant Rice Variety for Bangladesh Using Marker-Assisted Backcrossing. *Rice Sci.* **2015**, *22*, 16–26.

IRGSP. The Map Based Sequence of the Rice Genome. *Nature* **2005**, *436*, 793–800.

Ishimaru, K. Identification of a Locus Increasing Rice Yield and Physiological Analysis of Its Function. *Plant Physiol.* **2003**, *133*, 1083–1090.

Jain, S. M.; Brar, D. S.; Ahloowalia, B. S. *Molecular Techniques in Crop Improvement.* Kluwer Academic Publishers: The Netherlands, 2002.

Jantaboon, J.; Siangliw, M.; Im-mark, S.; Jamboonsri, W.; Vanavichit, A.; Toojinda, T. Ideotype Breeding for Submergence Tolerance and Cooking Quality by Marker-Assisted Selection in Rice. *Field Crop Res.* **2011**, *123*, 206–213.

Jena, K. K.; Mackill, D. J. Molecular Markers and Their Use in Marker-Assisted Selection in Rice. *Crop Sci.* **2008**, *48*, 1266–1276.

Jiang, W.; Zhou, H.; Bi, H.; Fromm, M.; Yang, B.; Weeks, D. P. Demonstration of CRISPR/Cas9/sgRNA-Mediated Targeted Gene Modification in Arabidopsis, Tobacco, Sorghum and Rice.*Nucleic Acids Res.* **2013**, *41*, e188. DOI: 10.1093/nar/gkt780

Jiao, Y. Q.; Wang, Y. H.; Xue, D. W.; Wang, J.; Yan, M. X.; Liu, G. F.; Dong, G. J.; Zeng, D. L.; Lu, Z. F.; Zhu, X. D.; Qian, Q.; Li, J. Y. Regulation of Osspl14byosmir156defines Ideal Plant Architecture in Rice. *Nat. Genet.* **2010**, *42*, 541–544.

Jin, L.; Lu, Y.; Shao, Y. F.; Zhang, G.; Xiao, P.; Shen, S. Q.; Corke, H.; Bao, J. S. Molecular Marker Assisted Selection for Improvement of the Eating, Cooking and Sensory Quality of Rice (Oryza Sativa L.). *J. Cereal Sci.* **2010**, *51*, 159–164.

Khan, M. H.; Dar, Z. A.; Dar, S. A. Breeding Strategies for Improving Rice Yield—A Review. *Agric. Sci.* **2015**, *6*, 467–478.

Khanna, A.; Sharma, V.; Ellur, R. K.; Shikari, A. B.; Gopala Krishnan, S.; Singh, U. D.; Prakash, G.; Sharma, T. R.; Rathour, R.; Variar, M.; Prashanthi, S. K.; Nagarajan, M.; Vinod, K. K.; Bhowmick, P. K.; Singh, N. K.; Prabhu, K. V.; Singh, B. D.; Singh, A. K. Development and Evaluation of Near-Isogenic Lines for Major Blast Resistance Gene(S) in Basmati Rice. *Theor. Appl. Genet.* **2015**, *128*, 1243–1259.

Khush, G. S. Strategies for Increasing the Yield Potential of Cereals: Case of Rice as an Example. *Plant Breed.* **2013**, *132*, 433–436.

Kottapalli, K. R.; Lakshmi, N. M.; Jena, K. K. Effective Strategy for Pyramiding Three Bacterial Blight Resistance Genes Into Fine Grain Rice Cultivar, Samba Mahsuri, Using Sequence Tagged Site Markers. *Biotechnol. Lett.* **2010**, *32*, 989–996.

Kottearachchi, N. S. Utility of Dna Markers in Rice Breeding. *Eur. Int. J. Sci. Technol.* **2013**, *2*, 111–122.

Kovach, M. J.; Calingacion, M. N.; Fitzgerald, M. A.; Mccouch, S. R. The Origin and Evolution of Fragrance in Rice (Oryza Sativa L.). *Proc. Natl. Acad Sci. U. S. A* **2009**, *106*, 14444–14449.

Kumar, V.; Rajvanshi, S. K.; Yadav, R. K. Potential Application of Molecular Markers in Improvement of Vegetable Crops. *Int. J. Adv. Biotechnol. Res.* **2014**, *5*, 690–707.

Lau, W. C. P.; Rafii, M. Y.; Ismail, M. R.; Puteh, A.; Latif, M. A.; Ramli, A. Review of Functional Markers for Improving Cooking, Eating, and the Nutritional Qualities of Rice. *Front Plant Sci.* **2015**, *6*(832), 1–11. DOI: 10.3389/fpls.2015.00832.

Li, J. M.; Thomson, M.; McCouch, S. R. A Fine Mapping of a Grain-Weight Quantitative Trait Locus in the Pericentromeric Region of Rice Chromosome 3. *Genetics* **2004**, *168*, 2187–2195.

Li, Y.; Fan, C.; Xing, Y.; Jiang, Y.; Luo, L.; Sun, L.; Shao, D.; Xu, C.; Li, X.; Xiao, J.; He, Y.; Zhang, Q Natural Variation in Gs5 Plays an Important Role in Regulating Grain Size and Yield in Rice. *Nat. Genet.* **2011**, *43*, 1266–1269.

Liu, E.; Liu, Y.; Wu, G.; Zeng, S.; Thi, T. G. T.; Liang, L.; Liang, Y.; Dong, Z.; She, D.; Wang, H.; Zaid, I.; Hong, D. Identification of a Candidate Gene for Panicle Length in Rice (Oryza Sativa L.) Via Association and Linkage Analysis. *Front Plant Sci.* **2016a**, *7*(596), 1–13. DOI: 10.3389/fpls.2016.00596.

Liu, Y.; Chen, L.; Liu, Y.; Dai, H.; He, J.; Kang, H.; Pan, G.; Huang, J.; Qiu, Z.; Wang, Q.; Hu, J.; Liu, L.; Chen, Y.; Cheng, X.; Jiang, L.; Wan, J. Marker Assisted Pyramiding of Two Brown Planthopper Resistance Genes, Bph3 and Bph27 (t), into Elite Rice Cultivars. *Rice* **2016b**, *9*(27), 1–7. DOI: 10.1186/s12284-016-0096-3.

Lou, J.; Chen, L.; Yue, G.; Lou, Q.; Mei, H.; Xiong, L.; Luo, L. Qtl Mapping of Grain Quality Traits in Rice. *J. Cereal Sci.* **2009**, *50*, 145–151.

Luo, Y.; Yin, Z. Marker-Assisted Breeding of Thai Fragrance Rice for Semi-Dwarf Phenotype, Submergence Tolerance and Disease Resistance to Rice Blast and Bacterial Blight. *Mol. Breed.* **2013**, *32*, 709–721.

Luo, Y.; Ma, T.; Zhang, A.; Ong, K. H.; Li, Z.; Yang, J.; Yin, Z. Marker-Assisted Breeding of the Rice Restorer Line Wanhui 6725 for Disease Resistance, Submergence Tolerance and Aromatic Fragrance. *Rice* **2016**, *9*, 66.

Luo, Y.; Zakaria, S.; Basyah, B.; Ma, T.; Li, Z.; Yang, J.; Yin, Z. Marker-Assisted Breeding of Indonesia Local Rice Variety Siputeh for Semi-Dwarf Phonetype, Good Grain Quality and Disease Resistance to Bacterial Blight. *Rice* **2014**, *7*(33), 1–8.

Mei, C.; Qi, M.; Sheng, G.; Yang, Y. Inducible Overexpression of a Rice Allene Oxide Synthase Gene Increases the Endogenous Jasmonic Acid Level, Pr Gene Expression, and Host Resistance to Fungal Infection. *Mol. Plant Microbe Interact.* **2006**, *19*, 1127–1137.

Meuwissen, T. H. E.; Hayes, B. J.; Goddard, M. E. Prediction of Total Genetic Value Using Genome-Wide Dense Marker Maps. *Genetics* **2001**, *157*, 1819–1829.

Miah, G.; Rafii, M. Y.; Ismail, M. R.; Puteh, A. B.; Rahim, H. A.; Asfaliza, R.; Latif, M. A. Blast Resistance in Rice: A Review of Conventional Breeding to Molecular Approaches. *Mol. Biol. Rep.* **2013**, *40*, 2369–2388.

Miao, L.; Wang, C.; Zheng, C.; Che, J.; Gao, Y.; Wen, Y. Molecular Mapping of a New Gene for Resistance to Rice Bacterial Blight. *Sci. Agric. Sin.* **2010**, *43*, 3051–3058.

Michael, J.; Kovach, M. N.; Calingacion, M.; Fitzgerald, A.; McCouch, S. R. The Origin and Evolution of Fragrance in Rice (Oryza Sativa L.). *Proc. Natl. Acad Sci.* **2009**, *106*(34), 144–149.

Miura, K.; Ashikari, M.; Matsuoka, M. The Role of QTLs in the Breeding of High-Yielding Rice. *Trends Plant Sci.* **2011**, *16*(6), 319–326.

Molla, K. A.; Debnath, A. B.; Ganie, S. A.; Mondal, T. K. Identification and Analysis of Novel Salt Responsive Candidate Gene Based Ssrs (Cgssrs) from Rice (Oryza Sativa L.). *BMC Plant Biol.* **2015**, *15*(122), 1–11.

Nandi, S.; Subudhi, P. K.; Senadhira, D.; Manigbas, N. L.; Sen-Mandi, S.; Huang, N. Mapping QTLs for Submergence Tolerance in Rice by AFLP Analysis and Selective Genotyping. *Mol. Gen. Genet.* **1997**, *255*, 1–8.

Neeraja, C. N.; Maghirang, R. R.; Pamplona, A.; Heuer, S.; Collard, B. C. Y.; Septiningsih, E. M.; Vergara, G.; Sanchez, D.; Xu, K.; Ismail, A. M.; Mackill, D. J. A Marker-Assisted Backcross Approach for Developing Submergence-Tolerant Rice Cultivars. *Theor. Appl. Genet.* **2007**, *115*, 767–776.

Ni, D.; Song, F.; Ni, J.; Zhang, A.; Wang, C.; Zhao, K.; Yang, Y.; Wei, P.; Yang, J.; Li, L. Marker-Assisted Selection of Two-Line Hybrid Rice for Disease Resistance to Rice Blast and Bacterial Blight. *Field Crop Res.* **2015**, *184*, 1–8.

Ni, D.; Zhang, S.; Chen, S.; Xu, Y.; Li, L.; Li, H.; Wang, Z.; Cai, X.; Li, Z.; Yang, J. Improving Cooking and Eating Quality of Xieyou57, an Elite Indica Hybrid Rice, by Marker-Assisted Selection of the Wx Locus. *Euphytica* **2011**, *179*, 355–362.

Nogoy, F. M.; Song, J. Y.; Ouk, S.; Rahimi, S.; Kwon, S. W.; Kang, K. K.; Cho, Y. G. Current Applicable Dna Markers for Marker Assisted Breeding in Abiotic and Biotic Stress Tolerance in Rice (Oryza Sativa L.). *Plant Breed. Biotech* **2016**, *4*, 271–284.

Pang, Y.; Ali, J.; Wang, X.; Franje, N. J.; Revilleza, J. E.; Xu, J.; Li, Z. Relationship of Rice Grain Amylose, Gelatinization Temperature and Pasting Properties for Breeding Better Eating and Cooking Quality of Rice Varieties. *PLoS One* **2016**, *11*(12), 1–14.

Parida, S. K.; Kumar, K. A. R.; Dalal, V.; Singh, N. K.; Mohapatra, T. Unigene Derived Microsatellite Markers for the Cereal Genomes. *Theor. Appl. Genet.* **2006**, *112*, 808–817.

Parida, S. K.; Kumar, K. A.; Dalal, V.; Singh, N. K.; Mohaptra, T. Unigene Derived Microsatellite Markers for the Cereal Genomes. *Theor. Appl. Genet.* **2006**, *112*, 808–817.

Peng, B.; Wang, L. Q.; Fan, C. C.; Jiang, G. H.; Luo, L. J.; Li, Y. B.; He, Y. Q. Comparative Mapping of Chalkiness Components in Rice Using Five Populations Across Two Environments. *BMC Genet.* **2014**, *15*, 49–63.

Pradhan, S. K.; Nayak, D. K.; Pandit, E.; Barik, S. R.; Mohanty, S. P.; Anandan, A.; Reddy, J. N. Characterization of Morpho-Quality Traits and Validation of Bacterial Blight Resistance in Pyramided Rice Genotypes Under Various Hotspots of India. *Aust. J. Crop Sci.* **2015**, *9*(2), 127–134.

Prince, S. J.; Beena, R.; Gomez, S. M.; Senthivel, S.; Babu, R. C. Mapping Consistent Rice (Oryza Sativa L.) Yield Qtls Under Drought Stress in Target Rainfed Environments. *Rice* **2015**, *8*(25), 1–13.

Qiu, Y. F.; Guo, J. P.; Jing, S. G.; Zhu, L. L.; He, G. C. High-Resolution Mapping of the Brown Planthopper Resistance Gene Bph6 in Rice and Characterizing Its Resistance in the 9311 and Nipponbare Near Isogenic Backgrounds. *Theor. Appl. Genet.* **2010**, *121*, 1601–1611.

Rabiei, B.; Valizadeh, M.; Ghareyazie, B.; Moghaddam, M.; Ali, A. J. Identification of Qtls for Rice Grain Size and Shape of Iranian Cultivars Using Ssr Markers. *Euphytica* **2004**, *137*, 325–332.

Rajpurohit, D.; Kumar, R.; Kumar, M.; Paul, P.; Awasthi, A. A.; Basha, P. O.; Puri, A.; Jhang, T.; Singh, K. Pyramiding of Two Bacterial Blight Resistance and a Semi Dwarfing Gene in Type 3 Basmati Using Marker-Assisted Selection. *Euphytica* **2011**, *178*, 111–126.

Saleem, M..; Akhtar, K.; Asghar, M.; Iqbal, Q.; Khan, A. Genetic Control of Late Blight, Yield and Some Yield Related Traits in Tomato (Lycopersicon Esculentum Mill). *Pak. J. Bot.* **2011**, *43*, 2601–2605.

Salgotra, R. K.; Gupta, D. B.; Millwood, R. J.; Balasubramaniam, M.; Stewart, C. N. Jr. Introgression of Bacterial Leaf Blight Resistance and Aroma Genes Using Functional Marker- Assisted Selection in Rice (Oryza Sativa L.). *Euphytica* **2012**, *187*, 313–323.

Salvi, S.; Bellotti, M.; Conti, S.; Frascaroli, E.; Giuliani, S.; Landi, P.; Maccaferri, M.; Natoli, V.; Sanguineti, C.; Sponza, G.; Talame, V.; Tuberosa, R. *The Art and Science of Cloning Qtls in Plants*. In Proceedings of the International Congress "In the Wake of the Double Helix: From the Green Revolution to the Gene Revolution", Bologna, Italy, May 27–31, 2003; Tuberosa, R., Phillips, R. L., Gale, M. Eds.; **2003**, pp 327–345.

Sandhu, N.; Singh, A.; Dixit, S.; Cruz, M. T. S.; Maturan, P. C.; Jain, R. K.; Kumar, A. Identification and Mapping of Stable Qtl with Main and Epistasis Effect on Rice Grain Yield Under Upland Drought Stress. *BMC Genet.* **2014**, *15*(63), 1–15.

Sattari, A.; Mahdinezhad, N.; Fakheri, B.; Noroozi, M.; Beheshtizadeh, M. Improvement of the Eating and Cooking Qualities of Rice: A Review. *Int. J. Farming Allied Sci.* **2015**, *4*(2), 153–160.

Septiningsih, E. M.; Pamplona, A. M.; Sanchez, D. L.; Neeraja, C. N.; Vergara, G. V.; Heuer, S.; Ismail, A. M.; Mackill, D. J. Development of Submergence-Tolerant Rice Cultivars: the Sub1 Locus and Beyond. *Ann. Bot.* **2009**, *103*(2), 151–160.

Shamsudin, N. A. A.; Swamy, B. P. M.; Ratnam, W.; Cruz, M. T. S.; Raman, A.; Kumar, A. Marker Assisted Pyramiding of Drought Yield QTLs into a Popular Malaysian Rice Cultivar, MR219. *BMC Genet.* **2016**, *17*(30), 1–16.

Shanti, M. L.; Shenoy, V. V.; Devi, G. L.; Kumar, V. M.; Premalatha, P.; Naveen, K. G.; Shashidhar, H. E.; Zehr, U. B.; Freeman, W. H. Marker Assistant Breeding for Resistance to Bacterial Leaf Blight in Popular Cultivar and Parental Lines of Hybrid Rice. *J. Plant Pathol.* **2010**, *92*, 495–501.

Shen, Y.-J.; Jiang, H.; Jin, J.-P.; Zhang, Z.-B.; Xi, B.; He, Y.-Y.; Wang, G.; Wang, C.; Qian, L.; Li, X.; Yu, Q.-B.; Liu, H.-J.; Chen, D.-H.; Gao, J.-H.; Huang, H.; Shi, T.-L.; Yang, Z.-N Development of Genome-Wide Dna Polymorphism Database for Map-Based Cloning of Rice Genes1[W]. *Plant Physiol.* **2004**, *135*, 1198–1205.

Shomura, A.; Izawa, T.; Ebana, K.; Ebitani, T.; Kanegae, H.; Konishi, S.; Yano, M. Deletion in a Gene Associated with Grain Size Increased Yields During Rice Domestication. *Nat. Genet.* **2008**, *40*, 1023–1028.

Singh, S.; Mackill, D. J.; Ismail, A. M, Physiological Basis of Tolerance to Complete Submergence in Rice Involves Genetic Factors in Addition to the Sub1 Gene. *AoB Plants* **2014**, *6*, 1–20.

Song, X. J.; Ashikari, M. Toward an Optimum Return from Crop Plants. *Rice* **2008**, *1*, 135–143.

Song, X. J.; Huang, W.; Shi, M.; Zhu, M. Z.; Lin, H. X. A Qtl for Rice Grain Width and Weight Encodes a Previously Unknown Ring-Type E3 Ubiquitin Ligase. *Nat. Genet.* **2007**, *39*, 623–630.

Spindel, J.; Begum, H.; Akdemir, D.; Virk, P.; Collard, B.; Redoña, E.; Atlin, G.; Jannink, J. L.; McCouch, S. R. Genomic Selection and Association Mapping in Rice (Oryza Sativa): Effect of Trait Genetic Architecture, Training Population Composition, Marker Number and Statistical Model on Accuracy of Rice Genomic Selection in Elite, Tropical Rice Breeding Lines. *PLoS Genet.* **2015**, *11*(2), 1–25 DOI: 10.1371/journal.pgen.1004982.

Su, Y.; Rao, Y.; Hu, S.; Yang, Y.; Gao, Z.; Zhang, G.; Liu, J.; Hu, J.; Yan, M.; Dong, G.; Zhu, L.; Guo, L.; Qian, Q.; Zeng, D. Map-Based Cloning Proves Qgc-6, a Major Qtl for Gel Consistency of Japonica/Indica Cross, Responds by Waxy in Rice (Oryza Sativa L.). *Theor. Appl. Genet.* **2011**, *123*, 859–867.

Sundaram, R. M.; Vishnupriya, R. M.; Biradar, S. K.; Laha, G. S.; Ashok Reddy, G.; Shobha Rani, N.; Sarma, N. P.; Sonti, R. V. Marker Assisted Introgression of Bacterial Blight Resistance in Samba Mahsuri, an Elite Indica Rice Variety. *Euphytica* **2008**, *160*, 411–422.

Sundaram, R. M; Vishnupriya, M. R; Shobha Rani, N.; Laha, G. S.; Viraktamath, B. C.; Balachandran, S. M.; Sarma, N. P.; Mishra, B.; Reddy, A. G.; Sonti, R. V. Ingr09070), a Paddy (Oryza Sativa) Germplasm with High Bacterial Blight Resistance, Yield and Fine-Grain Type. *Ind. J. Plant Genet. Resour.* **2010**, *23*, 327–328.

Swamy, B. P. M.; Kumar, A. Genomics-Based Precision Breeding Approaches to Improve Drought Tolerance in Rice. *Biotechnol. Adv.* **2013**, *31*, 1308–1318.

Tan, Y. F.; Xing, Y. Z.; Li, J. X.; Yu, S. B.; Xu, C. G.; Zhang, Q. Genetic Bases of Appearance Quality of Rice Grains in Shanyou 63, an Elite Rice Hybrid. *Theor. Appl. Genet.* **2000**, *101*, 823–829.

Tanksley, S. D.; Yong, N. D.; Paterson, A. H.; Bonierbale, M. W. Rflp Mapping in Plant Breeding: New Tools for an Old Science. *Biotechnol.* **1989**, *7*, 257–263.

Temnykh, S.; DeClerck, G.; Lukashova, A.; Lipovich, L.; Cartinhour, S.; McCouch, S. Computational and Experimental Analysis of Microsatellites in Rice (Oryza Sativa L.): Frequency, Length Variation, Transposon Associations, and Genetic Marker Potential. *Genomes Res.* **2001**, *11*(8), 1441–1452.

Thomson, M. J.; Ocampo, M.; Egdane, J.; Rahman, M. A.; Sajise, A. G.; Adorada, D. L.; Raiz, E. T.; Blumwald, E.; Seraj, J. I.; Singh, R. K.; Gregorio, G. B.; Ismail, A. M. Characterizing the Saltol Quantitative Trait Locus for Salinity Tolerance in Rice. *Rice* **2010**, *3*, 148–160.

Thomson, M. J.; Tai, T. H.; McClung, A. M.; Lai, X. H.; Hinga, M. E.; Lobos, K. B.; Xu, Y.; Martinez, C. P.; McCcouch, S. R. Mapping Quantitative Trait Loci for Yield, Yield Components and Morphological Traits in an Advanced Backcross Population Between Oryza Rufipogon and the Oryza Sativa Cultivar Jefferson. *Theor. Appl. Genet.* **2003**, *107*, 479–493.

Tian, Z. X.; Yan, C. J.; Qian, Q.; Yan, S.; Xie, H. L.; Wang, F.; Xu, J. F.; Liu, G. F.; Wang, Y. H.; Liu, Q. Q.; Tang, S. Z.; Li, J. Y.; Gu, M. H. Development of Gene-Tagged Molecular Markers for Starch Synthesis-Related Genes in Rice. *Chin. Sci. Bull* **2010**, *55*, 3768–3777.

Tong, H. H.; Chen, L.; Li, W. P.; Mei, H. W.; Xiong, Y. Z.; Yu, X. Q.; Xu, X. Y.; Zhang, S. Q.; Luo, L. J. Identification and Characterization of Quantitative Trait Loci for Grain Yield and Its Components Under Different Nitrogen Fertilization Levels in Rice (Oryza Sativa L.). *Mol. Breed.* **2011**, *28*, 495–509.

Toojinda, T.; Siangliw, M.; Tragoonrung, S.; Vanavichit, A. Molecular Genetics of Submergence Tolerance in Rice: Qtl Analysis of Key Traits. *Ann. Bot.* **2003**, *91*, 243–253.

Varshney, R. K.; Hoisington, D. A.; Tyagi, A. K. Advances in Cereal Genomics and Applications in Crop Breeding. *Trends Biotechnol.* **2006**, *24*, 490–499.

Venuprasad, R.; Impa, S.; Gowda, R. P. V.; Atlin, G. N.; Serraj, R. Rice Near-Isogenic-Lines (Nils) Contrasting for Grain Yield Under Lowland Drought Stress. *Field Crop. Res.* **2011**, *123*, 38–46.

Verma, D. K.; Srivastav, P. P. Extraction Technology for Rice Volatile Aroma Compounds. In: *Food Engineering: Emerging Issues, Modeling, and Applications*; Meghwal, M., Goyal, M. R., Eds.; as part of book series on Innovations in Agricultural and Biological Engineering; Apple Academic Press: USA, **2016**; Vol. 2, pp. 245–291.

Verma, D. K.; Srivastav, P. P. Proximate Composition, Mineral Content and Fatty Acids Analyses of Aromatic and Non-Aromatic Indian Rice. *Rice Sci. 2017*, *24*(1), 21–31.0

Verma, D. K.; Mohan, M.; Asthir, B. Physicochemical and Cooking Characteristics of Some Promising Basmati Genotypes. *Asian J. Food Agro.-Ind.* **2013**, *6*(2), 94–99.

Verma, D. K.; Mohan, M.; Prabhakar, P. K.; Srivastav, P. P. Physico-Chemical and Cooking Characteristics of Azad Basmati. *Int. Food Res. J.* **2015**, *22*(4), 1380–1389.

Verma, D. K.; Mohan, M.; Yadav, V. K.; Asthir, B.; Soni, S. K. Inquisition of Some Physico-Chemical Characteristics of Newly Evolved Basmati Rice. *Environ Ecol.* **2012**, *30*(1), 114–117.

Vu, H. T. T.; Le, D. D.; Ismail, A. M.; Le, H. Y. Marker-Assisted Backcrossing (MABC) for Improved Salinity Tolerance in Rice (Oryza sativa L.) to Cope with Climate Change in Vietnam. *Aust. J. Crop Sci.* **2012**, *6*(12), 1649–1654.

Wacera, H. R.; Safitri, F. A.; Lee, H. S.; Yun, B. W.; Kim, K. M. Genetic Mapping of Qtls That Control Grain Characteristics in Rice (Oryza Sativa L.). *J. Life Sci.* **2015**, *25*, 925–931.

Wan, X. Y.; Wan, J. M.; Jiang, L.; Wang, J. K.; Zhai, H. Q.; Weng, J. F.; Wang, H. L.; Lei, C. L.; Wang, J. L.; Zhang, X.; Cheng, Z. J.; Guo, X. P. Qtl Analysis for Rice Grain Length and Fine Mapping of an Identified Qtl with Stable and Major Effects. *Theor. Appl. Genet.* **2006**, *112*, 1258–1270.

Wang, L. Q.; Liu, W. J.; Xu, Y..; He, Y. Q.; Luo, L. J.; Xing, Y. Z.; Xu, C. G.; Zhang, Q. F. Genetic Basis of 17 Traits and Viscosity Parameters Characterizing the Eating and Cooking Quality of Rice Grain. *Theor. Appl. Genet.* **2007**, *115*, 463–476.

Wang, S.; Wu, K.; Yuan, Q.; Liu, X.; Liu, Z.; Lin, X.; Zeng, R.; Zhu, H.; Dong, G.; Qian, Q.; Zhang, G.; Fu, X. Control of Grain Size, Shape and Quality by OsSPL16 in Rice. *Nat. Genet.* **2012a**, *44*, 950–954.

Wang, S. K.; Wu, K.; Yuan, Q. B.; Liu, X. Y.; Liu, Z. B.; Lin, X. Y.; Zeng, R. Z.; Zhu, H. T.; Dong, G. J.; Qian, Q.; Zhang, G. Q.; Fu, X. D. Control of Grain Size, Shape and Quality by OsSPL16 in Rice. *Nat. Genet.* **2012b**, *44*, 950–954.

Weng, J. F.; Gu, S. H.; Wan, X. Y.; Gao, H.; Guo, T.; Su, N.; Lei, C. L.; Zhang, X.; Cheng, Z. J.; Guo, X. P.; Wang, J. L.; Jiang, L.; Zhai, H. Q.; Wan, J. M. Isolation and Initial Characterization of Gw5, a Major Qtl Associated with Rice Grain Width and Weight. *Cell Res.* **2008**, *18*, 1199–1209.

Wijerathna, Y. M. A. M. Marker Assisted Selection: Biotechnology Tool for Rice Molecular Breeding. *Adv. Crop Sci. Tech.* **2015**, *3*(4), 1–4.

Wu, H.; Liu, Y. Q.; He, J.; Liu, Y. L.; Jiang, L.; Liu, L. L.; Wang, C. M.; Cheng, X. N.; Wan, J. M. Fine Mapping of Brown Planthopper (Nilaparvata Lugens Stål) Resistance Gene Bph28(T) in Rice (Oryza Sativa L.). *Mol. Breed.* **2014**, *33*, 909–918.

Xing, Y. Z.; Zhang, Q. Genetic and Molecular Bases of Rice Yield. *Annu. Rev. Plant Biol.* **2010**, *61*, 421–442.

Xu, J.; Zhao, Q.; Du, P.; Xu, C.; Wang, B.; Feng, Q.; Liu, Q.; Tang, S.; Gu, M.; Han, B. Developing High Throughput Genotyped Chromosome Segment Substitution Lines Based on Population Whole-Genome Re-Sequencing in Rice (Oryza Sativa L.). *BMC Genomics* **2010**, *11*(1), 1–14.

Xu, K.; Mackill, D. J. A Major Locus for Submergence Tolerance Mapped on Rice Chromosome 9. *Mol. Breed.* **1996**, *2*, 219–224.

Xue, W. Y.; Xing, Y. Z.; Weng, X. Y.; Zhao, Y.; Tang, W. J.; Wang, L.; Zhou, H. J.; Yu, S. B.; Xu, C. G.; Li, X. H.; Zhang, Q. F. Natural Variation in Ghd7 is an Important Regulator of Heading Date and Yield Potential in Rice. *Nat. Genet.* **2008**, *40*, 761–767.

Yao, X. Y.; Li, Q.; Liu, J.; Jiang, S. K.; Yang, S. L.; Wang, J. Y. et al. Dissection of QTLs for Plant Height and Panicle Length Traits in Rice Under Different Environment. *Sci. Agric. Sin.* **2015**, *48*, 407–414.

Yoon, D. B.; Kang, K. H.; Kim, H. J.; Ju, H. G.; Kwon, S. J.; Suh, J. P.; Jeong, O. Y.; Ahn, S. N. Mapping Quantitative Trait Loci for Yield Components and Morphological Traits in an Advanced Backcross Population Between Oryza Grandiglumisand the O. Sativa Japonica Cultivar Hwaseongbyeo. *Theor. Appl. Genet.* **2006**, *112*, 1052–1062.

Yuan, L. Y.; Hong-jian, T.; Xiang-qian, Z.; Jie, X.; Geng-mi, L.; Shi-kai, H.; Guo-jun, D.; Zheng, S. Y.; Liwen, W.; Jiang, H.; Guo-you, Y.; Long, G. B. Molecular Improvement of Grain Weight and Yield in Rice by Using Gw6 Gene. *Rice Sci.* **2014**, *21*, 127–132.

Zeng, L.; Shannon, M. C. Salinity Effects on Seedling Growth and Yield Components of Rice. *Crop Sci.* **2000**, *40*, 996–1003.

Zhang, L.; Wang, J.; Wang, J.; Wang, L.; Ma, B.; Zeng, L.; Qi, Y.; Li, Q.; He, Z. Quantitative Trait Locus Analysis and Fine Mapping of the qPL6 Locus for Panicle Length in Rice. *Theor. Appl. Genet.* **2015**, *128*, 1151–1161.

Zhang, X. J.; Wang, J. F.; Huang, J.; Lan, H. X.; Wang, C. L.; Yin, C. F.; Wu, Y. Y.; Tang, H. J.; Qian, Q.; Li, J. Y.; Zhang, H. S. Rare Allele of Osppkl1 Associated with Grain Length Causes Extra-Large Grain and a Significant Yield Increase in Rice. *Proc. Natl. Acad. Sci. U. S. A.* **2012**, *109*, 21534–21539.

Zhou, L.; Chen, L.; Jiang, L.; Zhang, W.; Liu, L.; Liu, X.; Zhao, Z.; Liu, S.; Zhang, L.; Wang, J.; Wan, J Fine Mapping of the Grain Chalkiness Qtl Qpgwc-7 in Rice (Oryza Sativa L.). *Theor. Appl. Genet.* **2009**, *118*, 581–590.

Zong, G.; Wang, A.; Wang, L.; Liang, G.; Gu, M.; Sang, T.; Han, B. A Pyramid Breeding of Eight Grain-Yield Related Quantitative Trait Loci Based on Marker-Assistant and Phenotype Selection in Rice (Oryza Sativa L.). *J. Genet. Genomics* **2012**, *39*, 335–350.

Part IV
Molecular Advances and Trends for Quality Improvement in Rice

CHAPTER 9

THE CRISPR/CAS GENOME EDITING SYSTEM AND ITS APPLICATION IN RICE IMPROVEMENT

RUKMINI MISHRA[1,*], SATYABRATA NANDA[1,2], and
RAJ KUMAR JOSHI[1,3]

*1Centre of Biotechnology, Siksha O Anusandhan University,
Bhubaneswar, Odisha 751030, India, Mob.: +91-9556458058*

2sbn.satyananda@gmail.com, Mob.: +91-7751831025

3rajjoshi@soauniversity.ac.in, Mob.: +91-9437684176

Corresponding author. E-mail: rukmini.mishra@gmail.com

9.1 INTRODUCTION

Rice (*Oryza sativa* L.) is the major staple food crop in the world, feeding more than 50% of the world's population (Verma et al., 2012, 2013, 2015). Rapid increase in the human population together with acute weather conditions leading to biotic and abiotic stresses demand for a sustainable improvement of crop varieties including rice. Conventional rice improvement methods such as cross hybridization and mutation breeding techniques are gradually getting inhibited by the declination of the existing genetic variability thereby affecting the production for future supply (Chen and Gao, 2014). Molecular marker-assisted selection has the advantage of detecting constructive agricultural characters. However, large dependence on traditional crossing and backcrossing approaches for pyramiding of agriculturally important traits makes the process both times consuming and labor-intensive for rice improvement. Moreover, high variability in the stress factors often leads to rapid breakdown of stress response in the rice varieties with the course of time (Jia, 2003). Foreign gene transfer into plant genome has been a focal point of crop improvement for the last two decades. There are many examples of transgenic rice with improved traits such as enhancement in nutritional quality or resistance to insects, pathogens or herbicides (Chen et al., 2009). However, the concern of the general public and the environmental protection agencies over the cultivation of crop with foreign genes from distantly related organisms is a major hindrance to their widespread use. The adherence to the regulatory frameworks for environmental protection and addressing the public concerns has added to the cost of transgenic development (Lusser et al., 2012). As such, there are only a few transgenic crops such as soybean, corn and cotton with agriculturally valuable traits that are available for human use.

A rapid and inexpensive method is essential for bringing in high-yield, multi-stress resistant rice varieties. Plant Biotechnology is entering into a new phase where the genome engineering technologies are rapidly superseding the conventional mutagenesis methods. Customized site-specific nucleases (SSNs) are used as a common method to carry out induced targeted insertions, deletions and specific sequential manipulations in the genomes of model and crop plants (Voytas, 2013). Unlike random mutagenesis through EMS or gamma radiation, genome editing enables precise manoeuvring of specific genomic sequences for crop improvement. In the last decade, several SSNs including meganucleases, zinc-finger nucleases (ZFNs) and transcription activator-like effector nucleases (TALENs) have been extensively used in plants with a greater promise to revolutionize the strategy for crop improvement (Voytas and Gao, 2014). More recently, the invention of an efficient genome editing system based on bacterial monomeric DNA endonuclease, known as Cas9 [clustered regularly interspaced short palindromic repeats (CRISPR) associated protein 9] has made the process extremely simple, economical and versatile in many applications in both plants and animals (Doudna and Carpentier, 2014; Fichtner et al., 2014; Liang et al., 2015; Kumar and Jain, 2014; Khatodia et al., 2016). With this technique, it is now possible to first determine the desired DNA modification in the cultivars and then establish the genetic variations rapidly and precisely. This provides a sea change in our thinking toward generation of new crop varieties to feed the burgeoning populations of the developing countries. CRISPR/Cas system has already been introduced into the production pipelines of many crops including rice (Wang et al., 2016; Xu et al., 2016).

In this chapter, we provide a concise outline of the available genome editing systems. We describe the structure and machinery of CRISPR/cas9 system and its use in precise gene/genome editing. We also address the useful appliance of CRISPR/cas9 genome editing in the improvement of crop varieties in general and rice in particular. Finally, we are briefly overviewed the future usage of this system for rice improvement.

9.2   GENOME EDITING SYSTEMS-AN OVERVIEW

The genome editing processes relies on the SSNs which introduce double strand breaks (DSBs) at the specific target locus (Carroll, 2011). The breaks in the DNA can be repaired by either non-homologous end joining (NHEJ) or homologous recombination (HR) (Sander and Joung, 2014). The NHEJ and HR result in the introduction of indels of variable length or point mutations in the target DNA sequences, thereby altering the reading frames and generating gene knockouts, insertions, replacements and chromosomal rearrangements (Chen and Gao, 2014). The earliest known genome editing system made use of the homing endonucleases/meganucleases such as the yeast endonuclase I-*Sce*I enzyme and its homologous for activating the DSB repair system (Paques and Duchateau, 2007). Initially developed for the animal and human system, it has also been used for genome editing purposes in plants, including *Arabidopsis thaliana* and *Zea mays* (Gao et al., 2010; Antunes et al., 2012). However, it has an intricate process of redesigning for each new target site due to difficulty in engineering the overlapped DNA-binding and nuclease domains of the homing endonucleases (Paques and Duchateau, 2007). Further, it is also limited by the fact that the process requires expertise in protein engineering and artificial incorporation of recognition sites for homing endonucleases.

Hybrid nuclease proteins such as the ZFNs and TALENs constitute the next group of genome editing tools that provided a system to target precise genes. ZFNs are generated by a fusion of an artificial DNA binding domain (DBD) carrying a tandem array of zinc-finger motifs with the non-specific nuclease domain of the *Fok*I restriction endonucleases (Durai et al., 2005). The DBD consist of three or four zinc finger modules, each of which bind to a target nucleotide sequence of 9–12 base pairs (bp). Usually, two ZFNs are strung together with a 5–6 bp gap between them that can recognize a sequence of 18–24 bp. This results in the reconstitution of the *Fok*I dimer that induces the DSB (Durai et al., 2005). A wide variety of crop plants including tobacco, soybean, and maize has been subjected to efficient heritable mutagenesis using ZFNs (Townsend et al., 2009; Shukla et al., 2009; Curtin et al., 2011). The usage of ZFNs in a gene editing system is quite difficult because the multifaceted interactions between amino acid residues and the base pairs of the target sequence makes the designing of multi ZF containing DBD a complicated task. Further, the non-specific binding of the zinc-finger motifs is often results in off-target cleavage with ZFNs (Carrol, 2011; Voytas, 2013). Furthermore, the stiff licensing fees demanded by Sangamo Bioscience, the proprietary company for ZFNs limits the use of

ZFNs as genome editing tools in major public funded molecular genetic laboratories across the world.

Like ZFNs, TALENs are constructed by the fusion of an engineered array of DBD consisting of the transcription activator-like effector (TALE) protein from the bacterial genus *Xanthomonas* with the non-specific*Fok*I endonuclease domain (Christian et al., 2010). TALEs are chief virulence factors that are characterized by the presence a central DBD, N-terminal translocation signals and C-terminal acidic activation domain (AD) and nuclear localization signals (NLS) (Boch and Bonas, 2010; Bogdanove et al., 2010). TALEs are directly responsible for the transcriptional activation of the specific target genes determining the pathogenicity of the bacteria (Boch and Bonas, 2010). The central DBD consists of 14–20 tandem repeats of highly conserved 33–35 amino acid sequence. All repeats are nearly identical except for two residues at positions 12 and 13, labelled as repeat-variable di-residues (RVDs) that are responsible for pairing with a sense strand DNA base (Streubel et al., 2012). While residue 13 specifically recognizes the target DNA base, residue 12 helps to stabilize the repeat structure (Deng et al., 2012). Several recognition favourites have been demonstrated for the DNA base recognition by RVDs and therefore, TALENs can be tailored by assembling them according to required target DNA chain (Boch et al., 2009; Streubel et al., 2012). TALENs are often designed to target as many recognition sites as possible by inserting a thymine (T) before the first nucleotide of the target site, and numerous computer programs are available for competent designing of the TALEs and forecast of their targets (Shan et al., 2013; Feng et al., 2014). Likewise, several assembly methods, including standard cloning assembly (Sanders et al., 2011; Reyon et al., 2012a), Golden Gate assembly (Li et al., 2011; Weber et al., 2011) and solid-phase assembly (Reyon et al., 2012b) have been developed to facilitate the assembly of TALE arrays. Gene targeting using TALEN technology has been adopted in many plants, including rice, wheat, tobacco, barley and Arabidopsis (Li et al., 2012; Wendt et al., 2013; Christian et al., 2013; Wang et al., 2014). Although more efficient that ZFNs in terms of low off-target effects, the extensive repeat structure of the TALE DBD makes it a major challenge for assembly of individual DNA binding proteins to construct engineered TALENs.

The rapid innovation in the target genome modification area has led to the invention of the latest, most efficient and the extraordinary CRISPR/Cas genome editing system that have been effectively tested in diverse organisms with significant outcomes (Shan et al., 2013; Upadhyay et al., 2013; Butler et al., 2015; Jacobs et al., 2015; Čermák et al., 2015; Svitashev et al., 2015). These processes make use of a Cas9 endonuclease that can slice at a precise target location using two small RNA sequences, that is CRISPR RNA (crRNA) and trans-activating CRISPR RNA (tracrRNA). The two RNA sequences are combined together to form a dual RNA structure termed as single-guide RNA (sgRNA) (Qi et al., 2013). This sgRNA in turn guides the Cas9 endonuclease to create a DSB precisely at the specific target site of the genome (Jinek et al., 2012). As the sgRNA requires only ≈20 bp sequence for the recognition and precise cleavage of the target DNA sequence, they can be designed more easily as compared to ZFNs and TALENs. Not surprisingly, CRISPR/Cas system is being used to add, delete, activate or suppress target genes in a wide group of organisms. The detailed structure and mechanism of the CRISPR/Cas system is described in the next section.

9.3 CRISPR/CAS SYSTEM

The bacterial and archaeal genomes are often characterized by the presence of short repeat sequences interspaced with distinctive spacers, known as CRISPR loci. Sequence analysis has revealed that, the CRISPR loci exhibit high homology with the plasmid and viral genome sequences and together with Cas protein are responsible for regulating bacterial immunity against foreign genetic elements (Mojica et al., 2005; Pourcel et al., 2005; Bolotin et al., 2005). The CRISPR/Cas defense system is common in many archaeal and bacterial genomes and their ability to degrade foreign DNA has recently highlighted them as a tool of choice for genome editing. A typical CRISPR–Cas system targets either RNA or DNA of the pathogenic invaders (Hale et al., 2009; Wiedenheft et al., 2012). A variety of Cas proteins with DNase or RNase activity exist in the bacterial and archaeal genomes that play one or multiple important roles in the CRISPR/Cas system. While Cas1 and Cas2 metal-dependent DNAs are universally required for CRISPR–Cas systems, others have variable roles (Kumar and Jain, 2014). Cas3 protein possesses a histidine–aspartate (HD) domain and induces metal dependent cleavage of the double stranded sequence (Makarova et al., 2006). Cas4 are RecB endonucleases that recognize the CRISPR loci, while Cas5 and Cas6 are repeat associated mysterious proteins (RAMPs) with similar recognition motif as in case of Cas1 (Makarova et al., 2006). The most studied among the Cas proteins are the Cas9 which has a large globular recognition specificity (REC) domain and a small nuclease (NUC) domain that contains two nuclease sites, RuvC and HNH, and a PAM-interacting domain (PI domain) (Jinek at al., 2014). It is a multifunctional protein with significant role in the processing of pre-crRNA to crRNA and induces DSB at the specific target site. The CRISPR/Cas system is categorized into three distinct types based on the presence of specific Cas protein (Makarova et al., 2011). The type I system is found in both archaea and bacteria and target precise DNA chains using the cleaving properties of Cas1, Cas2 and Cas3 (Makarova et al., 2011). The type II system is entirely found in bacteria and make use of Cas9 along with Cas1, Cas2 and Cas4 which recognizes the sgRNA formed by the fusion of crRNA and the tracrRNA and cleaves the DNA within the confines of the protospacer adjacent motif (PAM) (Jinek et al., 2014). The type III CRISPR/Cas system involves Cas10, Cas1, Cas2, Cas6 and other RAMPs which bring about crRNA processing and target DNA cleavage (Makarova et al., 2011). The type III system has been commonly found in archae and only a few bacteria. Inclusively, the presence of three different kinds of CRISPR–Cas systems suggests a diversity in the functional mechanism of CRISPR/Cas mediated immunity in bacteria and archaea.

9.4　CRISPR/CAS9 MEDIATED GENOME EDITING

The specificity of the Cas9 endonuclease and the involvement of tracrRNA and PAM in the type II CRISPR/Cas system make it a promising component of genome editing. Typically, the type II CRISPR/Cas system executes three steps toward recognition and cleavage of pathogenic DNA in bacterial cell (Fig. 9.1). The first step involves the integration of the foreign DNA as spacer within the CRISPR locus. The PAM site consisting of a set of 2–5 conserved nucleotides that recognizes the DNA fragment and facilitates the introduction of a single copy of a ≈30 bp spacer sequence followed by its duplication at the leader end of the CRISPR repeat (Garneau et al., 2010). In the second step, the Cas protein facilitates the transcription of pre-crRNA from the CRISPR locus. Subsequently, the tracrRNA complementarily pair off with the duplicated region of crRNA thereby processing the pre-crRNAs into crRNA (Deltcheva et al., 2011) (Fig. 9.1). In the final step, crRNA binds with the CRISPR allied complex containing the Cas9 protein and guide it to the specific target for cleavage of foreign DNA (Garneau et al., 2010; Deltcheva et al., 2011). The Cas9-sgRNA scans for DNA double strand with PAMs using Watson-Crick pairing between the sgRNA and the target DNA (Fig. 9.1). Once attached to the PAM, the HNH nuclease domain of the cas9 cleaves the hybrid while RuvC results in a DSB by cleaving the other strand. Jinek et al., (2012) demonstrated in vitro cleavage of the target DNA using Cas9 and a sgRNA molecule generated by fusion of a tracrRNA and crRNA sequence. The revealing of this molecular mechanism of the CRISPR/Cas9 opened of new avenues for genome editing through the development of RNA-guided engineered nucleases (RGENs). The RGENs are artificially designed nucleases composed of two components, the Cas9 nuclease to induce DSB and an engineered sgRNA with specific 20 nucleotides that direct Cas9 to matching target site (Fig. 9.2). A DNA sequence with 20 nucleotides followed by NGG bases can be targeted by the first 20 nucleotides in the 5' end of the sgRNA to introduce a mutation at the specific site (Sander and Joung, 2014). The introduction of the DSB is followed by a cellular recombination repair mechanism such as NHEJ and HDR resulting in desired modification (Fig. 9.3). The simplicity in the designing of the CRISPR/Cas9 system together with the generation of efficient target mutations has resulted in major elemental discoveries in plant and animal biology (Doudna and Carpentier, 2014). It is now employed on a regular basis for targeted mutagenesis in bacteria, yeast, animals and a wide range of crop plants (reviewed in Kumar and Jain, 2014; Song et al., 2016). The availability of many Cas9 and sgRNA variants in the recent times will further add into the noble application of this tool in the field of crop breeding and biotechnology.

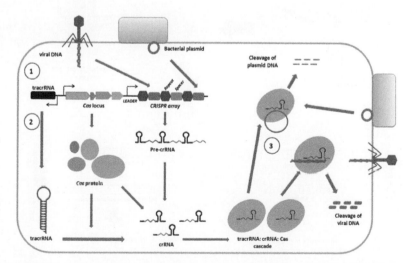

FIGURE 9.1   (See color insert.) CRISPR/Cas mediated bacterial defense against pathogen attack. The foreign DNA is recognized and integrated as spacer at the leader end of the CRISPR locus at the acquisition step (1). Transcription of the pre-CRISPR RNA (crRNA) which is processed into crRNA by Cas proteins and *trans*-activating crRNA (tracrRNA) at the expression step (2). The crRNA: tracrRNA: Cas9 cascade target and cleave a specific genetic element from the pathogen at the interference step (3).

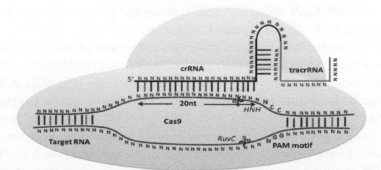

FIGURE 9.2 CRISPR/Cas9 mediated cleavage mechanism. sgRNA (crRNA: tracrRNA) target the protospacer adjacent motif (PAM) site and unzips the 20 bp complementary sequence of the target DNA resulting in sgRNA-target DNA heteroduplex. The nuclease domains RuvC and HNH each nicks one strand of the DNA resulting in double strand break (DSB). The blue line represents the backbone of target DNA while the black line and the folded sky line represent the sgRNA.

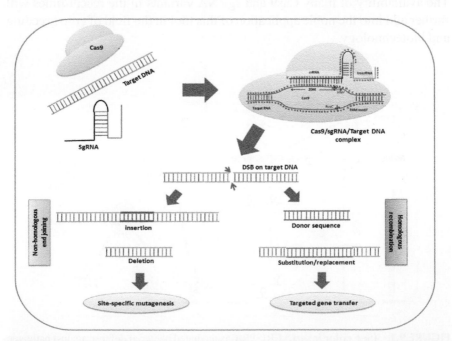

FIGURE 9.3 (See color insert.) Cas9/sgRNA mediated cleavage and DNA repair mechanism. The HNH and RuvC nuclease domain generate a double strand break at the 5′ of a protospacer adjacent motif (PAM) sequence. The DSB at the target site is subjected to cellular DNA repair through homologous recombination (HR) or non-homologous end joining (NHEJ). NHEJ results in site specific mutagenesis through insertion or deletion (indels) while HR results in target gene replacement.

9.5 APPLICATIONS OF CRISPR/CAS9 IN CROP IMPROVEMENT

A diminishing genetic base that depends on natural allelic variants of plants has become a major constraint in crop improvement and production (Chen and Gao, 2014). On the other hand, commercialization of genetically modified (GM) crops is frequently prohibited by public fear about health and environmental safety and by complex regulatory needs. With its simplicity and high efficiency, CRISPER/Cas9 has emerged as a promising strategy in meeting the food demands of growing population and providing sustainable productive agriculture (Liu et al., 2013; Khatodia et al., 2016). CRISPR/Cas9 editing in plants was first reported in 2013, when three independent research groups demonstrated the usage of CRISPR/Cas9in model plants like *Arabidopsis,* rice and *Nicotiana benthamiana* (Li et al., 2013; Nekrasov et al., 2013; Shan et al., 2013). During the last 3 years, CRISPER/Cas9 editing has made substantial progress in creating gene knockouts and induce targeted mutagenesis for several important crops including rice (Miao et al., 2013; Jiang et al., 2013b; Zhang et al., 2014), wheat (Upadhyay et al., 2013; Wang et al., 2014), maize (Liang et al., 2014), soybean (Sun et al., 2015), sorghum (Jiang et al., 2013b) and tomato (Brooks et al., 2014, Pan et al., 2016). In tomato, a CRISPR/Cas9 construct was designed to target the second exon of the tomato homolog of *Arabidopsis* ARGONAUTE7 (*SlAGO7*) (Brooks et al., 2014). The 35S promoter driven hCas9 was combined with two sgRNAs under *Arabidopsis thaliana* U6 control, aiming to produce a precise deletion in the *SlAGO7* gene for effortless detection of mutation. More recently, CRISPR/Cas9 gene editing was reported in two genes of tomato, namely *SlPDS* and *SlPIF4* through *Agrobacterium tumefaciens* mediated transformation method (Pan et al., 2016). A high mutation frequency of 83.56% was observed among the transgenic lines with 1–3 nucleotide deletions followed by 1 bp insertion. Upadhyay et al. (2013) reported the application of CRISPR/Cas9 mediated genome editing in wheat. The *inositol oxygenase (inox)* and *phytoene desaturase (pds)* genes from wheat were selected for inducing targeted mutations. Duplex cgRNA expression with Cas9 resulted in erasure of DNA region between the targets causing indel mutations in all the tested samples. Wang et al. (2014) used the type II CRISPR/Cas9 system and successfully edited three homeoalleles (*TaMLO-A, TaMLO-B and TaMLO-D*) of the *MLO* gene that confers resistance to powdery mildew in bread wheat. In Soybean, two genes, *Glyma01g38150* and *Glyma11g07220* representing the DNA methylation 1 (DDM1) from *Arabidopsis* were successfully edited using one sgRNA targeting both the genes (Jacobs et al. 2015). Two additional genes DD20 and DD43 were targeted with mutation rates of 59% and 76%, respectively (Li et al., 2015). Recently, a group of scientists from the Chinese Academy of Agricultural Sciences demonstrated the usage of six sgRNAs targeting two endogenous soybean genes, *GmFEI2* and *GmSHR* (Cai et al., 2015). Mutations were detected in soybean hairy roots and the results confirmed that the tailored CRISPR/Cas9 system exhibited the same level of competence for both exogenous and endogenous genes in soybean. The study indicated that CRISPR/Cas9 genome editing could be used as a principal means for studying root functional genomics in crop plants.

During the last couple decades, targeted mutagenesis has become the method of choice for functional genomics studies. Professor Zhengyan and his group from China were the first to demonstrate highly efficient targeted mutagenesis in *Arabidopsis* and rice using CRISPR/Cas9 (Feng et al., 2014). They introduced sgRNA expression cassette along with Cas9 for different target genes and detected a high mutation rate (88%) regardless of the chromatin arrangement and gene structure. CRISPR/Cas9 system has been successfully used for targeted mutagenesis in maize (*Zea mays*) (Liang et al., 2014; Feng et al., 2016). The sgRNA-Cas9 construct targeting a marker gene *Zmzb7* was transformed into maize protoplast resulting in significant insertions and deletions (Feng et al., 2016). The results revealed that CRISPR/Cas9 system can be effectively used for targeting genes in the heterochromatin as well as euchromatin regions of the maize genome irrespective of their expression. A more recent experiment demonstrated the use of *Streptococcus pyogenes* Cas9 endonuclease and sgRNA for targeting five different genomic regions within the maize genome (Svitashev et al., 2016). Mutations were detected at all the target sites and the progenies demonstrated an expected Mendelian segregation of mutations, edits and target insertions. Efficient targeted mutagenesis has also been reported in potato (Butler et al., 2015). In this study, CRISPR/Cas9 expressing sgRNA targeting the potato ACETOLACTATE SYNTHASE1 (*StALS1*) gene was evaluated in both diploid and tetraploids potato using the *Agrobacterium* mediated transformation method. Target mutation to the tune of 3–60% was reported for the *ALS* locus. Sun et al., (2015) also reported precise mutagenesis in the protoplasts and hairy roots of soybean. CRISPR/Cas9 gene editing involving three soybean genes *Glyma06g14180*, *Glyma08g02290* and *Glyma12g37050* resulted in 3.2–9.7% and 14.7–20.2% mutation efficiencies for two different vector-marker systems.

CRISPR/Cas9 system has been effectively used to accelerate the development of new crop varieties by modifying the genomic regions that control major agronomic traits (Wang et al., 2016). Recently, Wang et al. (2016) established a tRNA-based multiplex gene editing strategy for hexaploid wheat. CRISPR/Cas9 components were assessed using the transient expression in the wheat protoplasts followed by next-generation sequencing of the targeted genomic regions. Multiple single guide RNAs (gRNAs) were evaluated for the ability to edit the homeologous copies of four genes affecting important agronomic traits. The study also demonstrated that the tRNA processing system of wheat can generate mutagenic gRNA molecules from a poly-*cis*tronic gene construct assembled using multiple gRNAs separated by the tRNA spacers. Random, mutagenesis together with the existing natural genetic variation have been conventionally used on a regular basis to bring in new traits into cultivated varieties. The Cas9/sgRNA system is a modern plant breeding approach which could be faster than the conventional methods in producing a null segregant line without the transgenic insert (Araki and Ishii, 2015; Woo et al., 2015). The CRISPR/Cas system helps in introducing specific alterations into a plant genome, which gets inherited and the transgene region is simply removed after target gene modification to make transgene free plants during crop variety improvement (Mahfouz et al., 2014; Xu et al., 2015). Xu et al. (2015) segregated out the transgene with self-fertilization in the T_1 generation resulting in transgene free rice lines.

SSNs can directly target and cleave the viral genome and therefore offers a promising approach for engineering resistance to single and multiple viral infections in plants (Chaparro-Garcia et al., 2015; Zaidi et al., 2016). Many recent reports have demonstrated the application of CRISPR/Cas9 system to confer molecular immunity against plant viruses (Ali et al., 2015b, Ji et al., 2015; Baltes et al., 2015; Iqbal et al., 2016). Baltes et al. (2015) performed transient expression of multiple SSNs including CRISPR/Cas9 for targeting bean yellow dwarf virus (BeYDV) based replicons. Results demonstrated greater magnitude of gene targeting by the viral mediated replicons as compared to *A. tumefaciens* T-DNA. Further, the CRISPR/Cas9 resulted in significant introduction of target mutations within viral genome and subsequent reduction in the viral copy number (Baltes et al., 2015). Ji et al. (2015) identified 49 candidate target sites within the small genome of the Gemini virus, beet severe curly top virus (BSCTV) for target mutagenesis using CRISPR/Cas9. Transient assays performed in *N. benthamiana* demonstrated that sgRNA-Ca9 construct introduced mutations at the targeted sites leading to the inhibition of the viral accumulation. The over expression of sgRNA-Cas9 in transgenic Arabidopsis and *N. benthamiana* revealed durable resistance against BSCTV (Ji et al., 2015). Ali et al. (2015a) developed a CRISPR/Cas9 construct with sgRNAs targeting open reading frames for viral Rep and coat proteins as well as non-coding intergenic regions (IR) from the origin of replication (ori) point for three Gemini viruses, tomato yellow leaf curl virus (TYLCV), beet curly top virus (BCTV), and Merremia mosaic virus (MeMV). The results demonstrated the development of viral variants with mutations in both coding and non-coding regions. However, CRISPR/Cas9 targeting intergenic sequences provided significant viral interference activity and limited generation of viral variants with growth and replication. This suggests that CRISPR/Cas9 mediated editing of non-coding regions of the viral genome could be critical in developing durable resistance strategies against multiple Gemini viruses. These studies suggest that CRISPR/Cas9 has a huge potential for targeting several sites in the viral genome for developing broad spectrum resistance against multiple viruses in crop plants.

The availability of multiple variants of Cas9 and an equal number of sgRNA expression cassettes has led to the development of CRISPR/cas9 tool kit for multiple functions in crop plants. Qi-Jun Chen and his group from the China Agricultural University, Beijing, China has developed a CRISPR/Cas9 binary vector set based on pCAMBIA backbone, as well as a gRNA (guide RNA) module vector set, as a toolkit for multiplex genome editing (Xing et al., 2014). The tool kit carries a maize codon optimized Cas9 together with one or more sgRNA to induce target mutagenesis using a single cloning step. The tool kit has proved to be highly efficient and has been validated through transformation using maize protoplasts as well as transgenic lines from maize and *A. thaliana*. Based on the demonstrated outcomes, the tool kit could facilitate expression of CRISPR/Cas9 in a variety of plant systems and introduce mutations or knock out numerous plant genes simultaneously. Recently, Lowder et al. (2015) have developed a molecular toolbox for comprehensive application of CRISPR/Cas9 in crop plants. The toolbox is characterized by the assembly of functional CRISPR/cas9 constructs for both monocots and dicots using the Gateway cloning and Golden Gate cloning method. This is capable of performing multiplexed gene editing and transcriptional activation or repression of endogenous genes in a wide variety of plants. The effectiveness of the CRISPR/cas9 toolbox has already been demonstrated in model plants such as Arabidopsis and tobacco and crop plants including, rice and maize (Lowder et al., 2015). These developments could be of immense help in widening the applicability of Cas9/sgRNA system toward improvement of crop plants for a variety of traits (Polstein and Gersbach, 2015; Zetsche et al., 2015).

9.5.1   CRISPER/CAS9 GENOME EDITING AND RICE IMPROVEMENT

Rice is a major staple crop in the family Poaceae, feeding more than half of the world's population. Besides, it is used a model plant among the monocots for multiple biological studies because of its small genome size and ease of genetic transformation. The conventional method of rice improvement takes approximately a decade time to generate new varieties via crossing and backcrossing during which new types of biotic and abiotic stresses crops in. Although, the *Agrobacterium* mediated transformation is well established in rice, it is still recalcitrant to a wide number of rice varieties and often results in only a few transformed lines with improved traits. In recent years, the development of SSNs has demonstrated to be highly useful for genome editing in rice. TALENs were used to create a mutation in the promoter region of the rice sucrose-efflux transporter gene (OsSWEET14), thereby preventing the transcriptional induction and consequently enhancing resistance to bacterial blight (Li et al., 2012). Engineered TALENs were also used to target and disrupt the *Oryzasativa*betaine aldehyde dehydrogenase 2 (OsBADH2) gene resulting in enhanced fragrance of rice due to increase in the accumulation of the major fragrance compound, 2-acetyl-1-pyrroline (2AP) (Shan et al., 2015). More recently, Shan et al. (2015) used the TALEN-mediated genomic modification strategy to generate knockout mutations in four rice genes, including *OsDEP1*, *OsBADH2*, cytokininoxydase 2 (OsCKX2) and *OsSD1*. However, as the process of vector construction and the repetitive property of the binding domain in TALENS are complicated, there are greater concerns regarding vector stability in plant genomes (Ma et al., 2015). On the other hand, the CRISPR/Cas9 strategy of generating target mutations based on base-pairing of the engineered sgRNAs to the target DNA sites is more efficient in developing knockout lines with improved traits and characters. There have been many reports suggesting the substantial use of CRISPR/Cas9 strategy for rice improvement during the last 4 years (Miao et al., 2013; Xu et al., 2014; Xu et al., 2016; Hu et al., 2016; Liang et al., 2016). A detailed list of rice genes targeted by CRISPR/Cas9 system is represented in Table 9.1.

In 2013, three groups independently demonstrated for the first time, the efficiency of the engineered CRISPR/Cas9 system for inducing target mutagenesis in rice (Feng et al., 2013; Shan et al., 2013; Jiang et al., 2013a). Feng et al. (2013) made use of the *S. pyogenes* derived Cas9driven by the CaMV 35S promoter while the chimeric sgRNA was under the control of the OsU6–2 promoter of rice. CRISPR/Cas9 construct was used to target three rice genes, *Rice Outermost Cell-specific gene5* (*ROC5*), *Stromal Processing Peptidase*

TABLE 9.1   List of Genes Targeted by CRISPR/Cas9 System in Rice (*Oryza sativa* L.).

Target genes from rice	sgRNA promoter	Cas9 promoter	Version of Cas9	References
OsGW2, OsGW5, OsTGW6	OsU3, OsU6, TaU3	Ubi	Rice codon-optimized	Xu et al. 2016
OsALS	OsU6	2×P35S	Rice codon-optimized	Endo et al. 2016
Os ERF922	Os U6a	Ubi	Rice codon-optimized	Wang et al., 2016
LPA1, LG1, GL1-1	Os U6a	2×35S	Rice codon-optimized	Hu et al. 2016
OsMPK5	OsU3/U6	CaMV 35S	Human codon-optimized	Xie et al. 2015
OsMPKs	OsU3	Ubi	Rice codon-optimized	Xie et al. 2015
46 rice genomic targets	OsU3/U6	Ubi/35S	Plant codon-optimized	Ma et al. 2015
OsPDS, Oso2g23823, OsMPK2	OsU3	2×P35S	Rice codon-optimized	Wang et al. 2015
OsCDKB2	Os U3	CaMV 35S	Rice codon-optimized	Endo et al., 2015
OsMSH1, OsMYB1, OsROC5, OsSPP, OsYSA, OsDERF1, OsEPSPS, OsPDS, OsPMS3,	OsU3	CaMV 35S	Rice codon-optimized	Zhang et al. 2014
OsSWEET11, OsSWEET14	OsU6	Ubi	Rice codon-optimized	Zhou et al. 2014
OsBEL	AtU6	2×P35S	Plant codon-optimized	Xu et al. 2014
OsSWEET11, OsSWEET14	OsU6	CaMV 35S	Wild-type SpCas9	Jiang et al. 2013b
OsROC5, OsSPP, OsYSA	AtU6	CaMV 35S	Human codon-optimized	Feng et al., 2013
OsPDS, OsBADH2, Oso2g23823, OsMPK2	OsU3	2×P35S	Rice codon-optimized	Shan et al. 2013
OsCAO1, OsLAZY	OsU3	Ubi	Rice codon-optimized	Miao et al. 2013

(*SPP*) and *Young Seedling Albino* (*YSA*). The transformed rice plants demonstrated homozygous or bi-allelic mutants implying that the DSBs were created by CRISPR/Cas in the early meristematic cell of the transgenic rice calli. RFLP analysis together with cloning and sequencing of bands from the transgenic plants showed the incidence of sequence mutations in the target regions with mutation frequency as high as 84% in the T_0 and T_1 rice lines (Feng et al., 2013). Shan et al. (2013) made use of single customized sgRNA to introduce target mutations in three rice genes, *OsBADH2, Os02g23823* and *OsMPK2*. The mutation frequency of the sgRNA: Cas9 edited rice genes was considerably higher than that induced by TALENs (Shan et al., 2013). In the same year, Jiang et al. (2013b) reported the use of CRISPR/Cas9/sgRNA construct to modify the regulatory sequence upstream of the bacterial blight susceptibility genes, OsSWEET14 and OsSWEET11. PCR amplification and sequencing of the OsSWEET14 in transgenic lines revealed the deletion of 9 nt from the promoter region of the OsSWEET14 gene. Likewise, the sequence analysis of the OsSWEET11 promoter revealed seven distinct mutations (deletions and substitutions) in the promoter region of the transgenic lines. These reports suggest that Cas9/sgRNA system could be successfully used for target gene modification toward the development of rice lines with improved noble traits.

Xu et al. (2014) reported gene targeting in rice via the *A. tumefaciens*-mediated CRISPR/Cas9 system. Three sgRNAs were designed to target diverse sites in the second exon of the Bentazon Sensitive Lethal (BEL, LOC_Os03g0760200) gene that confers resistance to bentazon and sulfonylurea herbicides in rice. Plant codon-optimized SpCas9 (pSpCas9) coding region, driven by a double 35S promoter (pd35s) was used in the experiment. A binary vector containing the sgRNA and Cas9 expression cassette was transferred into rice tissue by *A. tumefaciens* mediated stable transformation system. The mutagenic efficiency ranged from ~2 to ~16% and the phenotypic analysis revealed that CRISPR/Cas9 induced transgenic plants were sensitive to bentazon. This study indicates that sgRNA/Cas9 induced gene targeting can make use of the conventional transformation protocols for purposeful modification of agricultural traits in rice.

Analysis of epistatic relationships in genetic pathways or functional characterization of redundant gene family members requires the generation of mutants with multiple gene modifications. A conventional mutation experiment in rice involves several rounds of inter-crossing among plants carrying single mutations over a long period of time. CRISPR/Cas9 system provides an alternative means to generate mutations in multiple rice genes in a short time. Ma et al. (2015) generated multiple sgRNA cassettes that were cloned into CRISPR/Cas9 binary vectors using the Golden gate assembly system. As many as 46 rice genomic targets were mutated using this system at a mutation rate of 84.5%. Recently, Wang et al. (2015) developed a multiplex editing system based on the conventional isocadamers for concurrent editing of several genes in rice. A single binary vector was designed which contained *S. pyogenes* encoded Cas9 fused with three sgRNAs targeting phytoenedesaturase (OsPDS), Os02g23823 and OsMPK2 genes from rice plant. A rice *Agrobacterium*-mediated transformation with the binary vector resulted in 110 T_0 transgenic plants, majority of which showed mutations at the designated target sites of the three genes. More recently, Liang et al. (2016) used a new strategy to facilitate the congregation of multiple sgRNA cassettes in a single binary construct. They developed two modules to construct CRISPR/Cas9/sgRNA system that can assemble up to 10 sgRNA cassettes by the use of 11 different restriction enzyme sites. Transformed rice plants were significantly edited with deletion, substitution, insertion and inversion at 82% of the desired target sites (Liang et al., 2016). All these reports indicate that, the CRISPR/Cas9 system is highly efficient to generate multiple gene mutations using conventional strategy that could be subsequently used for the acceleration of rice breeding in future.

Although CRISPR/Cas9 editing in plants occur at a single locus causing deletions and/or insertions as exemplified in many species, it also has the potential to allow more complex genome rearrangement such as modification of the cis-regulatory region controlling gene expression or alteration of the non-coding RNA genes controlling plant development. Such study require the removal or replacement of the entire large regions of the chromosomes in a targeted manner. Zhou et al. (2014) reported a Cas9/sgRNA induced large chromosome segment deletion in rice. They used a set of facile Cas9/sgRNA vectors to facilitate multiplex targeting of four sugar efflux transporter genes (OsSWEET11, OsSWEET12, OsSWEET13 and OsSWEET14) in rice. The results revealed di-allelic mutations in T_0 transgenic rice plants at a frequency of 87–100%. These constructs can also be used for knock-out screening of the entire rice genome with sgRNA libraries, that can result in mutant rice populations with greater heritable variability and precision as compared to T-DNA or transposons insertion libraries. These results provided the first evidence of large chromosomal deletions (115–245 kb) in rice protoplasts using CRISPR/Cas9 suggesting the potentiality of the platform for target genome editing involving basic research as well as agricultural applications (Zhou et al. 2014).

CRISPR/Cas9 genome engineering can be directly used for developing disease resistance rice germplasm (Wang et al. 2016). The blast disease of rice caused by *Magnaporthe grisea* is one of the most destructive diseases affecting rice worldwide. As the conventional methods of rice improvement against blast is highly time consuming and labour intensive, the enhancement of pathogen triggered immunity (PTI) is considered the most effective approach for the development of blast resistant rice varieties (Wang et al., 2007; Bundo and Coca, 2016). RNA interference based down regulation of defense responsive transcription factors is also used for enhancing rice resistant to blast (Yara et al., 2007) but RNAi lines exhibit variable expressions and requires the screening of large populations over multiple generations. Moreover, RNAi lines are often considered as transgenic and are subjected to regulatory procedures which add into their cost of development (McGinnis, 2010). The recent development of SSNs could be a suitable alternative toward control of blast disease in rice. A recent report has shown that CRISPR/Cas9 SSN targeting the rice ethylene responsive factor 922 (OsERF922) gene resulted in enhanced resistance to *M. grisea* (Wang et al., 2016). ERFs are a family of transcription factors that are involved in the modulation of both biotic as well as abiotic stress tolerance (Muller and Munne-Bosch, 2015). Transformation of the rice protoplast with the C-ERF922-expressing vector resulted in twenty-one C-ERF922-induced mutant lines (42%) out of 50 T0 plants. Comparison of the T_2 lines with the wild type plants revealed no significant differences with respect to agronomic traits. However, the pathogen infection assay demonstrated a significantly lower rate of blast lesion development in all the mutant lines at both the seedling and tillering stages. This suggest that, CRISPR/Cas9 based gene modification could be used as a suitable alternative for the development of broad spectrum resistance against blast as well as other pathogens affecting rice productivity.

The major agronomic traits in rice such as grain yield, quality and stress responses are quantitative and controlled by multiple genomic loci. The conventional breeding of rice toward detection and development of quantitative trait loci (QTLs) is highly laborious and time consuming. Besides, the close linkage between the positive and the undesirable traits within the QTL results in severe technical lacunae for their separation. The availability of multiple SSNs including the latest CRISPR/Cas9 system provides a practical method to manipulate genomic targets to improve the quantitative traits in rice. Recently, Xu et al. (2016) made use of the CRISPR/Cas9 mediated multiplex genome editing system to enhance the grain weight in a widely compatible rice variety, LH422. Three genes from the grain weight controlling QTL, *grain width 2* (GW2), *grain width 5 (GW5)* and *thousand grain weight* 6 (TGW6) that negatively regulate grain weight in rice were used as mutagenic targets. An OsCas9-contained binary vector pHUN412 containing sgRNA fused with Ubiquitin promoter was transferred into rice tissue via *Agrobacterium*-mediated transformation. Twenty-one di-allelic mutant lines were generated with multiple mutations in all the three target genes. Quantitative phenotypic assessment of the T_2 double (*gw5tgw6*) and triple (*gw2gw5tgw6*) mutant lines revealed significantly increased grain size and thousand-grain weight (TGW) as compared to the wild type LH422. These findings suggested that pyramiding via Cas9/sgRNA editing significantly improved the grain yield in rice and this strategy could be further extended toward modification of QTLs controlling other important agronomic traits in rice.

A major problem associated with SSNs including the CRISPR/Cas9 system is the introduction of unexpected mutations due to off-target mutations (Kumar and Jain, 2014). Improper concentration in the Cas9: sgRNA ratio, presence of promiscuous PAM sites and insufficient Cas9 codon optimization are some of the factors responsible for undesired cleavage of the DNA regions (Song et al., 2016). However, this low specificity of target cleavage by CRISPR/Cas9 system can also be exploited to induce knockouts of homeologous and paralogous sequences. Recently, Endo et al. (2015) designed a single sgRNA to target multiparalogous sequences of cyclin-dependent kinase (CDK) genes in rice. The 20 bpsgRNA sequence possessed similarity to four rice CDK genes, CDKA1, CDKA2, CDKB1 and CDKB2 with variable number of mismatches. The Cas9/sgRNA transformed rice calli resulted in single, double and triple mutants of CDKA2, CDKB1 and CDKB2 using the same sgRNA. As the multigene knockouts of these cell cycle regulators has not been developed earlier, these rice mutants could be a suitable material to study the function and redundancy of the CDKs in plant system. Moreover, the genome information of many crop plants especially the allopolyploids is either not available or partially available and therefore they lag behind with respect to diploid crops. In this circumstance, the disruption of duplicated genes is possible even from a plant with partial genome information using the off-target mutations of CRISPR/Cas9.

9.7 FUTURE PROSPECTS AND RESEARCH OPPORTUNITIES

In the last 3 years, CRISPER/Cas has emerged as a promising genome editing tool in plants due to its simplicity, efficiency, amenability to multiplexing, versatility and minimal off target effects. Cas9/sgRNA system holds significant in generating crop varieties with precious new agronomic and nutritional traits for the benefit of farmers and consumers. The technology has been successfully used for targeted mutagenesis in rice and enhancing disease resistance, yield and quality improvement. This system gives an absolute understanding of gene functions in plants and also has a possibility to generate mutations in genes where T-DNA insertion or EMS mutant is not available. The technology can be effectively used for knocking-out whole gene families as multiplesgRNAs can be easily assembled into a single delivery vector. Large chromosomal deletions (up to 245 kb) generated with CRISPR/Cas9 has also been reported opening up opportunities for deleting gene clusters. Even though the off-target activity is negligible in plants as compared to animal systems, still there is a need to address the off-target mutations systematically in major crops and other plant species. The Cas9 variants have helped widening the range of sequences for genome editing in plants and could be widely used in basic plant research and crop breeding. The employment of paired Cas9 nickase with D10A Cas9 and two sgRNAs has been suggested to be highly effective in reducing the risk of off-target mutations (Ran et al., 2013). Recently, Woo et al. (2015) transfected preassembled complexes of purified Cas9 nickase protein and guide RNA into plant protoplasts of rice and other model crops and achieved targeted mutagenesis in regenerated plants at a frequency of 46%. These plants with single engineered point mutations were indistinguishable from natural varieties suggesting that Cas9 mutagenized crop plants can be easily accepted for agricultural use by the general public. Although CRISPR/Cas9 SSN as a genetic scalpel has impressed the crop biotechnologist for the last couple of years, its sustenance largely depends upon how we can address to its problem related to mutation efficiency, heritability of the mutations mediated by CRISPR/Cas9, off-target probability, influence on local chromatin structure, side-effects on linked genes and precise molecular mechanism for efficient delivery in plants.

9.8 SUMMARY AND CONCLUSION

Genome editing is a new tool that can be used to introduce targeted modifications into the genome by adding, removing or replacing specific target DNA sequences. The artificially engineered SSNs such as ZFNs, TALENs, and CRISPR-associated endonuclease Cas9 (CRISPR/Cas9) have proven to be highly effective for genome editing in a wide variety of organisms including plants. The CRISPR/Cas9 system is the most recently developed targeted genome modification system and seems to be more efficient, inexpensive, easy, user friendly and rapidly adopted genome editing tool. Rice is being the staple food for more than half of the world population demands more efficient methods for increasing its productivity in a short period of time. The development of genetically edited rice plant similar to those developed by conventional mutation breeding using the CRISPR/Cas9 system makes it a highly potential tool for providing sustainable rice productivity for feeding the ever growing population. Recently, the large-scale genome editing in rice using the CRISPR/Cas9 system has not only improved the yield and quality but also has enhanced the disease resistance ability. In this chapter, we briefly overviewed the structure and mechanism of the CRISPR/Cas9 system and its adoption in crop plants for various applications. We also described the broad application of the CRISPR/Cas9 targeted genome editing toward the improvement of agronomic traits in rice. This suggest that CRISPR/Cas9 genome editing is an affordable and elegant technology that can accelerate basic and applied research to enhance a wide variety of agronomic traits in crops plants in general and rice in particular.

ACKNOWLEDGMENT

SN is thankful for the award of Junior Research Fellowship (JRF) from the Dept. of Biotechnology, Govt. of India. We are thankful to Prof. Manoj Ranjan Nayak, President, Siksha O Anusandhan University, Bhubaneswar, India for his guidance and support. We also thank DST-FIST, Govt. of India, for the research infrastructure facilities provided to Centre of Biotechnology, Siksha O Anusandhan University.

KEYWORDS

- Agrobacterium tumefaciens
- Cas9
- CRISPR/Cas9
- crop improvement
- crRNA
- DNA
- double-stranded break
- genome editing
- knockouts The CRISPR/Cas Genome Editing System 257
- loss-of-function
- multiplex editing
- mutation
- non-homologous end-joining
- off-targets
- protospacer adjacent motif
- sgRNA
- site-specific nucleases
- sustainable agriculture
- TALENs
- targeted genome modification
- tracrRNA
- transgenic plants
- type II CRISPR–Cas system
- ZFNs

REFERENCES

Ali, Z.; Abulfaraj, A.; Idris, A.; Ali, S.; Tashkandi, M.; Mahfouz, M. M. CRISPR/Cas9-Mediated Viral Interference in Plants. *Genome Biol.* **2015a,** *16,* 238–249.

Ali, Z.; Abul-Faraj, A.; Li, L.; Ghosh, N.; Piatek, M.; Mahjoub, A.; Aouida, M.; Piatek, A.; Baltes, N. J.; Voytas, D. F.; Dinesh-Kumar, S.; Mahfouz, M. M. Efficient Virus-Mediated Genome Editing in Plants Using the CRISPR/Cas9 System. *Mol. Plant* **2015b,** *8* (8), 1288–1291.

Antunes, M. S.; Smith, J. J.; Jantz, D.; Medford, J. Targeted DNA Excision in Arabidopsis by a Re-Engineered Homing Endonuclease. *BMC Biotechnol.* **2012,** *12,* 86.

Araki, M.; Ishii, T. Towards Social Acceptance of Plant Breeding Through Genome Editing. *Trends Plant Sci.* **2015,** *20,* 145–149.

Baltes, N. J.; Hummel, A. W.; Konecna E; Cegan R; Bruns, A. N.; Bisaro, D. M.; Voytas, D. F. Conferring Resistance to Geminiviruses with the CRISPR-Cas Prokaryotic Immune System. *Nat. Plants* **2015,** doi: 10.1038/nplants.2015.145.

Boch, J.; Bonas, U. Xanthomonas AvrBs3 Family-Type III Effectors: Discovery and Function. *Annu. Rev. Phytopathol.* **2010,** *48,* 419–436.

Boch, J.; Scholze, H.; Schornack, S.; Landgraf, A.; Hahn, S.; Kay, S.; Lahaye, T.; Nickstadt, A.; Bonas, U. Breaking the Code of DNA Binding Specificity of TAL-Type III Effectors. *Science* **2009,** *326,* 1509–1512.

Bogdanove, A. J.; Schornack, S.; Lahaye, T. TAL Effectors: Finding Plant Genes for Disease and Defense. *Curr. Opin. Plant Biol.* **2010,** *13,* 394–401.

Bolotin, A.; Quinquis, B.; Sorokin, A.; Ehrlich, S. D. Clustered Regularly Interspaced Short Palindrome Repeats (CRISPRs) Have Spacers of Extra-Chromosomal Origin. *Microbiology* **2005,** *151,* 2551–2561.

Brooks, C.; Nekrasov, V. Lippman, Z. B.;Van Eck, J. Efficient Gene Editing in Tomato in the First Generation Using the Clustered Regularly Interspaced Short Palindromic Repeats/CRISPR-Associated9 System. *Plant Physiol.* **2014,** *166,* 1292–1297.

Bundo, M.; Coca, M. Enhancing Blast Disease Resistance by Overexpression of the Calcium-Dependent Protein Kinase OsCPK4 in Rice. *Plant Biotechnol. J.* **2016,** *14,* 1357–1367.

Butler, N. M.; Atkins, P. A.; Voytas, D. F.; Douches. D. S. Generation and Inheritance of Targeted Mutations in Potato (*Solanum tuberosum* L.) Using the CRISPR/Cas System. *PLoS One* **2015,** *10* (12), 1–12.

Cai, Y.; Chen, L.; Liu, X.; Sun, S.; Wu, C.; Jiang, B.; Han, T.; Hou, W. CRISPR/Cas9-Mediated Genome Editing in Soybean Hairy Roots. *Plo One* **2015,** *10* (8), 1–13.

Carroll, D. Genome Engineering with Zincfinger Nucleases. *Genetics* **2011,** *188,* 773–782.

Čermák, T.; Baltes, N. J.; Čegan, R.; Zhang, Y.; Voytas, D. F. High-Frequency, Precise Modification of the Tomato Genome. *Genome Biol.* **2015,** *16,* 232. Doi: 10.1186/s13059–015–0796–9.

Chaparro-Garcia, A; Kamoun, S.;Nekrasov, V. Boosting Plant Immunity with CRISPR/Cas. *Genome Biol.* **2015,** *16,* 254. Doi: 10.1186/s13059–015–0829–4.

Chen, H.; Lin, Y. J.; Zhang, Q. F. Riview and Prospects of Transgenic Rice Research. *Chin. Sci. Bull.* **2009,** *54,* 4049.

Chen, K.; Gao, C. Targeted Genome Modification Technologies and their Applications in Crop Improvements. *Plant Cell Rep.* **2014,** *33,* 575–583.

Christian, M. L.; Cermak, T.; Doyle, E. L.; Schmidt, C.; Zhang, F. et al., Targeting Dnadouble-Strand Breaks with Tal Effector Nucleases. *Genetics* **2010,** *186,* 757–761.

Christian, M.; Qi, Y.; Zhang, Y.; Voytas, D. F. Targeted Mutagenesis of Arabidopsis Thaliana using Engineered TAL Effector Nucleases (TALENs). *G3 (Bethesda)* **2013**, *3*, 1697–1705.

Curtin, S. J.; Zhang, F.; Sander, J. D.; Haun, W. J.; Starker, C.; Baltes, N. J.; Reyon, D.; Dahlborg, E. J.; Goodwin, M. J.; Coffman, A. P.; Dobbs, D.; Joung, J. K.; Voytas, D. F.; Stupar, R. M. Targeted Mutagenesis of Duplicated Genes in Soybean with Zinc-Finger Nucleases. *Plant Physiol.* **2011**, *156*, 466–473.

Deltcheva, E.; Chylinski, K.; Sharma, C. M.; Gonzales, K.; Chao, Y.; Pirzada, Z. A.; Eckert, M. R.; Vogel, J.; Charpentier, E. CRISPR RNA Maturation by Trans-Encoded Small RNA and Host Factor RNase III. *Nature* **2011**, *471*, 602–607.

Deng, D.; Yan, C.; Pan, X.; Mahfouz, M.; Wang, J.; Zhu, J. K.; Shi, Y.; Yan, N. Structural Basis for Sequence-Specific Recognition of DNA by TAL Effectors. *Science* **2012**, *335*, 720–723.

Doudna, J. A.; andCharpentier, E. Genome Editing. The New Frontier of Genome Engineering with CRISPR-Cas9.*Science* **2014**, *346*,1258096 doi: 10.1126/science.1258096.

Durai, S.; Mani, M.; Kandavelou, K.; Wu, J.; Porteus, M. H.; Chandrasegaran, S. Zinc Finger Nucleases: Custom-Designed Molecular Scissors for Genome Engineering of Plant and Mammalian Cells. *Nucleic Acids Res.* **2005**, *33*, 5978–5990.

Endo, M.; Mikami, M.; Toki, S. Multigene Knock-Out Utilizing off-Target Mutation so of the Crisper/Cas9 System in Rice. *Plant Cell Physiol.* **2015**, *56* (1), 41–47.

Endo, M.; Mikami, M.; Toki, S. Biallelic Gene Targeting in Rice. *Plant Physiol.* **2016**, *170*, 667–677.

Feng, Z.; Zhang, B.; Ding, W.; Liu, X.; Yang, D.; Wei, P.; Cao, F.; Zhu, S.; Feng, Z.; Mao, Y.; Zhu, J. Efficient Genome Editing in Plants Using a CRISPR/Cas System. *Cell Res.* **2013**, *23*, 1229–1232.

Feng, Z.; Mao, Y.; Xu, N.; Zhang, B.; Wei, P.; Yang, D.;et al., Multigeneration Analysis Reveals the Inheritance, Specificity, and Patterns of CRISPR/ Cas-Induced Gene Modifications in Arabidopsis. *Proc. Natl. Acad. Sci. U. S. A.* **2014**, *111*, 4632–4637.

Feng, C.; Yuan, J.; Wang, R.; Liu, Y.; Birchler, J. A.; Han, F. Efficient targeted genome modification in maize using CRISPR/Cas9 system. *J. Genet. Genom.* **2016**, *43*, 37–43.

Fichtner, F.; Urrea, C. R.;Ülker, B. Precision Genetic Modifications: A New Era in Molecular Biology and Crop Improvement. *Planta* **2014**, *239*, 921–939.

Gao, H.; Smith, J.; Yang, H.; Jones, S.; Djukanovic, V.; Nicholson, M. G.; West, A.; Bidney, D.; Falco, S. C.; Jantz, D.; Lyznik, L. A. Heritable Targeted Mutagenesis in Maize Using a Designed Endonuclease. *Plant J.* **2010**, *61*, 176–187.

Garneau, J. E.; Dupuis, M. E.; Villion, M.; Romero, D. A.; Barrangou, R.; Boyaval, P.; Fremaux, C.; Horvath, P.; Magaádan, A. H.; Moineau, S. The CRISPR/Cas Bacterial Immune System Cleaves Bacteriophage and Plasmid DNA. *Nature* **2010**, *468*, 67–71.

Hale, C. R.; Zhao, P.; Olson, S.; Duff, M. O.; Graveley, B. R.; Wells, L.; Terns, R. M.; Tern, M. P. RNA-Guided RNA Cleavage by a CRISPR RNA–Cas Protein Complex. *Cell* **2009**, *139*, 945–956.

Iqbal, Z.; Sattar, M. N.; Shafiq, M. CRISPR/Cas9: A Tool to Circumscribe Cotton Leaf Curl Disease. *Front. Plant Sci.* **2016**, *7*, 475.

Jacobs, T. B.; LaFayette, P. R.; Schmitz, R. J.; Parrott, W. A. Targeted Genome Modifications in Soybean with CRISPR/Cas9. *BMC Biotechnol.* **2015**, *15*, 16.

Ji, X.; Zhang, H.; Zhang, Y.; Wang, Y.; Gao, C. Establishing a CRISPR-Cas-Like Immune System Conferring DNA Virus Resistance in Plants. *Nat. Plants* **2015**, *1*, 15144.

Jia, Y. Marker Assisted Selection for the Control of Rice Blast Disease. *Pestic. Outlook* **2003,** *14*(4), 150–152.

Jiang, W.; Bikard, D.; Cox, D.; Zhang, F.; Marraffini, L. A. RNA-guided Editing of Bacterial Genomes Using CRISPR–Cas Systems. *Nat. Biotechnol.* **2013a,** *31,* 233–239.

Jiang, W.; Zhou, H.; Bi, H.; Fromm, M.; Yang, B.; Weeks, D. P.Demonstration of CRISPR/Cas9/sgRNA-mediated targeted genemodification in Arabidopsis, tobacco, sorghum and rice. **2013b,** *Nucleic Acids Res. 41*(20), e188.

Jinek, M.; Chylinski, K.; Fonfara, I.; Hauer, M.; Doudna, J. A.; Charpentier, E. A Programmable Dual-RNA-Guided DNA Endonuclease in Adaptive Bacterial Immunity. *Science* **2012,** *337,* 816–821.

Jinek, M.; Chylinski, K.; Fonfara, I.; Hauer, M.; Doudna, J. A.; Charpentier, E. A Programmable Dual-RNA-Guided DNA Endonuclease in Adaptive Bacterial Immunity. *Science* **2012,** *337,* 816–821.

Jinek, M.; Jiang, F.; Taylor, D. W.; Sternberg, S. H.; Kaya, E.; Ma, E.; Anders, C.; Hauer, M.; Zhou, K.; Lin, S.; Kaplan, M.; Iavarone, A. T.; Charpentier, E.; Nogales, E.; Doudna, J. A. Structures of Cas9 Endonucleases Reveal RNA-Mediated Conformational Activation. *Science* **2014,** *343*(6176), 1247–1297.

Khatodia, S.; Bhatotia, K.; Passricha, N.; Khurana, S. M.; Tuteja, N. The CRISPR/Cas Genome-Editing Tool: Application in Improvement of Crops. *Front. Plant Sci.* **2016,** *7,* 1–13.

Kumar, V.; Jain, M. The CRISPR–Cas System for Plant Genome Editing: Advances and Opportunities. *J. Exp. Bot.* **2014,** *4,* 1–11.

Li, Y.; Fan, C.; Xing, Y.; Jiang, Y.; Luo, L.; Sun, L.; Shao, D.; Xu, C.; Li, X.; Xiao, J.; He, Y.; Zhang, Q. Natural variation in GS5 plays an important role in regulating grain size and yield in rice. *Nat. Genet.* **2011,** *43,* 1266–1269.

Li, T.; Liu, B.; Spalding, M. H.; Weeks, D. P.; Yang, B High-Efficiency TALEN-Based Gene Editing Produces Disease-Resistant Rice. *Nat. Biotechnol.* **2012,** *30,* 390–392.

Li, Z. S.; Liu, Z. B.; Xing, A. Q.; Moon, B. P.; Koellhoffer, J. P.; Huang, L. X.; Ward, R. T.; Clifton, E.; Falco, S. C.; Cigan, A. M. Cas9-guide RNA directed genome editing in soybean. *Plant Physiol.* **2015,** *169,* 960–970.

Li, J.; Norville, J. E.; Aach, J.; McCormack, M.; Zhang, D.; Bush, J. Multiplex and Homologous Recombination-Mediated Genome Editing in Arabidopsis and Nicotianabenthamianausing Guide RNA and Cas9. *Nat. Biotechnol.* **2013,** *31,* 688–691.

Liang, Z.; Zhang, K.; Chen, K.; Gao, C. Targeted Mutagenesis in Zeamays Using TALENS and the CRISPR/Cas System. *J. Genet. Genomics* **2014,** *41,* 63–68.

Liang, Q.; Huashan, L.; Yunhan, J.; Chunsheng, D. The Molecular Mechanism of CRISPR/Cas9 System and Its Application in Gene Therapy of Human Diseases. *Yi Chuan* **2015,** *37,* 974–982.

Liang, G.; Zhang, H.; Lou, D.; Yu, D. Selection of Highly Efficient Sgrnas for CRISPR/Cas9-Based Plant Genome Editing. *Sci. Rep.* **2016,** *6,* 21451.

Liu, W.; Yuan, J. S.; Stewart, C. N. Advanced Genetic Tools for Plant Biotechnology. *Nat. Rev. Genet.* **2013,** *14,* 781–793.

Lowder, L. G.; Zhang, D.; Baltes, N. J.; Paul, J. W.; Tang, X.; Zheng, X.; Voytas, D. F.; Hsieh, T. F.; Zhang, Y.; Qi, Y. A CRISPR/ Cas9 Toolbox for Multiplexed Plant Genome Editing and Transcriptional Regulation. *Plant Physiol.* **2015,** *169,* 971–985.

Lusser, M.; Parisi, C.; Plan, D.; Rodriguez-Cerezo, E. Deployment of New Biotechnologies in Plant Breeding. *Nat. Biotechnol.* **2012,** *30,* 231–239.

Ma, X. L.; Zhang, Q.; Zhu, Q.; Liu, W.; Chen, Y.; Qiu, R.; Wang, B.; Yang, Z.; Li, H.; Lin, Y.; Xie, Y.; Shen, R.; Chen, S.; Wang, Z.; Chen, Y.; Guo, J.; Chen, L.; Zhao, X.; Dong, Z.; Liu, Y. A Robust CRISPR/Cas9 System Forconvenient, High-Efficiency Multiplex Genome Editing in Monocot and Dicotplants. *Mol. Plant* **2015**, *8*(8), 1274–1284.

Ma, X.; Zhu, Q.; Chen, Y.; Liu, Y. G. CRISPR/Cas9 Platforms for Genome Editing in Plants: Developments and Applications. *Mol. Plant* **2016**, *9*(7), 961–974.

Mahfouz, M. M.; Piatek, A.; Stewart, C. N. Genome Engineering Via Talens and Crisper/ Cas9 Systems: Challenges and Perspectives. *Plant Biotechnol. J.* **2014**, *12*, 1006–1014.

Makarova, K. S.; Grishin, N. V.; Shabalina, S. A.; Wolf, Y. I.; Koonin, E. V. A Putative RNA-Interference-Based Immune System in Prokaryotes: Computational Analysis of the Predicted Enzymatic Machinery, Functional Analogies with Eukaryotic RNAi, and Hypothetical Mechanisms of Action. *Biol. Direct.* **2006**, *1*, 7.

Makarova, K. S.; Haft, D. H.; Barrangou, R.; Brouns, S. J. J.; Charpentier, E.; Horvath, P.; Moineau, S.; Mojica, F. J. M.; Wolf, Y. I.; Yakunin, A. F.; van der Oost, J.; Koonin, E. V. Evolution and classification of the CRISPR–Cas systems. *Nat. Rev. Microbiol.* **2011**, *9*, 467–477.

McGinnis, K. M. RNAi for Functional Genomics in Plants. *Brief Funct. Genomics* **2010**, *9*(2), 111–117.

Miao, J.; Guo, D.; Zhang, J.; Huang, Q.; Qin, G.; Zhang, X.; Wan, J.; Gu, H.; Qu, L. J. Targeted Mutagenesis in Rice Using CRISPR-Cas System. *Cell Res.* **2013**, *23*, 1233–1236.

Mojica, F. J.; Diez-Villasensor, C.; Garcia-Martinez, J.; Soria, E. Intervening Sequences of Regularly Spaced Prokaryotic Repeats Derive from Foreign Genetic Elements. *J. Mol. Evol.* **2005**, *60*, 174–182.

Muller, M.; Munne-Bosch, S. Ethylene Response Factors: A Key Regulatory Hub in Hormone and Stress Signaling. *Plant Physiol.* **2015**, *169*(1), 32–41.

Nekrasov, V.; Staskawicz, B.; Weigel, D.;Jones, J. D.; Kamoun, S. Targeted Mutagenesis in the Model Plant Nicotianabenthamianausing Cas9 RNA-Guided Endonuclease. *Nat. Biotechnol.* **2013**, *31*, 691–693.

Pan, C. C.; Ye, 1.; Qin, L.; Liu, X.; He, Y.; Wang, X.; Chen, L.; Lu, G; CRISPR/Cas9-Mediated Efficientand Heritable Targeted Mutagenesis in Tomato Plants in the First and Later Generations. *Sci. Rep.* **2016**, *6*, 24765.

Paques, F.; Duchateau, P. Meganucleases and DNA Double-Strand Break-Induced Recombi- nation: Perspectives for Gene Therapy. *Curr. Gene. Ther.* **2007**, *7*, 49–66.

Polstein, L. R.; Gersbach, C. A. A Light-Inducible CRISPR-Cas9system for Control of Endogenous Gene Activation. *Nat. Chem. Biol.* **2015**, *11*(3), 198–200.

Pourcel, C.; Salvignol, G.; Vergnaud, G. CRISPR Elements in Yersinia Pestis Acquire New Repeats by Preferential Uptake of Bacteriophage DNA, and Provide Additional Tools for Evolutionary Studies. *Microbiology* **2005**, *151*, 653–663.

Qi, L. S.; Larson, M. H.; Gilbert, L. A.; Doudna, J. A.; Weissman, J. S.; Arkin, A. P.; et al., Repurposing CRISPR as an RNA-Guided Platform for Sequence-Specific Control of Gene Expression. *Cell* **2013**, *152*, 1173–1183.

Ran, F. A.; Hsu, P. D.; Lin, C.-Y.; Gootenberg, J. S.; Konermann, S.; Trevino, A. E.; Scott, D A.; Inoue, A.; Matoba, S.; Zhang, Y.; Zhang, F. Double nicking by RNA-guided CRISPR Cas9 for enhanced genome editing specificity. *Cell* **2013**, *154*, 1380–1389.

Reyon, D.; Khayter, C.; Regan, M. R.; Joung, J. K.; Sander, J. D. Engineering Designer Transcription Activator-Like Effector Nucleases (TALENs) by REAL or REAL-Fast assembly. *Curr. Protoc. Mol. Biol.* **2012a**, *100*, 12151–121514

Reyon, D.; Tsai, S. Q.; Khayter, C.; Foden, J. A.; Sander, J. D.; Joung J. K. FLASH Assembly of TALENs for High-Throughput Genome Editing. *Nat. Biotechnol.* **2012b,** *30,* 460–465.

Sander, J. D.; Joung, J. K. CRISPR-Cas Systems for Editing, Regulating and Targeting Genomes. *Nat. Biotechnol.* **2014,** *32*(4), 347–355.

Sander, J. D.; Cade, L.; Khayter, C.; Reyon, D.; Peterson, R. T.; Joung, J. K.; Yeh, J. R. Targeted Gene Disruption in Somatic Zebrafish Cells Using Engineered Talens. *Nat. Biotechnol.* **2011,** *29,* 697–698.

Shan, Q.; Wang, Y.; Li, J.; Zhang, Y.; Chen, K.; Liang, Z. et al., Targeted Genome Modification of Crop Plants Using a CRISPR-Cas System. *Nat. Biotechnol.* **2013,** *31,* 686–688.

Shan, Q.; Zhang, Y.; Chen, K.; Zhang, K.; Gao, C. Creation of Fragrant Rice by Targeted Knockout of the Osbadh2 Gene Using Talen Technology. *Plant Biotechnol. J.* **2015,** *13,* 791–800.

Shukla, V. K.; Doyon, Y.; Miller, J. C.; DeKelver, R. C.; Moehle, E. A.; Worden, S. E.; Mitchell, J. C.; Arnold, N. L.; Gopalan, S.; Meng, X.; Choi, V. M.; Rock, J. M.; Wu, Y. Y.; Katibah, G. E.; Zhifang, G.; McCaskill, D.; Simpson, M. A.; Blakeslee, B.; Greenwalt, S. A.; Butler, H. J.; Hinkley, S. J.; Zhang, L.; Rebar, E. J.; Gregory, P. D.; Urnov, F. D. Precise Genome Modification in the Crop Species Zea Mays Using Zincfinger Nucleases. *Nature* **2009,** *459,* 437–441.

Song, G.; Jia, M.; Chen, K.; Kong, X.; Khattak, B.; Xie, C.; Li, A.; Mao, L. CRISPR/Cas9: A Powerful Tool for Crop Genome Editing. *Crop J.* **2016,** *4,* 75–82.

Streubel, J.; Blucher, C.; Landgraf, A.; Boch, J. Tal Effector Rvd Specificities and Efficiencies. *Nat. Biotechnol.* **2012,** *30,* 593–595.

Sun, X.; Hu, Z.; Chen, R.; Jiang, Q.; Song, G.; Zhang, H.; Yajun, X. Targeted Mutagenesis in Soybeanusing the CRISPR-Cas9 System. *Sci. Rep.* **2015,** *5,* 10342.

Svitashev, S.; Schwartz, C.; Lenderts, B.; Young, J. K.; Cigan, A. M. Genome editing in maize directed by CRISPR–Cas9 ribonucleoprotein complexes. *Nat. Commun.* **2016,** *7*(13274), 1–7.

Svitashev, S.; Young, J. K.; Schwartz, C.; Gao, H.; Falco, S. C.; Cigan A. M. Targeted Mutagenesis, Precise Gene Editing, and Site-Specific Gene Insertion in Maize Using Cas9 and Guide RNA. *Plant Physiol.* **2015,** *169,* 931–945.

Townsend, J. A.; Wright, D. A.; Winfrey, R. J.; Fu, F.; Maeder, M. L.; Joung, J. K.; Voytas, D. F. High-Frequency Modification of Plant Genes Using Engineered Zinc-Finger Nucleases. *Nature* **2009,** *459,* 442–445.

Upadhyay, S. K.; Kumar, J.; Alok, A.; Tuli, R. RNA-guided genome editing for target gene mutations in wheat. *G3 (Bethesda)* **2013,** *3*(12), 2233–2238.

Verma, D. K.; Mohan M; Yadav, V. K.; Asthir, B.; Soni, S. K. Inquisition of Some Physico-Chemical Characteristics of Newly Evolved Basmati Rice. *Environ. Ecol.* **2012,** *30* (1), 114–117.

Verma, D. K.; Mohan, M.; Asthir, B. Physicochemical and Cooking Characteristics of Some Promising Basmati Genotypes. *Asian J. Food Agro Ind.* **2013,** *6* (2), 94–99.

Verma, D. K.; Mohan, M.; Prabhakar, P. K.; Srivastav, P. P. Physico-Chemical and Cooking Characteristics of Azad Basmati. *Int. Food Res. J.* **2015,** *22* (4), 1380–1389.

Voytas, D. F. Plant Genome Engineering with Sequence-Specific Nucleases. *Annu. Rev. Plant Biol.* **2013,** *64,* 327–350.

Voytas, D. F.; Gao, C. Precision Genome Engineering and Agriculture: Opportunities and Regulatory Challenges. *PLoS Biol.* **2014,** *12* (6), e1001877.

Wang, H.; Hao, J.; Chen, X.; Hao, Z.; Wang, X.; Lou, Y.; Peng, Y.; Guo, Z. Overexpression of Rice WRKY89 Enhances Ultraviolet B Tolerance and Disease Resistance in Rice Plants. *Plant Mol. Biol.* **2007,** *65*(6), 799–815.

Wang, Y.; Cheng, X.; Shan, Q.; Zhao, Y.; Liu, J.; Gao, C.; Qiu, J. L. Simultaneous Editing of Three Homoeoalleles in Haxaploid Bread Wheatconfers Heritable Resistance to Powdery Mildew. *Nat Biotechnol.* **2014,** *32,* 947–951.

Wang, S.; Zhang, S.; Wang, W.; Xiong, X.; Meng, F.; Cui, X. Efficient Targeted Mutagenesis in Potato by the CRISPR/Cas9 System. *Plant Cell Rep.* **2015,** *34*(9), 1473–1476.

Wang, F.; Wang, C.; Liu, P.; Lei, P.; Hao, W.; Gao, Y.; Guang, Y.; Zhao, K. Enhanced Rice Blast Resistance by CRISPR/ Cas9-Targeted Mutagenesis of the ERF Transcription Factor Gene OsERF922. *PLoS One* **2016,** *11*(4), e0154027.

Weber, E.; Gruetzner, R.; Werner, S.; Engler, C.; Marillonnet, S. Assembly of Designer TAL Effectors by Golden Gate Cloning. *PLoS One* **2011,** *6*(5), 1–5.

Wendt, T.; Holm, P. B.; Starker, C. G.; Christian, M.; Voytas, D. F.; Brinch-Pedersen, H.; Holme, I. B. TAL Effector Nucleases Induce Mutations At a Pre-Selected Location in the Genome of Primary Barley Transformants. *Plant Mo.l Biol.* **2013,** *83,* 279–285.

Wiedenheft, B.; Sternberg, S. H.; Doudna, J. A. RNA Guided Genetic Silencing Systems in Bacteria and Archaea. *Nature* **2012,** *482,* 331–338.

Woo, J. W.; Kim, J.; Kwon, S. I.; Corvalán, C.; Cho, S. W.; Kim, H. DNA-Free Genome Editing in Plants with Preassembled CRISPR-Cas9 Ribonucleoproteins. *Nat. Biotechnol.* **2015,** *33*(11), 1162–1164.

Xie, K.; Minkenberg, B.; Yang, Y. Boosting CRISPR/Cas9 Multiplexediting Capability with the Endogenous tRNA-Processing System. *Proc. Nat. Acad. Sci. U. S. A.* **2015,** *112,* 3570–3575.

Xing, H. L.; Dong, L.; Wang, Z. P.; Zhang, H. Y.; Han, C. Y.; Liu, B.; Wang, X. C.; Chen, Q. J. A CRISPR/Cas9 Toolkit for Multiplex Genome Editing in Plants. *BMC Plant Biol.* **2014,** *14,* 327.

Xu, R.; Li, H.; Qin, R.; Wang, L.; Li, L.; Wei, P.; Yang, J. Genetargeting Using the Agrobacterium Tumefaciens-Mediated CRISPR-Cassystemin Rice. *Rice* **2014,** *7* (5), 1–4.

Xu, R. F.; Li, H.; Qin, R. Y.; Li, J.; Qiu, C. H.; Yang, Y. C.; Ma, H.; Li, L.; Wei, P. C.; Yang, J. B. Generation of Inheritable and "Transgene Clean" Targeted Genome-Modified Rice in Later Generations Using the CRISPR/Cas9 System. *Sci. Rep.* **2015,** *5,* 11491.

Xu, R.; Yang, Y.; Qin, R.; Li, H.; Qiu, C.; Li, L.; Wei, P.; Yang, J. Rapid Improvement of Grain Weight Via Highly Efficient CRISPR/Cas9-Mediated Multiplex Genome Editing in Rice. *J. Genet. Genomics* **2016,** *43,* 529–532.

Yara, A.; Yaeno, T.; Hasegawa, M.; Seto, H.; Montillet, J. L.; Kusumi, K.; Seo, S.; Iba, K. Disease Resistance Against Magnaporthegrisea is Enhanced in Transgenic Rice with Suppression of Omega-3 Fatty Acid Desaturases. *Plant Cell Physiol.* **2007,** *48*(9), 1263–1274.

Zaidi, S. S. A.; Mansoor, S.; Ali, Z.; Tashkandi, M.; Mahfouz, M. M. Engineering Plants for Geminivirus Resistance with CRISPR/Cas9 System. *Trends Plant Sci.* **2016,** *21*(4), 279–281.

Zetsche, B.; Volz, S. E.; Zhang, F. A Split-Cas9 Architecture for Inducible Genome Editing and Transcription Modulation. *Nat. Biotechnol.* **2015,** *33,* 139–142.

Zhang, H.; Zhang, J.; Wei, P.; Zhang, B.; Gou, F.; Feng, Z.; Mao, Y.; Yang, L.; Zhang, H.; Xu, H.; Zhu, J. K. The CRISPR/Cas9 System Produces Specific and Homozygous Targeted Gene Editing in Rice in One Generation. *Plant Biotechnol. J.* **2014,** *12,* 797–807.

Zhou, H.; Liu, B.; Weeks, D. P.; Spalding, M. H.; Yang B Large Chromosomal Deletions and Heritable Small Genetic Changes Induced by CRISPR/Cas9 in Rice. *Nucleic Acids Res.* **2014,** *42,* 10903–10914.

CHAPTER 10

ROLE OF GENETIC ENGINEERING AND BIOTECHNOLOGY AS ADVANCED MOLECULAR TOOL AND TREND IN QUALITY IMPROVEMENT OF RICE CROP

AZIZ AHMAD[1,*], DEEPAK KUMAR VERMA[2,*], and PREM PRAKASH SRIVASTAV[2,3]

[1]School of Fundamental Sciences, Universiti Malaysia Terengganu, 21030, Kuala Nerus, Terengganu, Malaysia, Tel.: +609 6683176, Mob.: +6 19 3475443, Fax: +609 6683434

[2]Department of Agricultural and Food Engineering, Indian Institute of Technology, Kharagpur, West Bengal 721302, India, Tel.: +91 3222281673, Mob.: +91 7407170259, Fax: +91 3222282224

[3]pps@agfe.iitkgp.ernet.in, Mob.: +91 9434043426, Tel.: +91 3222-283134, Fax: +91 3222–282224

*Corresponding author. E-mail: aaziz@umt.edu.my; deepak.verma@agfe.iitkgp.ernet.in, rajadkv@rediffmail.com

10.1 INTRODUCTION

The global rice demand is continuously increased. Asia is the largest rice consumer with more than 90% of the world's rice consuming population. It was estimated that by 2035 the milled rice production should be increased up to 105 million t to satisfy the growing population of world rice consumers (IRRI, 2017). Nonetheless, the statistical data from 1990 to 2000s shows a trend of decline, from 2.73 to 0.88% (Zhu et al., 2010).

This phenomenon has been contributed by the urbanization, rapid decline of arable land, and climate change (Zhao and Fitzgerald, 2013). Despite yield, rice grain quality is an important trait that influence consumer's acceptance and marketable rice (Verma et al., 2012; 2013, 2015). Generally, rice quality refers to the milling quality of rice, the components of milling outputs, physical appearances before and after cooking, and nutritional quality (Bhattacharya, 2011; Verma et al., 2012, 2013, 2015; Verma and Srivastav, 2017). Rice breeding programs now require a new paradigm. Knowledge in advance biotechnology, genomics, and techniques in molecular biology permit researchers to identify and characterize many valuable traits in rice. For instance, "Super Rice" is developed by a rice plant breeder at International Rice Research Institute (IRRI) in Philippines which produces more grain and less stem in comparison to traditional rice varieties as well as first high-yieldng type rice variety (Fig. 10.1; Malcolm, 1998). The genotyping of target genes, genetic markers linked to genes of interest have been extensively exploited to facilitate the understanding of complex genetic traits and selection during rice breeding. IRRI maintains the collections of rice germplasm at International Rice Genebank (IRG), established in 1977, just after foundation of IRRI in 1960 (Jackson, 1997). The germplasms of rice are most genetically diverse and complete (Table 10.1). The markers are based on sizes of DNA products, such as cleaved amplified polymorphic sequences (CAPS) markers, marker-assisted selection (MAS), simple sequence repeats (SSR), and small insertions and deletions (InDels). MAS allows precise marker-assisted backcrossing (MABC) of transfer desirable loci into breeding lines (Xu et al., 2012), enhances genetic pool of environmental dependent-phenotypes (Moose and Mumm, 2008), and the selection accuracy (Zhou et al., 2016). Functional markers are generated on the basis of functional characterization of polymorphisms of genes that target trait variation (Andersen and Lubberstedt, 2003); it offers an efficient fixation of alleles in populations. In addition over the years, quantitative trait loci (QTL) mapping and association analysis utilizing a varieties of mapping populations; F_2, recombinant inbred lines (RILs), backcross and doubled haploid (DH) have generated a lot of information about rice traits.

Genetic engineering is an important instrument in plant breeding programs. The system allows the insertion of traits encoding gene (s) without disturbing the required features of an elite genotype (Fig. 10.2). The advances and efficiency of *Agrobacterium*-mediated transformation method in Chapter 7 in this book entitled as "*Agrobacterium*-mediated genetic transformation practices for improvement of rice quality and production" displays great development on rice genetic engineering and biotechnology. Despite the advances in scientific and plant genetic transformation technologies, the drawback of many transgenic plants is that they fail to display the expected transgene expression. Gene expression is largely influenced by the epigenetic effects and post-transcriptional gene silencing (Hoang et al., 2016). The presence of foreign gene or additional copies of endogenous genes may trigger gene silencing. Transgene silencing could be overcome by artificial chromosomes, expression of extrachromosomal episomal and viral silenicng suppressors, or enclosure of intron of related gene. The genome editing technology permits the transformation process to be integration free or at the directed integration location for targeted mutagenesis (Khan et al., 2016; Xu et al., 2016a, b; Zhou et al., 2017).

FIGURE 10.1 (See color insert.) Biotechnology in the development of rice variety A comparative view of grain and stem of (A) Traditional type rice plant, and (B) High-yieldng type, *vis-n-vis* (C) Super Rice plant.

Source: Modified from Khush (2001).

TABLE 10.1   Origin of the Rice Accessions in the International Rice Genebank Collection at IRRI.

Country	Accessions
Bangladesh	5 923
Cambodia	4 908
China	8 507
India	16 013
Indonesia	8 993
Lao PDR	15 280
Malaysia	4 028
Myanmar	3 335
Nepal	2 545
Philippines	5 515
Sri Lanka	2 123
Thailand	5 985
Viet Nam	3 039
7 countries with > 1000 and < 2000 accessions	10 241
105 countries < 1000 accessions	11 821
Total	108 256

Sources: From GOI (Government of India). Taxonomy, Geographic Origin and Genomic Evolution. In: Biology of *Oryza saiva* L. (Rice). Series of Crop Specific Biology Documents. Department of Biotechnology, Minister of Science and Technology and Minister of Environment and Forests, Government of India, India, 2011, pages 3–10.

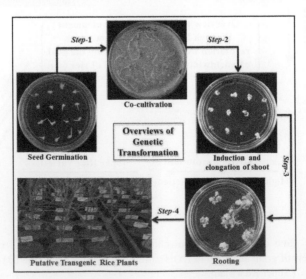

FIGURE 10.2   (See color insert.) Genetic Transformation in the developmemt of transgenic rice plant.

Source: Adapted from Kim et al. (2012).

10.2   CHEMICALS AND IRRADIATIONS-INDUCED MUTANT

The implementation of induced mutant technology on rice unfastens new opportunity such as enrichment of the genetic resources (Wu et al., 2005), gene mapping (Yang et al., 2003, 2009; Liu et al., 2007; Xiao et al., 2011), isolation of isogenic lines (Xiao et al., 2009), and producing mutant population. Worldwide, more than 500 rice varities have been developed from direct induce mutant or derivative of mutant breeding lines (Jeng et al., 2012). IRRI alone generates more than 60,000 IR64 mutants using chemicals and irradiations (Wu et al., 2005). Table 10.2 summarizes the physionomies, physological, grain quality, and nutritional properties of few induced mutant rice varities. These mutant populations are ready for manipulation and genetic improvement breeding.

The most widely cultivated mutant rice cultivar is the imidazolinone herbicide resistance (Andres et al., 2014). It is the breeding line of imidaz-olinone-tolerant rice line developed in early 1990s (Croughan, 1994). This cultivar was patented as the "Clearfield®" rice technology by BASF Agro-chemical Products. This technology is now widely adopted to control the weedy rice that causes up 60% yield losses for inpuddled transplanted rice and up to 80% in direct seeding rice (Dass et al., 2016). The primary target site of the imidazolinones (IMIs) class of herbicides is the acetohydroxyacid synthase (AHAS) or acetolactate synthase (ALS) enzyme that catalyzes the first step in the biosynthetic pathways of amino acid leucine, isoleucine, and valine (Tan et al., 2005; Livore et al., 2010, 2015).

Point mutation is target systems that have been widely implemented on rice either by random or specific target gene sequence. The random point mutation is usually induced using the mutagenic agents, chemicals, or irradiations. Chemical mutagenesis efficiently generates phenotypic distinc-tions in homogeneous genetic backgrounds permitting genes functional analysis. The commonly used mutagenic agents are the sodium azide (NaN_3) (Tai, 2007; Gruszka et al., 2012; Jeng et al., 2012; Chi et al., 2014), ethyl methanesulfonate (EMS) (Jeng et al., 2011; Mohapatra et al., 2014; Lima et al., 2015), diepoxybutane (Wu et al., 2005), and fast neutron and gamma ray (cobalt-60 or cesium-137) irradiations (Wu et al., 2005; Radhamani et al., 2015a,b; Joshi et al., 2016). Other chemicals that used as mutagens including the antibiotics (streptonigrin), alkylating agents (acridines, azide, epoxides, ethylenamines, hydroxylamine, lactones, nitrogen mustards, nitrous acid, sulfates, sulfonates, sulfones, sulfur mustards), 5-bromo-uracil and 8-ethoxy caffeine. Other irridations are the x-rays, neutrons from uranium-235, β-radiation from phosphorus 32 or carbon 14, and ultraviolet

TABLE 10.2   Physionomies and Phytochemical Varitions of Induced Mutant Rice.

Agents	Characteristics	References
NaN$_3$	The SA-586 mutants have variations in the levels of lipid, protein, total anthocyanins, total flavonoids, total phenolics, total proanthocyanidins, total tocopherols, total tocotrienols and total γ-oryzanol varied among. The brans of mutants M-18, M-50 and M-56 contains higher γ-oryzanol, proanthocyanidins, and vitamin E than the wildtype (SA-586), while M-54 is rich in Fe.	Jeng et al., (2011)
	The calcium (Ca), copper (Cu), iron (Fe), magnesium (Mg), manganese (Mn), phosphorus (P), potassium (K), and zinc (Zn) contents varied among the mutants. The mutants M-IR-58 and M-IR-75 accumulates high content of Fe, while the M-IR-49, M-IR-175 and M-IR-180, accumulates high content of Zn.	Jeng et al., (2012); Gruszka et al. (2012)
	The mutant developed exhibits herbicide resistancy toward an effective amount of an imidazolinone (C$_3$H$_4$N$_2$O) herbicide, a sulfonylurea herbicide, a pyrimidinyloxybenzoate herbicide, a triazolopyrimidine herbicide, and a sulfonylamino-carbonyltriazolinone herbicide, or mixture there of to the weeds and rice plant	Rizwan et al. (2015); Livore et al. (2015; 2010)
	The mutant SA419 exhibits rapid grain filling than it wildtype the Tainung 67.	Jeng et al. (2003a,b)
Ethyl methane sulfonate (EMS)	The "ewst1" a mutant of Nagina22 (an upland rice variety) exhibits water stress tolerance with increasing the root length, relative ewater content, membrane cell stability and chlorophyll content. The mutant seed also exhibits higher germination rate in PEG inclusion medium.	Lima et al. (2015); Mohapatra et al. (2014)
γ-ray irradiation	The IR64-mutant exhibits a salinity tolerance from EC6 dS/m to EC14 ds/m. The mutant also exhibits better harvest index than the wild type, panicle length, plant height, shoot and root weight, and tiller number, total biomass. Mutants D100–209 and D100–211 produces higher yield (18 and 34%) than the WT plants, when grow on salinity EC10 dS/m	(Joshi et al., 2016).
	A total of 652 mutants were generated from the variety of ADT (R)-47. Mutant shows very high genetic variability; high heritability coupled with high genetic advance in the number of productive tillers per plant and single plant yield, high heritability with moderate genetic advance on the panicle weight and 100 grain weight and high heritability coupled with low genetic advance were on the days to 15% flowering, plant height, panicle length.	Radhamani et al. (2015a,b)

TABLE 10.2 *(Continued)*

Agents	Characteristics	References
	A mutant developed from the cv. "Dongan," exhibits resistancy to a tryptophan analogue, the 5-methyltryptophan. The irradiation had altered the negative feedback by anthranilate synthase (AS) in tryptophan biosynthesis, causes tryptophan over-accumulation. Thus producing rice lines with high tryptophan.	(Chun et al. 2012)
	M4 of mutant of IET 1412 and IET14143 derived from Tulaipanja exhibits genetic variability for yield related traits. Four mutants had increased grain yield, which was contributed by the number of panicle, spikelet, grains, and weight of grains. The IET13541 a mutant of aromatic rice "Gobindabhog" after treated with highest dose of 450 Gy exhibits alteration in the flowering time. A moderate dose of 350 Gy formed two clusters (Clusters II and IV). Cluster III comprising one mutant of Pigmented Mutant and two mutants of genotype IET13541.	Kole and Chacraborty (2012)

at 2500–2900 nm (Livore et al., 2010). Chemical and irradirion mutagenesis generates a large numbers of rare mutations that enable for selection of valuable traits, phenotypes and characterization of functional genes. The commonly observed morphological dissimilarities are the plant architecture, growth habit, pigment development and physiological traits at vegetative or reproductive stages. In most cases, the NaN_3-induced mutant exhibits decrases in plant heights or generates semi-dwaft traits (Jeng et al., 2011).

The mutation of target DNA may be detected using the next-generation sequencing (NGS)-based target enrichment methods (Jeng et al., 2012) or integration of multiplex, semi-nested PCR combined with GNS library construction (Che et al., 2014). The targeting-induced local lesions in genomes (TILLING) is used to exploit the variations in induced mutant for gene discovery, functional studies, and crop improvement (Casella et al., 2013). The efficiency of induced mutant is comparable with those of insertional mutagenesis. To further increase it efficacy a cataloguing systematic for mutations should be developed. For example, Henry et al. (2014) utilised the multiplexed global exome capture and sequencing coupled with bioinformatics tools to identify large-scale-induced mutation resources. From 72 independent M2 individuals, they managed to identify 18,000 induced mutations and recover more than 2600 genes potentially deleted (Henry et al., 2014).

10.3 TARGETED GENOME ENGINEERING

Targeted genome engineering or targeted gene mutagenesis is a robust technology that useful to facilitate genetic improvement of rice. The system permits direct manipulation of the genetic sequence either delete, insert or replace of the DNA (Li et al., 2014a; 2016a,b). The genome editing was first popularized in early 2000s, with homologous recombination (HR)-mediated targeting to generate knock-in or knock-out mutants (Capecchi et al., 1989). Among the earlier reports on the uses of HR-mediated targeting on rice was by Johzuka-Hisatomi et al. (2008). They managed in obtaining a fertile transgenic rice plant with the *Waxy* or *Adh2* gene distrupted by the insertion of the *hpt* gene on embryogenic rice calli using *Agrobacterium*-mediated transformation. Nonetheless, HR-mediated targeting has low efficiency, requires an entensive work and time consuming for mutant selection (Gaj et al., 2013). As alternative, a rapid, inexpensive and high-throughput gene knockdown by RNA interference (RNAi) targeted have been introduced (McManus and Sharp, 2002). However, knockdown by RNAi is incomplete

and inpractical due to unpredictable off-target effects, and offers a temporary inhibition of gene function, unable to link phenotype to genotype (Gaj
et al., 2013). Comparatively, the SSN-based genome editing can achieve
complete knockout without incorporating exogenous DNA (Wang et al.,
2016a). Advances in genome editing technology had released the Zinc-
finger nucleases (ZFNs) in year 1998 (Beerli et al., 1998), transcription-like
effector nucleases (TALENs) in year 2009 (Boch et al., 2009; Moscou and
Bogdanove, 2009) and recently the clustered regulatory interspaced short
palindromic repeats -associated (CRISPR/Cas system) in year 2013 (Shan
et al., 2013).

TALENs are fusion protein of TALE or TAL effector as DNA binding
domain and a nonspecific *Fok*I endonuslease (Chen et al., 2014; Li et al.,
2014a; 2012a). The system permits the insertion of determined chromosomal loci with DNA double-strand breaks (DSBs). This results in error-
prone non-homologous end joining (NHEJ) lead to loss or modification
of gene function (Chen et al., 2014). A large variety of heritable mutations
in rice was regenerated by TALEN-mediated targeted mutagenesis (Zhang
et al., 2015). In Figure 10.3, overview of TALEN-mediated targeted mutagenesis is well elaborated (Chen et al., 2014). The protocol for rice genome
modification, based on TALEN, has been reviewed (Khan et al., 2016; Li
et al., 2014a). Table 10.2 shows target genes and rice traits that have been
modified using TALENs.

Rice genetic engineering requires a robust technology such as the
CRISPR/Cas system. The complex of Cas9 nucleases and RNA-guided
(gRNA) allows a specific cleavage of genomic DNA and complimentary
binding of targeted sequence (Xie and Yang, 2013; Jiang et al., 2013). Off-
target editing can be avoided by locating the target sequence (TS) adjacent
to the 5'-NGG protospacer adjacent motif (PAM) sequence. In 2014, Naito
et al. (2014) had developed the CRISPRdirect, a soft ware that aid users to
locate the target sequence and reduce the off-target sites (Naito et al., 2014).
CRIPR/Cas has few advantgaes over ZFns and TALENs; increases number
of potential recombination sites (Zheng et al., 2016a), improved delivery due
to the small size of the guide RNAs, enables multiple different genomic loci
or multiplexing genom editing due to target specificity (Endo et al., 2014;
Zhou et al., 2014; Lowder et al., 2015; Mikami et al., 2016; Wang et al.,
2015a; Xu et al., 2016b), efficiency of selection by complex quantitative
traits, and full-genome genotyping (Khlestkina and Shumny, 2016). In addition, it is a promising tools in the production of inheritable and "transgene
clean" targeted genome-modified rice in the T_1 generation (Xu et al., 2015).

The main challenge of the CRISPR-Cas technology in targeting non coding gene in plants are the quest for an effective delevery system, off-target mutations and targeting miRNA and lncRNAs (Basak and Nithin, 2015).

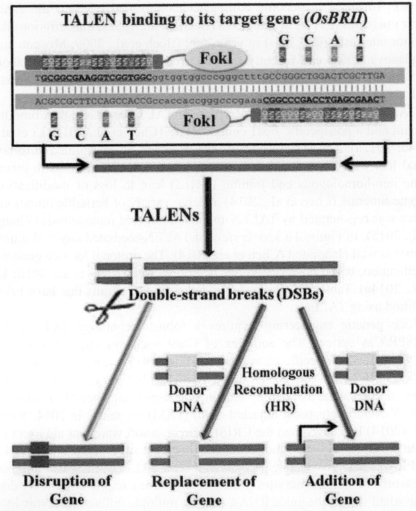

FIGURE 10.3 (See color insert.) Overview of TALEN-mediated targeted mutagenesis (Adapted from Chen et al., 2014) (Note: For details on targeted mutagenesis mediated by TALEN, reader should refer to paper of Chen, K.; Shan, Q. and Gao, C. An efficient TALEN mutagenesis system in rice. *Methods*, 2014, 64 (1): 2–8).

Efforts have been made to improve the efficiency of CRISPR/Cas9 genome editing tools to edit endogenous rice genes (Mazumdar et al., 2016).

These includ the selection of highly efficient sgRNAs (Liang et al., 2016) and single-strand conformational polymorphism (SSCP) (Zheng et al., 2016a) in delivering the targeted mutagenesis in rice, *Agrobacterium*-mediated transformation of SSN expression constructs in practical (Endo et al., 2016). The cas9 expression cassette also influence the successful of CRISPR/Cas9-mediated targeted mutagenesis (Mikami et al., 2015a) and culture period of rice callus (Mikami et al., 2015b). The *Agrobacterium tumefaciens*-mediated CRISPR/Cas9 system enables to generate an efficient target rice gene and creates a specific rice mutant (Xu et al., 2016a; Xu et al., 2014; Zhang et al., 2014; Jiang et al., 2013). Among rice traits that have been explored using these genome editing system are herbicide resistant, fungus resistant, yield and grain quality, and plant architecture. Nevertheless, protein coding genes are the focus of all these experiments (Table 10.3).

10.4 GENETIC TRANSFORMATION OF C$_4$ PHOTOSYNTHESIS IN RICE

The rice growth and yield is mainly depends on the photosynthesis. Photosynthesis is a major process that generated most of the organic compounds, which assimilates CO_2 from environment. Theoretically, plants that efficiently assimilated CO_2 will exhibit better growth and yield produces. Rice is a C$_3$ photosynthesis (C$_3$) plants that utilised the ribulose-5-phosphate decarboxylase (Rubisco) for CO_2 assimilation in the chloroplast of mesophyll cell. On the other hand, the C$_4$ photosynthesis (C$_4$) plants employed the phosphoenol-pyruvate carboxylase (PEPC) to fix the CO_2 in the cytosol of mesophyll cells. Subsequently, fixed CO_2 is transported to the Benson-Calvin cycle proceed in the chloroplast of bundle-sheath cells. Furthermore, PEPC is very sensitive event to the lower concentration of CO_2. To increase the photosynthesis and finally lead to yield improvement, establishment of C$_4$-engineered rice is a strategic approach. Since, the conventional breeding methods are impossible to transfer the C$_4$ photosynthesis characteristics to rice. Advance in genetic engineering and biotechnology has made this dream possible and achievable. The first engineered C$_4$ rice was reported by Ku and co-workers (Ku et al., 1999). The gene encoding PEPC from maize was successfully transformed and highly expressed in the *japonica* rice cultivar "Kitaake". Subsequently, many engineered rice with C4 photosynthesis related genes have been developed (Wang et al., 2016b; Lian et al., 2014; Miyao et al., 2011; Kajala et al., 2011) and comparative studies on traits compared to the wild-type have been reports (Table 10.4).

TABLE 10.3 The Improved Rice Traits Using the Targeted Genome Engineering or Targeted Gene Mutagenesis System.

Genes	Traits	System	References
OsALS	Herbicide resistant	TALEN	Li et al (2016c).
OsEPSPS	Glyphosate resistance	TALEN	Wang et al. (2015b)
OsSWEET13/Os11N3	bacterial blight susceptible	TAL	Zhou et al. (2015a); Li et al. (2012a)
waxy gene	Grain quality	TALEN	Nishizawa-Yokoi et al. (2016)
LOXs	Grain quality/seed shelf-life	TALEN	Ma et al. (2015)
OsBADH2	Grain quality/fragrance rice	TALEN	Shan et al. (2015)
BADH2 CKX2 and DEP1	Grain quality/grain weight	(CRISPR)/Cas9	Xu et al. (2016)
TGMS	Thermo sensitive	(CRISPR)/Cas9	Zhou et al. (2016)
GRAIN SIZE3 (GS3) and Gn1a	Grains yield	CRISPR/Cas9	Shen et al. (2016)
Gn1a, DEP1, GS3, and IPA1	Grain quality and plant architecture	CRISPR/Cas9	Li et al. (2016d)
DROOPING LEAF (DL) gene	Plant architecture	CRISPR/Cas9	Ikeda et al. (2016)
OsPDS	Plant architecture/albino	CRISPR/Cas9	Ishizaki, (2016)
OsEPSPS	Herbicide resistance	CRISPR/Cas9	Li et al. (2016c)
OSSPS and OsSPS	Grain quality	CRISPR/Cas9	Hashida et al. (2016)
OsALS	Herbicide resistant	Cas9 nickase with paired gRNAs	Mikami et al. (2016); Sun et al. (2016)
OsROC5 and *OsDEP1*	Plant architecture	CRISPR/Cas9	Zheng et al. (2016a)
OsBEIb	Grain quality	CRISPR/Cas9	Baysal et al. (2016)
SDH	Blast resistance	CRISPR/Cas9	Arazeo et al. (2015)
NAL1	Architecture	CRISPR/Cas9	Hu et al. (2016)
JAZ9	Jasmonic signalling cascade	CRISPR/Cas9	Jang et al. (2016)
OsERF922	Blast resistance	CRISPR/Cas9	Wang et al. (2016a)

TABLE 10.4 Effects of Overexpression of Phosphoenolpyruvate Carboxylase Gene Family on Physiognomies of C_4-Engineered Rice Lines.

Improved Traits	Description	References
Photosynthetic rate, high temperature tolerance	Investigation on immature embryos (8–12 days) of *indica* rice cultivar line IR68899B transformed with maize *pepc* gene, driven by its own promoter using the particle bombardment method. Four selection of transgenic embryogenic cali on 50-mg/L hygromycin with a duration of 14-days per cycle. 9% out of 300 transgenic plants shows irregularity in morphology and physiology traits. The engineered rice exhibited higher photosynthetic rate and drought tolerance.	Bandyopadhya et al. (2007)
Grain-filling rates, total grain numbers	Investigation on *Oryza sativa* cv. Kitaake transformed with maize C_4pepc (PC) by Ku et al (1999). The transgenic (PC) exhibits higher dry weights of the leaf, panicle, stem and sheath per plant. The starch synthesis enzyme; ADPGPase during the flowering stage is higher in PC than in the WT. The transgenic rice demonstrats a small difference in starch content among them	Lian et al. (2014); Ding et al. (2013)
Resistance towards nitric oxide, photosynthetic rate	Investigation on *Oryza sativa* cv. Kitaake transformed with maize C_4pepc (PC) by Ku et al (1999). Both transgenic and control (WT) were treated with NO donors, an NO scavenger, phospholipase inhibitors, a Ca^{2+} chelator, a Ca^{2+} channel inhibitor, and a hydrogen peroxide (H_2O_2) inhibitor. The NO donors increased the photosynthetic rate (P_N), phospholipase D and Ca^{2+} levels in PC than wild-type (WT). Phospholipase inhibitors and a Ca^{2+} chelator decrease the photosynthesis in WT stronger than in PC. NO donors' increases endogenous levels of NO in WT higher than in PC. NO is involved in the complex control of PEPC activity in PC leaves.	Chen et al. (2014);
Photosynthetic rate, yield and oxidative stress	Investigation on *japonica* rice (*Oryza sativa* L.) cultivar Zhonghua 8 transformed with P_{rbcS} rice Rubisco small sub-unit promoter using the *Agrobacterium*-mediated transformation protocol. Selection was based on *GUS* β-glucuronidase and *PMI* phosphomannose isomerase. The transgenic lines (PC) display higher net photosynthetic rates than the wild-type lines. PC also show different phenotype under upland conditions, but same in wetland fields. Photosynthesis and yield are improved under drought situation.	Ding et al. (2012; 2013); Shen et al. (2015)
Photosynthetic rate, oxidative stress	Investigation on transgenic rice was previously developed by Ku et al (1999). The activities of superoxide dismutase, catalase, and ascorbate peroxidase of the transgenic plants higher than in wild-types, both in intact plants and cell suspension.	Ren et al (2014); Qian et al. (2015b); Huo et al. (2016)

TABLE 10.4 (Continued)

Improved Traits	Description	References
Grain yield, biomass, drought tolerance, nitrogen-use and water-use efficiency	Investigation on rice cultivar Kitaake transformed with combination with maize C_4-specific phosphoenolpyruvate carboxylase (PCK) or maize-specific pyruvate orthophosphate dikinase (PPDK) singly. Transgenic plants (PCs) exhibits higher photosynthetic rate and produces higher grain yields in three moisture conditions; moderate drought (MD), severe drought (SD) and well-watered (WW). Increases in the phosphoenolpyruvate carboxylase (PEPC) and carbonic anhydrase (CA) activity in transgenic plants. PCs also exhibits higher stomatal conductance, leafwater content and oxidative activity in all moisture conditions. MDA content increases with increasing drought level and leaf age, the volumes of root exudates, root oxidation activity (ROA), biomass, and grain yield of transgenic plants were greater than WT plants.	Gu et al. (2013); Qian et al. (2015a)
Photosynthesis rates, grain number, high-temperature tolerance	Investigation on indica rice variety Hang2 transformed with phosphoenolpyruvate carboxylase gene from sugar by *Agrobacterium tumefaciens*-mediated protocols. PEPC activity, photosynthetic rates and total grain numbers were higher in transgenic compared to the WT.	Lian et al. (2014)
Phosphoenol pyruvate carboxylase	Investigation on *japonica* rice cultivar Tainung 67 (TNG67) transformed with maize C_4 *pepc* gene (pBIH2) using *Agrobacterium*-mediated techniques. After antibiotic selection, 29 transgenics were obtained. The PEPC enzyme activity in transgenic rice plants was higher than wild type. Two T_5 homozygous rice lines, exhibits higher PEPC enzyme activity between 7 and 9.4 folds higher than WT.	Chen et al. (2016a)
Agronomic traits; root length, biomass and leaf length and width	Investigation on indica rice cultivar "Khitish" transformed with *PEPC-CA, PPDK* and pGL2 using particle bombarded method. Only two transgenic plant lines exhibits higher than control. Two line exhibits greater leaf length and breadth than the control plants, and all transgenic plant lines longer root length and biomass than the controls.	Shen et al. (2016)

Generally, the C_4 photosynthesis engineered rice has one major advantage, which improves the drought tolerant traits (Liu et al., 2016). The novel C_4 rice is better than the C_3 plants to be planted on soil with water scarcity problems or under drought prone condition and high temperature (Karki et al., 2013). Rice with improve physiological traits; high photosynthetic rate, higher nitrogen-use, water efficiencies high light intensity acclimation (Shen et al., 2015), lower CO_2 concentrations and tolerant to water scarcity, the C_4 rice is very versatile future crops (Liu et al., 2013).

10.5 INSIGHT OF RICE GRAIN QUALITY

Advance in genetic engineering also allows scientist and researchers to improve rice quality. The rice market value is determined by the grain appearance quality. Colour, odours, hardness, percentage of cracks rice, grains shape and chalkiness are among the physical appearance of rice. Off it, grain shape and endosperm chalkiness are two focal grain quality traits in rice improvement breeding programs (Verma et al., 2015). Furthermore, grain size is the main trait determining the rice yield. The grain hardness is closely associated with the milling quality. If the grains hardness is increases then the milling breakage is decreases, thus the head rice yield is also reduce. Grain shape is dignified as grain length, width, length-to-width ratio and thickness (Bai et al., 2010; Verma et al., 2012, 2013, 2015). It is widely accepted as a complex quantitative trait and closely associated with grain weight, the rice yield. In 2012, Tanabata and co-workers had introduces an automation methods using a computer to assists the aqurate measurement of rice shape (Tanabata et al., 2012). Since, grain shape or size is polygenic traits and and involves multiple signaling pathways (Zheng et al., 2016b). Two main groups of the gene are those controlling cell elongation and those controlling cell division (Segami et al., 2016; 2012). To date, an extensive stidies have been carried out to determine QTLs and gene related to grains quality. QTLs and genes related to grain shape that have been identified from mapping population derived from two contrasting parents of japonica, indica and/or cross between japonica and indica rice (Table 10.5).

A short-grain phenotype causes by in the loss function of cell elongation gene the *SRS3* that encode a kinesin-13 protein (Kitagawa et al., 2010; Segami et al., 2012). Those *srs* mutant exhibites a short-grain phenotypes similar to the mutant of brassinosteroid (BR)-related gene (Segami et al., 2016; 2012). Meanwhile, cell division is regulated by *D1* and *TUD1* genes that encode the α-subunit of a heterotrimeric G-protein and U-box ubiquitin

TABLE 10.5 QTL and Grain Quality-Related Genes from Mapping Population of Japonica, Indica and/or Cross Between Japonica and Indica Parents Rice.

QTL	Traits	Encode protein	References
GS3	Grain length and weight	encodes the γ-subunit 3 of a heterotrimeric G-protein	Chen et al. (2016b); Fan et al. (2006); Mao et al. (2010); Li et al. (2012b)
GW2 and qSW5/ GW5	Grain width and weight	GW2 encodes a RING-type protein domain with E3 ubiquitin ligase activity; qSW5/GW5 encodes a nucleoprotein that interacts with polyubiquitin	Shomura et al. (2008); Song et al. (2007); Weng et al. (2008)
GS5	Grain width, filling, and weight	encodes a putative serine carboxypeptidase and acts as a positive regulatory factor in the cell cycle	Li et al. (2011)
GW8	Grain width	encodes the transcription factor Squamosa promoter-binding protein-like 16 containing the miR156-targeted site	Wang et al. (2012)
qGL3/GL3.1 (OsPPKL1)	Grain length	encodes a Ser/Thr phosphatase of the protein phosphatase of the kelch-like family	Qi et al. (2012); Zhang et al. (2012); Xu et al. (2015); Wan et al. (2006)
SRS3 and SR5	Grain length	Kinesin-13 protein	Kitagawa et al. (2010); Segami et al. (2016; 2012)
TGW6	Grain weight	encodes a protein with indole-3-acetic acid-glucose hydrolase activity	Ishimaru et al. (2013)
GW6a	Grain weight	encodes a functional GCN5-related N-acetyl-transferase-like protein that harbors intrinsic histone acetyl-transferase activity	Song et al. (2015)
GL7/GW7	Grain length and width	encodes a homolog of the Arabidopsis thaliana TONNEAU1 recruiting motif (TRM) protein	Wang et al. (2015b,c); Bai et al. (2010); Kato et al. (2011)
qSS7	Grain length	LONGIFOLIA1 (LNG1)	Qiu et al. (2012); Shao et al. (2012)

TABLE 10.5 (*Continued*)

QTL	Traits	Encode protein	References
GIFI	Grain filling	GRAIN INCOMPLETE FILLING 1, Encode for cell-wall invertase for carbon portioning during early grain filling	Wang et al. (2008)
FLO2 or	Grains size and starch quality	*FLOURY ENDOSPERM 2*	She et al. (2010)
qPGWC5 (*qDEC5*) locus	Waxy grains	the bidirectional sugar transporter SWEET3a, the UDP-glucuronosyl/UDP-glucosyltransferase family protein and the class III peroxidase 73	Chen et al. (2016b)
qPGWC7 (*qDEC7*,	Starch biosynthesis	the UDP-arabinopyranose mutase 3 gene	Chen et al. (2016b)
qPGWC8 (*qDEC8*)	Starch biosynthesis	beta-glucosidase, GBA2 type domain containing protein and fructose-bisphosphate aldolase	Chen et al. (2016b)

ligase, respectively (Hu et al., 2013; Izawa et al., 2010). Two conditions determines the grain weight are the cell division in outer glumes that contributes to the hull size and the grain filling rate, accumulation of storage starch in the hull (Xing and Zhang, 2010). *GW2* gene for grain width and weight that encode for RING-type protein with E3 ubiquitin ligase activity (Song et al., 2007). Carbon portioning during early grain-filling of rice is regulates by *GIF1* (Wang et al., 2008). Grain size and starch quality is regulates by *FLO2,* which affecting the storage substances accumulation in the endosperm (She et al., 2010). Mutant with loss-of-function of *GW2* promotes cell numbers, resulting in a larger (wider) spikelet hull, thus directly encourages grain filling grain width, weight and yield (Song et al., 2007).

Chalkiness is traits that increasing postharvest loss due to weakens the endosperm and (Zhao et al., 2016). Chalkiness is characterised by the white or opaque areas or "chalky" formed in the endosperm (Bhattacharya, 2011). The type of chalkiness refers to the position of the opaque areas; "white-belly" is the chalky on the germ side or ventral edge of the grain. It is called "chalky-center" or "white-center" if the chalkiness in the centre of the grain. "White-back" refers to any chalky area on the dorsal side, but its uncommon (Ikehashi and Khush, 1979). Usually, high percentage of grains with chalkiness or opaque parts dignified the grains break during the milling process (Cooper et al., 2008; Lisle et al., 2000). Increases in chalky is associated decreases in head rice yield. Chalky grains have a lower starch granules density, loosely packed, of than vitreous ones thus more susceptible to breakage during milling (Zhao and Fitzgerald, 2013). This contributes to the inferior cooking and eating qualities, thus it will decreases the acceptable and marketable rice (Fitzgerald and Resurrecvion 2009). Generally, rice with more than 20% chalky kernels is unacceptable in the world's markets; the slender and translucent kernels are preferred by the majority consumers, but preferences vary among consumer in different countries (Calingacion et al., 2014). Rice grain chalkiness is a complex trait, regulates by multiple genes and highly influences by environmental factors (Zhao et al., 2016; Sun et al., 2015; Zhou et al., 2015b; Li et al., 2014b; Zhao and Fitzgerald, 2013; Wan et al., 2005). The percentage of chalkiness is closely related with the temperature and/or humidity during grain filling, high temperature has increases the occurrence of chalkiness (Zhao and Fitzgerald 2013; Fitzgerald and Resurreccion 2009; Cooper et al., 2008; Yamakawa et al., 2007; Counce et al., 2005; Lisle et al., 2000). Numerous genes and/or QTLs for grain chalkiness have been mapped in various rice populations across the rice-growing environments. Twelve chromosomes have been found and mapped with quantitative traits and QTL for chalkiness traits in rice. These huge informations provide a high possible marker-assisted selection for rice improvement breeding. Although many QTLs have been mapped, only one chalkiness regulating gene *Chalk5*, encodes a vacuolar H^+-translocating pyrophosphatase (V-PPase) with PPi hydrolysis and H^+-translocation activity was successfully isolated (Li et al., 2014b). It was shown that overexpression of Chalk5 increases chalkiness of the endosperm by disturbing the pH homeostasis in the endomembrane trafficking system in developing seeds (Li et al., 2014b). There is an increases in small vesicle-like structure couple with protein bodies' biogenesis causes the formation of air spaces among endosperm storage substances. Nonetheless, the interaction mechanisms and regulation pathway of rice chalkiness associated genes remain unclear (Chen et al., 2016b,c).

The physical characteristics are highly related with the physicochemical properties. The rates of water uptake, hydration and swelling, volume expansion, pasting properties, cooked rice texture, loss of solids rice during cooking, and gel consistency and gelatinisation temperature. The water uptake and hydration is refered to the water absorption by rice during cooking. Swelling or bulk swelling is the physical expansion of individual rice grain during cooking. The expansion of rice upon cooking is called as "volume expansion". Volume of expansion determines the stickiness of cooked rice. Excess of water during cooking may causes insoluble of rice grains or loss of rice-solids. Cooked rice textures are identified as soft, sticky, hard or non-sticky and paste viscosity or pasting properties. Storage factor contributes to the gel consistency, harder if stored for long period. In general, amylose content, gel consistency and gelatinisation temperature are three physicochemical characters linked to rice quality criterion. Nonetheless, all these physicochemical properties are contributes by the compositions of grain starch, the amylose (Zhang et al., 2016; Hasjim et al., 2012).

Starch encompasses about 90% of total dry matter of the storage substance of rice grains. The endosperm starch consists of amylose (20–30%) and amylopectin (70–80%). Amylose content (AC) is a complex trait that controlled by many genes and determines by the granule-bound starch synthase (Umemoto and Terashima, 2002) code by the *Waxy* gene. Two functional alleles are the Wx^b is mainly found in *japonica* cultivars, and Wx^a is found in *indica* cultivars and various wild rice species It was shows that the transcripts levels of the Wx^a was ten-fold higher than those of Wx^b. Thus, endosperms of indica rice have higher AC (25–30%) than *japonica* rice (15–18%) (Chen et al., 2008). Takemoto-Kuno et al. (2015) had identified a QTL, *qAC2* that interacts with *Wx* and controls the low amylose content in rice. *The qAC2* annotated three genes that encode the glucose/

ribitol dehydrogenase family, glycosyl transferase, and auxin responsive proteins (Takemoto-Kuno et al. 2015). A transgenic rice with the antisense *W*x gene has reduced AC in mature seeds, up to 7% in the T2 generation (Chen et al., 2002). The cooking and eating quality (GC and GT) of the transgenic lines were also improved. Furthermore, AC is influenced by the heading date (Tanaka et al., 2006; Wada et al., 2006).

Cooking and eating quality are the key characters for rice grain quality that driving market demand, by way of consumers prefer good palatable rice (Kwon et al., 2011; Verma et al., 2012, 2013, 2015). These qualities are determines by the physicochemical characteristics such as amylose content (AC), gel consistency (GC) and gelatinization temperature (GT) of rice (Table 10.6), alkali digestion value and protein content (Bao et al., 2004; Kobayashi et al., 2008; Wada et al., 2008; Verma et al., 2012, 2013, 2015).

TABLE 10.6 The Characterization of Cooking and Eating Quality of Rice Grain

Amylose classes	Gelatinisation temperature	Gel consistency
High (>25%)	High (>74°C)	Hard (length of gel <40 mm)
Intermediate (20–25%)	Intermediate (70–74°C)	Medium (length of gel 41–60 mm)
Low (10–19%)	Low (<70°C)	Soft texture (length of gel >61 mm)
Very low (3–9%)	–	–
Waxy (0–2%)	–	–

Source: Kobayashi et al. (2008); Wada et al. (2008); Verma et al. (2012, 2013, 2015); Bhat and Riar, (2016).

10.6 SUMMARY AND CONCLUSIONS

This proposed chapter emphasized on the current trends of genetic engineering and biotechnology employed in quality improvement of rice crop. The focus of genetic engineering of rice is vey wide from the growth and development until the grain quality. Thousands of mutants have been regenerated wordwide in various research laboratories through randomly induced-mutant or direct specific target gene mutation. The genetic transformation techniques also facilate the mutation process. Insertation of gene of interest into rice genome has produced a large varity of rice with the desired traits, which contribute to enrichment of germplasm and material for rice improvement breeding. Molecular breeding through genetic engineering and biotechnology helps to reduce the duration for production of quality traits with specific targeted physiognomies.

ACKNOWLEDGMENT

Deepak Kumar Verma and Prem Prakash Srivastav are indebted to Department of Science and Technology, Ministry of Science and Technology, Govt. of India for an individual research fellowship (INSPIRE Fellowship Code No.: IF120725; Sanction Order No. DST/INSPIRE Fellowship/2012/686 and Date: 25/02/2013).

KEYWORDS

- 5-bromo-uracil
- 8-ethoxy caffeine
- AGP glucose pyrophosphorylase
- auxin responsive proteins
- BASF Agrochemical Products
- bound starch synthase
- Class III peroxidase 73 Cytosol
- fructose-bisphosphate aldolase
- glucose/ribitol dehydrogenase family
- indole-3-acetic acid-glucose hydrolase
- N-acetyl-transferase-like protein
- phosphoenol-pyruvate carboxylase
- pyrimidinyloxybenzoate herbicide
- streptonigrin
- UDP-glucuronosyl/UDPglucosyltransferase family protein
- Uranium 235

REFERENCES

Andersen, J. R.; Lubberstedt, T. Functional Markers in Plants. *Trends Plant Sci.* **2003**, *8*, 554–560.

Andres, A.; Fogliato, S.; Ferrero, A.; Vidotto, F. Susceptibility to Imazamox in Italian Weedy Rice Populations and Clearfield® Rice Varieties. *Weed Res.* **2014**, *54*(5). 492–500. DOI: 10.1111/wre.12099.

Bai, X.; Luo, L.; Yan, W.; Kovi, M.; Zhan, W.; Xing, Y. Z. Genetic Dissection of Rice Grain Shape Using a Recombinant Inbred Line Population Derived from Two Contrasting Parents and Fine Mapping a Pleiotropic Quantitative Trait Locus qGL7. *BMC Genet.* **2010**, *11*, 1–11. DOI: 10.1186/1471-2156-11-16 pmid:20184774.

Bandyopadhyay, A.; Datta, K.;Zhang, J.; Yang, W.; Raychaudhuri, S.; Datta, S. K. Enhanced Photosynthesis Rate in Genetically Engineered Indica Rice Expressing *pepc* Gene Cloned from Maize. *Plant Sci.* **2007**, *172*(6), 1204–12049.

Bao, J. S.; Sun, M.; Zhu, L. H.; Corke, H. Analysis of Quantitative Trait Loci for Some Starch Properties of Rice (*Oryza sativa* L.) Thermal Properties, Gel Texture and Swelling Volume. *J. Cereal Sci.* **2004**, *39*, 379–385.

Basak, J.; Nithin, C. Targeting Non-Coding RNAs in Plants with the CRISPR-Cas Technology is a Challenge yet Worth. *Front. Plant Sci.* **2015**, 19;6:1001. doi: 10.3389/fpls.2015.01001

Baysal, C.; Bortesi, L.; Zhu, C.; Farré, G.; Schillberg, S.; Christou, P. CRISPR/Cas9 Activity in the Rice *OsBEIIb* Gene Does not Induce Off-Target Effects in the Closely Related Paralog *OsBEIIa. Mol.* Breed. **2016**, *36*, 108. DOI: 10.1007/s11032-016-0533-4.

Beerli, R. R.; Segal, D. J.; Dreier, B.; Barbas, C. F. Toward controlling gene expression at will: specific regulation of the erbB-2/HER-2 promoter by using polydactyl zinc finger proteins constructed from modular building blocks. *Proc. Natl Acad. Sci. USA*, 3rd Ed., **1998**, *95*, 14628–14633.

Bhat, F. M.; Riar, C. S. Physicochemical, Cooking, and Textural Characteristics of Grains of Different Rice (*Oryza sativa L.*) Cultivars of Temperate Region of India and Their Interrelationships. *J. Texture Studies* **2016**, DOI: 10.1111/jtxs.12227.

Bhattacharya, K. R. *Rice Quality: A Guide to Rice properties and Analysis*. Woodhead Publ. Ltd.: Oxford/Cambridge/Phladelphia/New Delhi, 2011.

Boch, J.; Scholze, H.; Schornack, S.; Landgraf, A.; Hahn, S.; Kay, S.; Lahaye, T.; Nickstadt, A. Bonas, U. Breaking the Code of DNA Binding Specificity of TAL-Type III Effectors. *Science.* **2009**, *326*(5959), 1509–1512. DOI: 10.1126/science.1178811.

Calingacion, M.; Laborte, A.; Nelson A, Resurreccion, A.; Concepcion, J. C.; Daygon, V. D.; Mumm, R.; Reinke, R.; Dipti, S.; Bassinello, P. Z.; Manful, J.; Sophany, S.; Lara, K. C.; Bao, J.; Xie, L.; Loaiza, K.; El-hissewy, A.; Gayin, J.; Sharma, N.; Rajeswari, S.; Manonmani, S.; Rani, N. S.; Kota, S.; Indrasari, S. D.; Habibi, F.; Hosseini, M.; Tavasoli, F.; Suzuki, K.; Umemoto, T.; Boualaphanh, C.; Lee, H. H.; Hung, Y. P.; Ramli, A.; Aung, P. P.; Ahmad, R.; Wattoo, J. I.; Bandonill, E.; Romero, M.; Brites, C. M.; Hafeel, R.; Lur, H.-S.; Cheaupun, K.; Jongdee, S.; Blanco, P.; Bryant, R.; Thi Lang, N.; Hall, R. D.; Fitzgerald, M. Diversity of Global Rice Markets and the Science Required for Consumer-Targeted Rice Breeding. *PLoS One* **2014**, 9, e85106.

Capecchi, M. R. Altering the Genome by Homologous Recombination. *Science* **1989**, *244*, 1288–1292.

Casella, L.; Greco, R.; Bruschi, G.; Wozniak, B.; Dreni, L.; Kater, M.; Cavigiolo, S.; Lupotto, E.; Piffanelli, P. TILLING in European Rice: Hunting Mutations for Crop Improvement. *Crop Sci.* **2013**, *53*(6), 2550–2562. DOI: 10.2135/cropsci2012.12.0693.

Chen, M. J.; Bergman, C.; Pinson, S.; Fjellstrom, R. Waxy gene haplotypes: Associations with apparent amylase content and the effect by the environment in an international rice germplasm collection. *J. Cereal Sci.* **2008,** *47*, 536–545.

Chen, K.; Shan, Q.; Gao, C. An Efficient TALEN Mutagenesis System in Rice. *Methods* **2014**, *64*(1), 2–8.

Chen, L.; Gao, W.; Chen, S.; Wang, L.; Zou, J.; Liu, Y.; Wang, H.; Chen, Z.; Guo, T. High-Resolution QTL Mapping for Grain Appearance Traits and Co-Localization of

Chalkiness-Associated Differentially Expressed Candidate Genes in Rice. *Rice* **2016b**, *9*, 48. DOI: 10.1186/s12284-016-0121-6.

Chen, P. Y.; Tsai, Y. T.; Ng, C. Y.; Ku, M. S. B.; Po, K. Y. Transformation and Characterization of Transgenic Rice and *Cleome Spinosa* Plants Carrying the Maize Phospho*enol*pyruvate Carboxylase Genomic DNA. *Plant Cell, Tissue Organ Cult.* **2016a**, 1–11. DOI: 10.1007/s11240-016-1128-9.

Chen, L.; Gao, W.; Guo, T.; Huang, C.; Huang, M.; Wang, J.; Xiao, W.; Yang, G.; Liu, Y.; Wang, H.; Chen, Z. A Genotyping Platform Assembled with High-Throughput DNA Extraction, Codominant Functional Markers, and Automated CE System to Accelerate Marker-Assisted Improvement of Rice. *Mol. Breed.* **2016c**, *36*, 123. DOI: 10.1007/s11032-016-0547-y.

Chen, X.; Liu, Q.; Wang, Z.; Wang, X.; Cai, X.; Zhang, J.; Gu, M. Introduction of an antisense Waxy gene into the main parent lines of indica hybrid rice. *Chinese Sci. Bull.* **2002**, *47*, 1192-1195.

Chi, X.; Zhang, Y.; Xue, Z.; Feng, L.; Liu, H.; Wang, F, Qi, X. Discovery of Rare Mutations in Extensively Pooled DNA Samples Using Multiple Target Enrichment. Plant Biotechnol. J. **2014**, DOI: 10.1111/pbi.12174.

Chun, J. B.; Ha, B. K.; Jang, D. S.; Song, M.; Lee, K. J, Kim, J. B.; Kim, S. H.; Kang, S. Y.; Lee, G. J. Seo, Y. W.; Kim, D. S. Identification of Mutations in OASA1 Gene from a Gamma-Irradiated Rice Mutant Population. *Plant Breed.* **2012**, *131*(2), 276–281. DOI: 10.1111/j.1439–0523.2011.01933.x.

Cooper, N. T. W.; Siebenmorgen, T. J.; Counce, P. A. Effects of Night Time Temperature During Kernel Development on Rice Physicochemical Properties. *Cereal Chem.* **2008**, *85*, 276–282.

Counce, P. A.; Bryant, R. J.; Bergman, C. J.; Bautista, R. C.; Wang, Y. J.; Siebenmorgen, T. J.; Moldenhauer, K. A.; Meullenet, J. F. C. Rice Milling Quality, Grain Dimensions, and Starch Branching as Affected by High Night Temperatures. *Cereal Chemi.* **2005**, *82*, 645–648.

Croughan, T. P. Application of Tissue Culture Techniques to the Development of Herbicide-Resistant Rice. *La. Agric.* **1994**, *37*, 25–26.

Dass, A.; Shekhawat, K.; Choudhary, A. K.; Sepat, S.; Rathore, S. S.; Mahajan, G.; Chauhan, B. S. Weed Management in Rice Using Crop Competition-a Review. *Crop Prot.* **2016**, pp 1–8. http://dx.doi.org/10.1016/j.cropro.2016.08.005.

Ding, Z. S.; Huang, S. H.; Zhou, B. Y.; Sun, X. F.; Zhao, M. Over-Expression of Phosphoenolpyruvate Carboxylase cDNA from C_4 Millet (*Seteria italica*) Increase Rice Photosynthesis and Yield Under Upland Condition but Not in Wetland Fields. *Plant Biotechnol. Rep.* **2013**, *7*, 155–163.

Ding, Z. S.; Zhou, B. Y.; Sun, X. F.; Zhao, M. High Light Tolerance is Enhanced by Overexpressed PEPC in Rice Under Drought Stress. *Acta Agrono. Sin.* **2012**, *38*, 285–292.

Endo, M.; Mikami, M.; Toki, S. Multigene Knockout Utilizing off-Target Mutations of the CRISPR/Cas9 System in Rice. *Plant Cell Physiol.* **2014**, *56*(1), 41–47. https://doi.org/10.1093/pcp/pcu154.

Endo, M.; Nishizawa-Yokoi, A.; Toki, S. Targeted Mutagenesis in Rice Using TALENs and the CRISPR/Cas9 System. *Methods Mol. Biol.* **2016**, *1469*, 123-135.

Fan, C.; Xing, Y.; Mao, H.; Lu, T.; Han, B.; Xu, C.; Li, X.; Zhang, Q. GS3, a Major QTL for Grain Length and Weight and Minor QTL for Grain Width and Thickness in Rice, Encodes a Putative Transmembrane Protein. *Theor. Appl. Genet.* **2006**, *112*, 1164–1171.

Fitzgerald, M. A.; Resurreccion, A. P. Maintaining the Yield of Edible Rice in a Warming World. *Funct. Plant Biol.* **2009**, *36*, 1037–1045.

Gaj, T.; gersbach, C. A.; barbas C. F. ZFN, TALEN and CRISPR/Cas-Based Methods for Genome Engineering. *Trends Biotechnol.* **2013**, *31*(7), 397–405. DOI: 10.1016/j. tibtech.2013.04.004.

GOI (Government of India). Taxonomy, Geographic Origin and Genomic Evolution. In: Biology of *Oryza saiva* L. (Rice). Series of Crop Specific Biology Documents. Department of Biotechnology, Minister of Science and Technology and Minister of Environment and Forests. Government of India: India, 2011, pp 3–10.

Gruszka, D.; Szarejko, I.; Maluszynski, M. Sodium Azide as a Mutagen In *Plant Mutation Breeding and Biotechnology;* Shu, Q.Y., Forster, B. P., Nakawaga, H.; eds.; Poland. 2012, pp 159. DOI: 10.1079/9781780640853.0159.

Gu, J. F.; Qiu, M.; Yang, J. C. Enhanced Tolerance to Drought in Transgenic Rice Plants Over-Expressing C$_4$ Photosynthesis Enzymes. *Crop J.* **2013**, 1, 105–114.

Hashida, Y.; Hirose, T.; Okamura, M.; Hibara, K.; Ohsugi, R.; Aoki, N. A Reduction of Sucrose Phosphate Synthase (SPS) Activity Affects Sucrose/Starch Ratio in Leaves but Does not Inhibit Normal Plant Growth in Rice. *Plant Sci.* **2016**, *253*, 40–49. http://dx.doi. org/10.1016/j.plant sci.2016.08.017.

Hasjim J.; Li E.; Dhital S. Milling of Rice Grains: The Roles of Starch Structures in the Solubility and Swelling Properties of Rice Flour. *Starch* **2012**, *64*(8), 631–645. DOI: 10.1002/star.201100204.

Henry, I. M.; Nagalakshmi, U.; Lieberman, M. C.; Ngo, K. J.; Krasileva, K. V.; Vasquez-Gross, H.; Akhunova, A.; Akhunov, E.; Dubcovsky, J.; Tai, T. H.; Comai, L. Efficient Genome-Wide Detection and Cataloging of EMS-Induced Mutations Using Exome Capture and Next-Generation Sequencing. *Plant Cell* **2014**, *26*, 1382–1397. DOI: http://dx.doi.org/10.1105/tpc.113.121590.

Hoang, T. M. Linh.; Tran, T. N.; Nguyen, T. K. T.; Williams, B.; Wurm, P.; Bellairs, S.; Mundree, S. Improvement of Salinity Stress Tolerance in Rice:Challenges and Opportunities. *Agronomy* **2016**, *6*, 54. DOI: 10.3390/agronomy6040054.

Hu, X.; Wang, C.; Fu, Y.; Liu, Q.; Jiao, X.; Wang, K. Expanding the Range of CRISPR/Cas9 Genome Editing in Rice. *Mol. Plant* **2016**, *9*(6), 943–945. http://dx.doi.org/ 10.1016/j. molp.2016.03.003.

Hu, W.; Wen, M.; Han, Z.; Tan, C.; Xiong, Y. Scanning QTLs for Grain Shape Using a Whole Genome SNP Array in Rice. *J. Plant Biochem. Physiol.* **2013**, *16*, 104.

Huo, K.; Li, X.; He, Y. F.; Wei, X. D.; Lu, W.; Zhao, C. F.; Wang. C. L. Exogenous ATP Enhance Signal Response of Suspension Cells of Transgenic Rice (*Oryza sativa* L.) Expressing Maize *C$_4$-pepc* Encoded Phosphoenolpyruvate Carboxylase Under PEG Treatment. *Plant Growth Regul.* **2016**, DOI: 10.1007/s10725-016-0238-z.

Ikeda, T.; Tanaka, W.; Mikami, M.; Endo, M.; Hirano, H. Y. Generation of Artificial *Drooping Leaf* Mutants by CRISPR-Cas9 Technology in Rice. *Genes Genet. Syst.* **2016**, *90*(4), 231–235. http://doi.org/10.1266/ggs.15-00030.

Ikehashi, H.; Khush, G. S. Methodology of Assessing Appearance of the Rice Grain, Including Chalkiness and Whiteness. *Chemical Aspects of Rice Grain Quality*, Los Baños, Laguna, Philippines, International Rice Research Institute, 1979, pp. 223–229.

IRRI. (International Rice Research Institute). International Rice Research Institute. Philippines. URL: http://www.irri.org. (Accessed Jan 01, 2017).

Ishimaru, K.; Hirotsu, N.; Madoka, Y.; Murakami, N.; Hara, N.; Onodera, H.; Kashiwagi, T.; Ujiie, K.; Shimizu, B.; Onishi, A.; Miyagawa, H.; Katoh, E. Loss of function of the IAA-glucose hydrolase gene TGW6 enhances rice grain weight and increases yield. *Nat Genet.* **2013**, *45*(6), 707–711.

Ishizaki, T. CRISPR/Cas9 in Rice Can Induce New Mutations in Later Generations, Leading to Chimerism and Unpredicted Segregation of the Targeted Mutation. *Mol. Breed.* 2016, *36*, 165. DOI: 10.1007/s11032-016-0591-7.

Izawa, Y.; Takayanagi, Y.; Inaba, N.; Abe, Y.; Minami, M.; Fujisawa, Y. *Function and Expression Pattern of the α Subunit of the Heterotrimeric G Protein in Rice. Plant Cell Physiol.* **2010**, *51*, 271–281.

Jackson, M. T. Conservation of Rice Genetic Resources: the Role of the International Rice Genebank at RRI. *Plant Mol. Biol.* **1997**, *35*, 61–67.

Jang, G.; Lee, S.; Um, T. Y.; Chang, S. H.; Lee, H. Y.; Chung, P. J.; Kim, J.; Choi, Y. D. Genetic Chimerism of CRISPR/Cas9-Mediated Rice Mutants. *Plant Biotechnol. Rep.* **2016**, *10*(6), 425–435. DOI: 10.1007/s11816-016-0414-7.

Jeng, T. L.; Ho, P. T.; Shih, Y. J.; Lai, C. C.; Wu, M. T.; Sung, J. M. Comparisons of Protein, Lipid, Phenolics, γ-Oryzanol, Vitamin E, and Mineral Contents in Bran Layer of Sodium Azide-Induced Red Rice Mutants. *J. Sci. Food Agric.* **2011**, *91*(8), 1341–1526. DOI: 10.1002/jsfa.4333.

Jeng, T. L.; Lin, Y. W.; Wang, C. H.; Sung, J. M. Comparisons and Selection of Rice Mutants with High Iron and Zinc Contents in Their Polished Grains that were Mutated from the *indica* type Cultivar IR64. *J. Food Compos. Anal.* **2012**, *28*(2), 149–154.

Jeng, T. L.; Tseng, T. H.; Wang, C. S.; Chen, C. L.; Sung J. M Starch Biosynthesizing Enzymes in Developing Grains of Rice Cultivar Tainung 67 and its Sodium Azide-Induced Rice Mutant. *Field Crops Res.* 2003b, *84*(3), 261–269. http://dx.doi.org/10.1016/S0378-4290(03)00094-7.

Jeng, T. L.; Wang, C. S.; Chen, C. L.; Sung, J. M. Effects of Grain Position on the Panicle on Starch Biosynthetic Enzyme Activity in Developing Grains of Rice Cultivar Tainung 67 and its NaN $_3$-Induced Mutant. *J. Agric. Sci.* **2003a**, *141*(3–4), 303–311.

Jiang, W.; Zhou, H.; Bi, H.; Fromm, M.; Yang, B.; Weeks, D. P. Demonstration of CRISPR/Cas9/sgRNA-Mediated Targeted Gene Modification in Arabidopsis, Tobacco, Sorghum and Rice. *Nucleics Acids Res.* **2013**, *41*(20), e188. https://doi.org/10.1093/nar/gkt780.

Joshi, R.; Prashat, R.; Sharma, P. C.; Singla-Pareek, S. L.; Pareek, A. Physiological Characterization of Gamma-Ray Induced Mutant Population of Rice to Facilitate Biomass and Yield Improvement Under Salinity Stress. *Indian J. Plant Physiol.* **2016**, *21*(4), 545–555. DOI: 10.1007/s40502-016-0264-x.

Jozuka-Hisatomi, Y.; Terada, R.; Lida S. Efficient Transfer of Base Changes from a Vector to the Rice Genome by Homologous Recombination: Involvement of Heteroduplex Formation and Mismatch Correction. *Nucliec Acids Res.* **2008**, *36*(14), 4727–4735. https://doi.org/10.1093/nar/gkn451.

Kajala, K.; Covshoff, S.; Karki, S.; Woodfield, H.; Tolley, B. J.; Dionora, M. J.; Mogul, R. T.; Mabilangan, A. E.; Danila, F. R.; Hibberd, J. M.; Quick, W. P. Strategies for Engineering a Two-Celled C$_4$ Photosynthetic Pathway into Rice. *J. Exp. Bot.* **2011**, *62*, 3001–3010.

Karki, S.; Rizal, G.; Quick, W. P. Improvement of Photosynthesis in Rice (*Oryza sativa* L.) by Inserting the C$_4$ Pathway. *Rice* **2013**, *6*, 28. DOI: 10.1186/1939-8433-6-28.

Kato, T.; Segami, S.; Toriyama, M.; Kono, I.; Ando, T.; Yano, M. et al. Detection of QTLs for Grain Length from Large Grain Rice (*Oryza sativa* L.). *Breed. Sci.* **2011**, *61*, 269–274. DOI: 10.1270/jsbbs.61.269.

Khan, Z.; Khan, S. H.; Mubarik, M. S.; Sadia, B.; Ahmad, A. Use of TALEs and TALEN Technology for Genetic Improvement of Plants. *Plant Mol. Biol. Rep.* **2016**, 1–19. DOI: 10.1007/s11105-016-0997-8.

Khlestkina, E. K.; Shumny, V. K. Prospects for Application of Breakthrough Technologies in Breeding: The CRISPR/Cas9 System for Plant Genome Editing. *Russ. J. Genet.* **2016**, *52*(7), 676–687. DOI: 10.1134/S102279541607005X.

Khush, G. S. Green Revolution: The Way Forward. *Nat. Rev. Genet.* **2001**, *2*, 815–822.

Kim, B.-G.; Han, S.-Y.; Shin, D.; Moon, S.-J.; Jeon, S.-A.; Byun, M.-O. Optimization of *Agrobacterium*-Mediated Transformation in Japonica-Type Rice *Oryza sativa* L. cv. Dongjin for high Efficiency. *Korean J. Breed. Sci.* **2012**, *44*(3), 221–228.

Kitagawa, K.; Kurinami, S.; Oki, K.; Abe, Y.; Ando, T.; Kono, I.; Yano, M.; Kitano, H.; Iwasaki, Y. A Novel Kinesin 13 Protein Regulating Rice Seed Length. *Plant Cell Physiol.* **2010**, *51*, 1315–1329.

Kobayashi, A.; Tomita, K.; Yu, F.; Takeuchi, Y.; Yano, M. Verification of Quantitative Trait Locus for Stickiness of Cooked Rice and Amylose Content by Developing Near-Isogenic Lines. *Breed. Sci.* **2008**, *58*, 25–242.

Kole, P. C.; Chakraborty, N. R. Assessment of Genetic Divergence in Induced Mutants of Short Grain Aromatic Non-Basmati Rice (*Oryza sativa L.*). *J. Trop. Agric.* **2012**, *89*(4), 211–215.

Ku, M. S. B.; Agarie, S.; Nomura, M.; Fukayama, H.; Tsuchida, H.; Ono, K.; Hirose, S.; Toki, S.; Miyao, M.; Matsuoka, M. High-Level Expression of Maize Phosphoenolpyruvate Carboxylase in Transgenic Rice Plants. *Nat. Biotechnol.* **1999**, *17*, 76–80. DOI: 10.1038/5256.

Kwon, S.; Cho, Y.; Lee, J.; Kim, J.; Kim, M.; Choi, I.; Hwang, H.; Koh, H.; Kim, Y. Identification of Quantitative Trait Loci Associated with Rice Eating Quality Traits Using a Population of Recombinant Inbred Lines Derived from a Cross Between Two Temperate Japonica Cultivars. *Mol. Cells* **2011**, *31*, 437–445.

Li, D.; Tang, N.; Fang, Z.; Xia, Y.; Cao, M. Co-Transfer of TALENs Construct Targeted for Chloroplast Genome and Chloroplast Transformation Vector into Rice Using Particle Bombardment. *J. Nanosci. Nanotechnol.* **2016a**, *16*(12), 12194–12201(8).

Li, J.; Meng, X.; Zong, Y.; Chen, K.; Zhang, H.; Liu, J.; Li, J.; Gao, C. Gene Replacements and Insertions in Rice by Intron Targeting Using CRISPR–Cas9. *Nat. Plants* **2016b**, *2*, Article number 16139. DOI: 10.1038/nplants.2016.139.

Li, M.; Li, X.; Zhou, Z.; Wu, P.; Fang, M.; Pan, X.; Lin, Q.; Luo, W.; Wu, G.; Li, H. Reassessment of the Four Yield-Related Genes *Gn1a*, *DEP1*, *GS3*, and *IPA1* in Rice Using a CRISPR/Cas9 System. *Front. Plant Sci.* **2016d**, *7*, 377. https://doi.org/10.3389/fpls.2016.00377.

Li, S.; Liu, W.; Zhang, X.; Liu, Y.; Li, N.; Li, Y. Roles of the Arabidopsis G Protein γ Subunit AGG3 and its Rice Homologs GS3 and DEP1 in Seed and Organ Size Control. *Plant Signal Behav.* **2012b**, *7*, 1357–1359.

Li, T.; Liu, B.; Chen, C. Y.; Yang, B. TALEN-Mediated Homologous Recombination Produces Site-Directed DNA Base Change and Herbicide-Resistant Rice. *J. Genet. Genomics* **2016c**, *43*(5), 297–305. http://dx.doi.org/10.1016/j.jgg.2016.03.005.

Li, T.; Liu, B.; Chen, C.; Yang, B. TALEN Utilization in Rice Genome Modifications. *Methods* **2014a**, *69*(1), 9–16. http://dx.doi.org/10.1016/j.ymeth.2014.03.019.

Li, T.; Liu, B.; Spalding, M. H.; Weeks, D. P.; Yang, B. High-Efficiency TALEN-Based Gene Editing Produces Disease-Resistant Rice Nat. Biotechnol. 2012a, *30*, 390–392.

Li, Y.; Fan, C.; Xing, Y.; Yun, P.; Luo, L.; Yan, B.; Peng, B.; Xie, W.; Wang, G.; Li, X.; Xiao, J.; Xu, C.; He, Y. Chalk5 Encodes a Vacuolar H^+−Translocating Pyrophosphatase Influencing Grain Chalkiness in Rice. *Nat. Genet.* **2014b**, 46, 398–404.

Li, Y.; Fan, C. C.; Xing, Y. Z.; Jiang, Y. H.; Luo, L. J.; Sun, L.; Shao, D.; Xu, C.; Li, X.; Xiao, J. et al. Natural Variation in GS5 Plays an Important Role in Regulating Grain Size and Yield in Rice. *Nat. Genet.* **2011**, *43*, 1266–1269. DOI: 10.1038/ng.977 pmid:22019783.

Lian, L.; Wang, X. W.; Zhu, Y. S.; He, W.; Cai, Q. H.; Xie, H. A.; Zhang, M. Q.; Zhang, J. F. Physiological and Photosynthetic Characteristics of Indica Hang2 Expressing the Sugarcane *PEPC* Gene. *Mol. Biol. Rep.* **2014**, *41*, 2189–2197.

Liang, G.; Zhang, H.; Lou, D.; Yu, D. Selection of Highly Efficient sgRNAs for CRISPR/Cas9-Based Plant Genome Editing. *Sci. Rep.* **2016**, *6*, Article number: 21451. DOI: 10.1038/srep21451.

Lima, J. M.; Nath, M.; Dokku, P.; Raman, K.V.; Kulkarni, K. P.; Vishwakarma, C.; Sahoo, S. P.; Mohapatra, U. B.; Mithra, S. V. A.; Chinnusamy, V.; Robin, S.; Sarla, N.; Seshashayee, M.; Singh, K.; Singh, A. K.; Singh, N.K.; Sharma, R. P.; Mohapatra, T. Physiological, Anatomical and Transcriptional Alterations in a Rice Mutant Leading to Enhanced Water Stress Tolerance. *AoB Plants* **2015**, *7*, plv023. DOI: 10.1093/aobpla/plv023.

Lisle, A. J.; Martin, M.; Fitzgerald, M. A. Chalky and Translucent Rice Grains Differ in Starch Composition and Structure and Cooking Properties. *Cereal Chem.* **2000**, *77*, 627–632.

Liu, X.; Li, X.; Zhang, C.; Dai, C.; Zhou, J.; Ren, C.; Zhang, J. Phosphoenolpyruvate Carboxylase Regulation in C_4-*PEPC*-Expressing Transgenic Rice During Early Responses to Drought Stress. *Physiol. Plant.* **2016**, DOI: 10.1111/ppl.12506.

Liu, X.; Lin, F.; Wang, L.; Pan, Q. The in Silico Map-Based Cloning of *Pi36*, a Rice Coiled-Coil Nucleotide-Binding Site Leucine-Rich Repeat Gene that Confers Race-Specific Resistance to the Blast Fungus. *Genetics* **2007**, *176*, 2541–2549.

Liu, Z.; Sun, N.; Yang, S.; Zhao, Y.; Wang, X.; Hao, X.; Qiao, Z. Evolutionary Transition from C_3 to C_4 Photosynthesis and the Route to C_4 Rice. *Biologia.* **2013**, *68*(4), 577–586. DOI: https://doi.org/10.2478/s11756-013-0191-5.

Livore, A. B.; Prina, A. R.; Birk, I.; Singh, B. Rice Plants Having Increased Tolerance to Imidazolinone Herbicides. EP 2,294,913 B1 Patent, 2015.

Livore, A. B.; Prina, A. R.; Singh, B. K. Ascenzi, R.; Whitt, S. R. Herbicide-Resistant Rice Plants, Polynucleotides Encoding Herbicide-Resistant Acetohydroxyacid Synthase Large Subunit Proteins, and Methods of Use. U.S. 20,100,029,485 A1 Patent, 2010.

Lowder, L. G.; Zhang, D.; Baltes, N. J.; Paul, J. III; W.; Tang, Xu.; Zheng, X.; Voytas, D. F.; Hsieh, T.; Zhang, Y.; Qi, Y. A CRISPR/Cas9 Toolbox for Multiplexed Plant Genome Editing and Transcriptional Regulation. *Plant Physiol.* **2015**, *169*(2), 971–985. DOI: http://dx.doi.org/10.1104/pp.15.00636.

Ma, L.; Zhu, F.; Li, Z.; Zhang, J.; Li, X.; Dong, J.; Wang, T. TALEN-Based Mutagenesis of Lipoxygenase LOX3 Enhances the Storage Tolerance of Rice (*Oryza sativa*) Seeds. PLOS One. **2015**. http://dx.doi.org/10.1371/journal.pone.0143877.

Malcolm, L. Super Rice. 1998, http://www.abc.net.au/science/slab/rice/story.htm#super. (Accessed Jan 24, 2017).

Mao, H.; Sun, S.; Yao, J.; Wang, C.; Yu, S.; Xu, C.; Li, X.; Zhang, Q. Linking Differential Domain Functions of the GS3 Protein to Natural Variation of Grain Size in Rice. *Proc. Natl. Acad. Sci. U. S. A.* **2010**, *107*, 19579–19584.

Mazumdar, S.; Quick, W. P.; Bandyopadhyay, A. CRISPR-Cas9 Mediated Genome Editing in Rice, Advancements and Future Possibilities. *Indian J. Plant Physiol.* **2016**, *21*(4), 437–445. DOI: 10.1007/s40502-016-0252-1.

McManus, M. T.; Sharp, P. A. Gene Silencing in Mammals by Small Interfering RNAs. *Nat. Rev. Genet.* **2002**, *3*, 737–747.

Mikami, M.; Toki, S.; Endo, M. Comparison of CRISPR/Cas9 Expression Constructs for Efficient Targeted Mutagenesis in Rice. *Plant Mol. Biol.* **2015a**, *88*(6), 561–572. DOI: 10.1007/s11103-015-0342-x.

Mikami, M.; Toki, S.; Endo, M. Parameters Affecting Frequency of CRISPR/Cas9 Mediated Targeted Mutagenesis in Rice *Plant Cell Rep.* **2015b**, *34*(10), 1807–1815. DOI: 10.1007/ s00299-015-1826-5.

Mikami, M.; Toki, S.; Endo, M. Precision Targeted Mutagenesis Via Cas9 Paired Nickases in Rice. *Plant Cell Physiol.* **2016**, *57*(5), 1058–1068. DOI: https://doi.org/10.1093/pcp/pcw049.

Miyao, M.; Masumoto, C.; Miyazawa, S.; Fukayama, H. Lessons from Engineering a Single-cell C_4 Photosynthetic Pathway into Rice. *J. Exp. Bot.* **2011**, *62*, 3021–3029.

Mohapatra, T.; Robin, S.; Sarla, N.; Sheshashayee, M.; Singh, A. K.; Singh, K.; Singh. N. K.; Mithra, S. V. A.; Sharma, R. P. MS Induced Mutants of Upland Rice Variety Nagina22: Generation and Characterization. *Proc. Indian Natl. Sci. Acad.* **2014**, *80*. pp. 163–172. 10.16943/ptinsa/2014/v80i1/55094

Moose, S. O.; Mumm, H. Molecular Plant Breeding as the Foundation for 21st Century Crop Improvement. *Plant Physiol.* **2008**, *147*, 969–977.

Moscou, M. J.; Bogdanove, A. J. A. Simple Cipher Governs DNA Recognition by TAL Effectors. *Science* **2009**, *326*(1501). DOI: 10.1126/science.1178817.

Naito, Y.; Hino, K.; Bono, H.; Ui-Tei, K. CRISPRdirect: Software for Designing CRISPR/ Cas Guide RNA with Reduced off-Target sites. *Bioinformatics* **2014**, *31*(7), 1120–1123. https://doi.org/10.1093/bioinformatics/btu743.

Nishizawa-Yokoi, A.; Cermak, T.; Hoshino, T.; Sugimoto, K.; Saika, H.; Mori, A.; Osakabe, K.; Hamada, M.; Katayose, Y.; Starker, C.; Voytas, D. F.; Toki, S. A Defect in DNA Ligase4 Enhances the Frequency of TALEN-Mediated Targeted Mutagenesis in Rice. *Plant Physiol.* **2016**, *170*(2), 653–666. DOI: 10.1104/pp.15.01542.

Qi, P.; Lin, Y. -S.; Song, X. -J.; Shen, J.-B.; Huang, W.; Shan, J.-X.; Zhu, M.-Z.; Jiang, L.; Gao, J. -P.; Lin, H.-X. The Novel Quantitative Trait Locus GL3.1 Controls Rice Grain Size and Yield by Regulating Cyclin-T1;3. *Cell Res.* **2012**, *22*, 1666–1680.

Qian, B. Y.; Li, X.; Liu, X. L.; Wang, M. Improved Oxidative Tolerance in Suspension Cultured Cells of C_4-Pepctransgenic Rice by H_2O_2 and Ca^{2+} Under PEG-6000. *J. Integr. Plant Biol.* **2015b**, *57*(6), 534–549.

Qian, B. Y.; Li, X.; Liu, X. L.; Chen, P. B.; Ren, C. G.; Dai, C. C. Enhanced Drought Tolerance in Transgenic Rice Over-Expressing of Maize C_4 Phosphoenolpyruvate Carboxylase Gene Via NO and Ca^{2+}. *J. Plant Physiol.* **2015a**, *175*, 9–20.

Qiu, X. J.; Gong, R.; Tan, Y. B.; Yu, S. B. Mapping and Characterization of the Major Quantitative Trait Locus qSS7 Associated with Increased Length and Decreased Width of Rice Seeds. *Theor. Appl. Genet.* **2012**, *125*, 1717–1726. DOI: 10.1007/s00122-012-1948-x pmid:22864386.

Radhamani, T.; Sassikumar, D.; Packiaraj, D.; Saraswathi, R. Analysis of Variability in Induced Mutants of ADT (R) 47 Rice *(Oryza sativa. L.). Int. J. Plant Res.* **2015b**, *28*(4), 111–113. DOI: 10.5958/2229-4473.2015.00092.0.

Radhamani, T.; Sassikumar, D.; Packiaraj, D.; Saraswathi, R.; Ravi, V. Principal Component Analysis of Variation Among the Induced Mutants of Rice *(Oryza sativa L.). Res. Crops.* **2015a**, *16*(4), 698–703. DOI: 10.5958/2348–7542.2015.00098.4.

Ren, C. G.; Li, X.; Liu, X. L.; Wei, X. D.; Dai, C. C. Hydrogen Peroxide Regulated Photosynthesis in C_4-pepc Transgenic Rice. *Plant Physiol. Biochem.* **2014**, *74*, 218–229.

Rizwan, M.; Akhtar, S.; Aslam, M.; Asghar, M. J. Development of Herbicide Resistant Crops Through Induced Mutations. *Adv. Life Sci.* **2015**, *3*(1), 1–8.

Segami, S.; Kono, I.; Ando, T.; Yano, M.; Kitano, H.; Miura, K.; Iwasaki, Y. Small and Round Seed 5 Gene Encodes Alpha-Tubulin Regulating Seed Cell Elongation in Rice. *Rice* **2012**, *5*, 4.

Segami, S.; Yamamoto, T.; Oki, K.; Noda, T.; Kanamori, H.; Sasaki, H.; Mori, S.; Ashikari, M.; Kitano, H.; Katayose, Y.; Iwasaki, Y.; Miura, K. Detection of Novel QTLs Regulating Grain Size in Extra-Large Grain Rice (*Oryza sativa* L.) Lines. *Rice* **2016**, *9*, 34. DOI: 10.1186/s12284-016-0109-2.

Shan, Q.; Wang, Y.; Chen, K.; Liang, Z.; Li, J.; Zhang, Y.; Zhang, K.; Liu, J.; Voytas, D. F.; Zheng, X.; Zhang, Y.; Gao, C. Rapid and efficient gene modification in rice and Brachypodium using TALENs. *Mol. Plant* **2013**, 6, 1365–1368.

Shan, Q. Zhang, Y. Chen, K. Zhang, K. Gao C. Creation of Fragrant Rice by Targeted Knockout of the *OsBADH2* Gene Using TALEN Technology. *Plant Biotechnol. J.* **2015**, *13*, 791–800.

Shao, G. N.; Wei, X. J.; Chen, M. L.; Tang, S. Q.; Luo, J.; Jiao, G. A. et al. Allelic Variation for a Candidate Gene for GS7, Responsible for Grain Shape in Rice. *Theor. Appl. Genet.* **2012**, *125*, 1303–1312. DOI: 10.1007/s00122-012-1914-7 pmid:22772587.

She, K. C.; Kusano, H.; Koizumi, K.; Yamakawa, H.; Hakata, M.; Imamura, T.; Fukuda, M.; Naito, N.; Tsurumaki, Y.; Yaeshima, M. A Novel Factor *FLOURY ENDOSPERM 2* is Involved in Regulation of Rice Grain Size and Starch Quality. *Plant Cell* **2010**, 22, 3280–3294.

Shen, W.; Chen, G.; Xu, J.; Jiang, Y.; Liu, L.; Gao, Z.; Ma, J.; Chen, X.; Chen, T.; Lv. C. Overexpression of Maize Phospho*enol*pyruvate Carboxylase Improves Drought Tolerance in Rice by Stabilization the Function and Structure of Thylakoid Membrane. *Photosynthetica* **2015**, *53*(3), 436–446. DOI: 10.1007/s11099-015-0111-8.

Shen, L.; Wang, C.; Fu, Y.; Wang, J.; liu, Q.; Zhang, X.; yan, C.; Qian, Q.; Wang, K. QTL Editing Confers Opposing Yield Performance in Different Rice Varieties. *J. Integr. Plant Biol.* **2016**. DOI: 10.1111/jipb.12501.

Shomura, A.; Izawa, T.; Ebana, K.; Ebitani, T.; Kanegae, H.; Konishi, S.; Yano, M. Deletion in a Gene Associated with Grain Size Increased Yields During Rice Domestication. *Nat. Genet.* **2008**, *40*, 1023–1028.

Song, X.; Huang, W.; Shi, M.; Zhu, M.; Lin, H. A QTL for Rice Grain Width and Weight Encodes a Previously Unknown RING-type E3 Ubiquitin Ligase. *Nat. Genet.* **2007**, *39*, 623–630.

Song, X. J.; Kuroha, T.; Ayano, M.; Furuta, T.; Nagai, K.; Komeda, N.; Segami, S.; Miura, K.; Ogawa, D.; Kamura, T.; Suzuki, T.; Higashiyama, T.; Yamasaki, M.; Mori, H.; Inukai, Y.; Wu, J.; Kitano, H.; Sakakibara, H.; Jacobsen, S. E.; Ashikaria, M. Rare allele of a previously unidentified histone H4 acetyltransferase enhances grain weight, yield, and plant biomass in rice. *Proc Natl Acad Sci U S A.* **2015**, *112*(1), 76–81.

Sun, W. Q.; Zhou, Q. L.; Yao, Y.; Qiu, X. J.; Xie, K.; Yu, S. B. Identification of Genomic Regions and the Isoamylase Gene for Reduced Grain Chalkiness in Rice. *PLoS One* **2015**, *10*, e0122013.

Sun, Y.; Zhang, X.; Wu, C.; He, Y.; Ma, Y.; Hou, H.; Guo, X.; Du, W.; Zhao, Y.; Xia, L. Engineering Herbicide-Resistant Rice Plants Through CRISPR/Cas9-Mediated Homologous Recombination of Acetolactate Synthase. *Mol. Plant.* **2016**, *9*(4), 628–631. http://dx.doi.org/ 10.1016/j.molp.2016.01.001.

Tai, T. H. Induced Mutations in Rice (*Oryza sativa* L.). *Isr. J. Plant Sci.* **2007**, *55*, 137–145.

Takemoto-Kuno, Y.; Mitsueda, H.; Suzuki, K.; Hirabayashi, H.; Ideta, O.; Aoki, N.; Umemoto, T.; Ishii, T.; Ando, I.; Kato, H.; Nemoto, H.; Imbe, T.; Takeuchi, Y. qAC2, a novel QTL that interacts with Wx and controls the low amylose content in rice (*Oryza sativa* L.). *Theor Appl Genet.* **2015**, *128*(4), 563-573.

Tan, S.; Evans, R. R.; Dahmer, M. L.; Singh, B. K.; Shaner, D. L. Imidazolinone-Tolerant Crops: History, Current Status and Future. *Pest Manage. Sci.* **2005**, *61,* 246–257.

Tanaka, I.; Kobayashi, A.; Tomita, K.; Takeuchi, Y.; Yamagishi, M.; Yano, M.; Sasaki, T.; Horiuchi, H. Detection of quantitative trait loci for stikiness and appearance based on eating quality test in japonica rice cultiva. *Breed. Res.* **2006**, *8*, 39–47.

Tanabata, T.; Shibaya T.; Hori, K.; Ebana, K.; Yano, M. SmartGrain: High-Throughput Phenotyping Software for Measuring Seed Shape Through Image Analysis. *Plant Physiol.* **2012**, *160*, 1871–1880.

Umemoto, T.; Yano, M.; Satoh, H.; Shomura, A.; Nakamura, Y. Mapping of a Gene Responsible for the Difference in Amylopectin Structure Between Japonica-Type and Indica-Type Rice Varieties. *Theor. Appl. Genet.* **2002**, *104*, 1–8.

Verma, D. K.; Mohan, M.; Prabhakar, P. K.; Srivastav, P. P. Physico-Chemical and Cooking Characteristics of Azad Basmati. *Int. Food Res. J.* **2015**, *22*(4), 1380–1389.

Verma, D. K.; Mohan, M.; Asthir, B. Physicochemical and Cooking Characteristics of Some Promising Basmati Genotypes. *Asian J. Food Agro Ind.* **2013**, *6*(2), 94–99.

Verma, D. K.; Mohan, M.; Yadav, V. K.; Asthir, B.; Soni, S. K. Inquisition of Some Physico-Chemical Characteristics of Newly Evolved Basmati Rice. *Environ. Ecol.* **2012**, *30*(1), 114–117.

Verma, D. P.; Srivastav, P. P. Proximate Composition, Mineral Content and Fatty Acids Analyses of Aromatic and Non-Aromatic Indian Rice. *Rice Sci.* **2017**, *24*(1), 21–31.

Wada, T.; Ogata, T.; Tsubone, M.; Uchimura, Y.; Matsue, Y. Mapping of QTLs for Eating Quality and Physicochemical Properties of the Japonica Rice "Koshihikari". *Breed. Sci.* **2008**, *58*, 427–435.

Wan, X. Y.; Wan, J. M.; Weng, J. F.; Jiang, L.; Bi, J. C.; Wang, C. M.; Zhai, H. Q. Stability of QTLs for Rice Grain Dimension and Endosperm Chalkiness Characteristics Across Eight Environments. *Theor. Appl.Genet.* **2005**, *110*, 1334–1346.

Wan, X. Y.; Wan, J. M.; Jiang, L.; Wang, J. K.; Zhai, H. Q.; Weng, J. F. et al. QTL Analysis for Rice Grain Length and Fine Mapping of an Identified QTL with Stable and Major Effects. *Theor. Appl. Genet.* **2006**, *112*, 1258–1270. DOI: 10.1007/s00122-006-0227-0 pmid:16477428.

Wang, E.; Wang, J.; Zhu, X.; Hao, W.; Wang, L.; Li, Q.; Zhang, L.; He, W.; Lu, B.; Lin, H.; Ma, H.; Zhang. G.; He, Z. Control of Rice Grain-Filling and Yield by a Gene with a Potential Signature of Domestication. *Nat. Genet.* **2008**, *40*, 1370–1374.

Wang, F.; Wang, C.; Liu, P.; Lei, C.; Hao, W.; Gao, Y.; Liu, Y.; Zhao, K. Enhanced Rice Blast Resistance by CRISPR/Cas9-Targeted Mutagenesis of the ERF Transcription Factor Gene *OsERF922*. *PloS One* **2016a**, http://dx.doi.org/10.1371/journal.pone.0154027.

Wang, M.; Liu, Y.; Zhang, C.; Liu, J.; Liu, X.; Wang, L.; Wang, W.; Chen, H.; Wei, C.; Ye, X.; Li, X.; Tu, J. Gene Editing by Co-Transformation of TALEN and Chimeric RNA/DNA Oligonucleotides on the Rice *OsEPSPS* Gene and the Inheritance of Mutations. *PLoS One* **2015a**, 10, e0122755.

Wang, S.; Li, S.; Liu, Q.; Wu, K.; Zhang, J.; Wang, S.; Wang, Y.; Chen, X.; Zhang, Y.; Gao, C.; Wang, F.; Huang, H.; Fu, X. The OsSPL16-GW7 Regulatory Module Determines Grain Shape and Simultaneously Improves Rice Yield and Grain Quality. *Nat. Genet.* **2015b**, *47*, 949–954.

Wang, S.; Tholen, D.; Zhu, X. C_4 Photosynthesis in C_3 Rice: A Theoretical Analysis of Biochemical and Anatomical Factors. *Plant Cell Environ.* **2016b**, *40*(1), 80–94. DOI: 10.1111/pce.12834.

Wang, S.; Wu, K.; Yuan, Q.; Liu, X.; Liu, Z.; Lin, X.; Zeng, R.; Zhu, H.; Dong, G.; Qian, Q. Control of grain size, shape and quality by OsSPL16 in rice. *Nat. Genet.* **2012,** *44*, 950–954.

Wang, Y.; Xiong, G.; Hu, J.; Jiang, L.; Yu, H.; Xu, J.; Fang, Y.; Zeng, L.; Xu, E.; Xu, J.; Ye, W.; Meng, X.; Liu, R.; Chen, H.; Jing, Y.; Wang, Y.; Zhu, X.; Li, J.; Qian, Q. Copy Number Variation at the GL7 Locus Contributes to Grain Size Diversity in Rice. *Nat. Genet.* **2015c**, *47*, 944–948.

Weng, J.; Gu, S.; Wan, X.; Gao, H.; Guo, T.; Su, N.; Lei, C.; Zhang, X.; Cheng, Z.; Guo, X.; Wang, J.; Jiang, L.; Zhai, H.; Wan, J. Isolation and Initial Characterization of GW5, a Major QTL Associated with Rice Grain Width and Weight. *Cell Res.* **2008**, *18*, 1199–1209.

Wu, J.; Wu, C.; Lei, C.; Baraoidan, M.; Bordeos, A.; Madamba, M. R. S.; Ramos-Pamplona, M.; Mauleon, R.; Portugal, A.; Ulat, v. J.; Bruskiewich, R.; Wang, G.; Leach, J.; Khush, G.; Leung, H. Chemical- and Irradiation-Induced Mutants of Indica Rice IR64 for Forward and Reverse Genetics. *Plant Mol. Biol.* **2005**, *59*(1), 85–97. DOI: 10.1007/s11103-004-5112-0.

Xiao, W.; Yang, Q.; Chen, Z.; Wang, H.; Guo, T.; Liu, Y.; Zhu, X. Blast-Resistance Inheritance of Space-Induced Rice Lines and Their Genomic Polymorphism by Microsatellite Markers. *Agric. Sci. China* **2009**, 8(4), 101–105.

Xiao, W.; Yang, Q.; Wang, H.; Guo, T.; Liu, Y.; Zhu, X.; Chen, Z. Identification and Fine Mapping of a Resistance Gene to *Magnaporthe Oryzae* in a Space-Induced Rice Mutant. *Mol. Breed.* **2011**, *28*(3), 303–312. DOI: 10.1007/s11032-010-9481-6.

Xie, K.; Yang, Y. RNA-Guided Genome Editing in Plants Using a CRISPR–Cas System. *Mol. Plant* **2013**, *6*(6), 1975–1983. http://dx.doi.org/10.1093/mp/sst119.

Xing, Y.; Zhang, Q. Genetic and Molecular Bases of Rice Yield. *Annu. Rev. Plant Biol.* **2010**, *61*, 421–442.

Xu, R.; Li, H.; Qin, R.; Li, J.; Qiu, C.; Yang, Y.; Ma, H.; Li, L.; Wei, P.; Yang, J. Generation of Inheritable and "Transgene Clean" Targeted Genome-Modified Rice in Later Generations Using the CRISPR/Cas9 System. *Sci. Rep.* **2015**, *5*, Article number: 11491. DOI: 10.1038/srep11491.

Xu, R.; Li, H.; Qin, R.; Wang, L.; Li, L.; Wei, P.; Yang, J. Gene Targeting Using the *Agrobacterium tumefaciens*-Mediated CRISPR-Cas System in Rice. *Rice* **2014**, *7*, 5. DOI: 10.1186/s12284-014-0005-6.

Xu, R.; Wei, P.; Yang J. Use of CRISPR/Cas Genome Editing Technology for Targeted Mutagenesis in Rice. *Methods Mol. Biol.* **2016a**, *1498*, 33–40.

Xu, R.; Yang, Y.; Qin, R.; Li, H.; Qiu, C.; Li, L.; Wei, P.; Yang, J. Rapid Improvement of Grain Weight *via* Highly Efficient CRISPR/Cas9-Mediated Multiplex Genome Editing in Rice. *J. Genet. Genomics* **2016b**, *43*(8), 529–532.

Yamakawa, H.; Hirose, T.; Kuroda, M.; Yamaguchi, T. Comprehensive Expression Profiling of Rice Grain Filling-Related Genes Under High Temperature Using DNA Microarray. *Plant Physiol.* **2007**, *144*, 258–277.

Yang, Q.; Lin, F.; Wang, L.; Pan, Q. Identification and Mapping of Pi41, a Major Gene Conferring Resistance to Rice Blast in the Oryza Sativa Subsp. Indica Reference Cultivar, 93–11. *Theor. Appl. Genetics* **2009**, *118*, 1027–1034.

Yang, Z.; Sun, X.; Wang, S.; Zhang, Q. Genetic and Physical Mapping of a New Gene for Bacterial Blight Resistance in Rice. *Theor. Appl. Genetic* **2003**, *106*, 1467–1472.

Zhang, C.; Zhou, L.; Zhu, Z.; Lu, H.; Zhou, X.; Qian, Y.; Li, Q.; Lu, Y.; Gu, M.; Liu, Q. Characterization of Grain Quality and Starch Fine Structure of Two Japonica Rice (*Oryza sativa*) Cultivars with Good Sensory Properties at Different Temperatures During the Filling Stage. *J. Agric. Food Chem.* **2016**, *64*, 4048–4057 DOI:10.1021/acs.jafc.6b00083.

Zhang, X. J.; Wang, J. F.; Huang, J.; Lan, H. X.; Wang, C. L.; Yin, C. F.; Wu, Y. Y.; Tang, H. J.; Qian, Q.; Li, J. Y.; Zhang, H. S. Rare Allele of OsPPKL1 Associated with Grain Length Causes Extra-Large Grain and a Significant Yield Increase in Rice. *Proc. Natl. Acad. Sci. U. S. A.* **2012**, *109*, 21534–21539.

Zhang, H.; Gou, F.; Zhang, J.; Liu, W.; li, Q.; Mao, Y.; Botella, J. R.; Zhu, J. TALEN-Mediated Targeted Mutagenesis Produces a Large Variety of Heritable Mutations in Rice. *Plant Biotechnol. J.* **2015**, *14*(1), 186–194. DOI: 10.1111/pbi.12372.

Zhang, H.; Zhang, J.; Wei, P.; Zhang, B.; Gou, F.; Feng, Z.; Mao, Y.; Yang, L.; Zhang, H.; Xu, N.; Zhu, J. The CRISPR/Cas9 System Produces Specific and Homozygous Targeted Gene Editing in Rice in One Generation. *Plant Biotechnol. J.* **2014**, *12*(6), 797–807. DOI: 10.1111/pbi.12200.

Zhao, X.; Fitzgerald, M. Climate Change: Implications for the Yield of Edible Rice. *PLoS One* **2013**, *8*, e66218.

Zhao, X.; Daygon, V. D.; McNally, K. L.; Hamilton, R. S.; Xie, F.; Reinke, R. F.; Fitzgerald, M. A. Identification of Stable QTLs Causing Chalk in Rice Grains in Nine Environments. *Theor. Appl. Genetics* **2016**, *129*, 141–153.

Zheng, X.; Li, L.; Li, J.; Zhao D. Constitutive Expression of *McCHITI–PAT* Enhances Resistance to Rice Blast and Herbicide, But Does not Affect Grain Yield in Transgenic Glutinous Rice. *Biotechnol. Appl. Biochem.* **2016b**, *63*(1), 77–85. DOI: 10.1002/bab.1342.

Zheng, X.; Yang, S.; Zhang, D.; Zhong, Z.; Tang, Xu.; Deng, K.; Zhou, J.; Qi, Y.; Zhang, Y. Effective Screen of CRISPR/Cas9-Induced Mutants in Rice by Single-Strand Conformation Polymorphism. *Plant Cell Rep.* **2016a**, *35*(7), 1545–1554. DOI: 10.1007/s00299-016-1967-1.

Zhou, B.; Lin, J. Z.; Peng, D.; Yang, Y. Z.; Guo, M.; Tang, D. Y.; Tan, X.; Li, X. M. Plant Architecture and Grain Yield are Regulated by the Novel DHHC-Type Zinc Finger Protein Genes in Rice (*Oryza sativa* L.). *Plant Sci.* **2017**, *254*, 12–21.

Zhou, H.; He, M.; Li, J.; Chen, L.; Huang, Z.; Zheng, S.; Zhu, L.; Ni, E.; Jiang, D.; Zhao, B.; Zhuang, C. Development of Commercial Thermo-Sensitive Genic Male Sterile Rice Accelerates Hybrid Rice Breeding Using the CRISPR/Cas9-Mediated *TMS5* Editing System. *Sci. Rep.* **2016**, *6*, Article number: 37395. DOI: 10.1038/srep37395.

Zhou, H.; Liu, B.; Weeks, D. P.; Spalding, M. H.; Yang, B. Large Chromosomal Deletions and Heritable Small Genetic Changes Induced by CRISPR/Cas9 in Rice. *Nucleic Acids Res.* **2014**, *42*(17), 10903–10914. DOI: 10.1093/nar/gku806.

Zhou, J.; Peng, Z.; Long, J.; Sosso, D.; Liu, B.; Eom, J.; Huang, S.; Liu, S.; Cruz, C. V.; Frommer, W. B.; White, F. F.; Yang, B. Gene Targeting by the TAL Effector PthXo2 Reveals Cryptic Resistance Gene for Bacterial Blight of Rice. *Plant J.* **2015a**, *82*(4), 632–643 DOI: 10.1111/tpj.12838.

Zhou, L.; Liang, S.; Ponce, K.; Marundon, S.; Ye, G.; Zhao, X. Factors Affecting Head Rice Yield and Chalkiness in *indica* Rice. *Field Crops Res.* **2015b**, *172*, 1–10.

Zhu, D.; Cheng, S.; Zhuang, Y.; Lin, X.; Chen, H. Analysis of Status and Constraints of Rice Production in the World. *Sci. Agric. Sin.* **2010**, *43*, 474–479.

CHAPTER 11

ASSOCIATION MAPPING IN RICE: A HIGH-RESOLUTION MAPPING TECHNIQUE FOR COMPLEX TRAITS

PARMESHWAR KUMAR SAHU[1,*], DEEPAK SHARMA[1,2,]
VIKASH KUMAR[3], SUVENDU MONDAL[2,4,]
GAUTAM VISHWAKARMA[2,5], and B. K. DAS[2,6]

[1]Department of Genetics and Plant Breeding, Indira Gandhi Krishi Vishwavidyalaya, Raipur, Chhattisgarh 492012, India, Mob.: +00-91-8103795885

[2]deepakigkv@gmail.com, Mob.: +00-91-9826647509

[3]Nuclear Agriculture and Biotechnology Division (NA&BTD), Bhabha Atomic Research Centre, Mumbai, Maharashtra 400085, India, Tel.: +91-22-25592331, vikash007barc@gmail.com

[4]suvenduhere@yahoo.co.in, Tel.: +91-22-25590779

[5]gtmvish@barc.gov.in, Tel.: +91-22-25593632

[6]bkdas@barc.gov.in, Tel.: +91-22-25592640

*Corresponding author. E-mail: parmeshwarsahu1210@gmail.com

11.1 INTRODUCTION

We can say that plant breeding is an art, science, and technology to change the genetic architecture of crop plants by using the principles of genetics in order to produce desired characteristics as per high economic values. The basic aim of genetics and plant breeding is to connect genotype to phenotype to improve the performance of crop plants. Association mapping (AM) is one of the techniques of modern plant breeding to connect plant phenotype with their genotype to define some new alleles/genes/quantitative trait loci

(QTLs) in large natural population. AM seeks specific alleles or loci linked to phenotyping differences in a trait. In short, AM can be defined as the technique of gene localization by linkage disequilibrium (LD), without cloning. Although this technique has some similarities with QTL mapping, they differ in terms of nature of population and statistical approaches. Identification and utilization of QTLs for desirable and important traits needs its tagging and mapping using molecular markers in the genome. QTLs/genes identification and mapping for desired trait is valuable for map-based cloning of the tagged genes and integrating these QTLs/genes into desirable cultivars through marker-assisted breeding (Asins, 2002; Khaing et al., 2014; Sehgal et al., 2016). Techniques of AM situate the QTLs/genes by correlating the phenotype of desired trait with genotypic data obtained using molecular markers in unstructured or loosely structured population. AM has the ability to identify and map the QTLs/genes with higher accuracy as compared to linkage-based QTL mapping therefore it is more powerful technique for marker-assisted selection, gene discovery, gene tagging, gene pyramiding, and sequence diversity with heritable phenotypic differences (Mackay and Powell, 2007).

AM stands on the principle that over several generations of recombination, correlations of linked molecular markers with trait of interest remain stable. Therefore, counterfeit associations between landrace/lines/genotype and phenotype can be perceived due to the degree of structure within the population, requiring the development of different statistical technique to explain for population structure (Balding, 2006; Vinod, 2011). AM give surety of high-resolution mapping by utilization of historical recombination events at the population level that may facilitate genetic fine mapping on non-model individuals where family-based linkage based approaches would not be possible. AM exploits ancestral recombination and natural genetic diversity within a population to scrutinize quantitative traits (Sahu et al., 2015). This technique could be potentially utilized for identifying desirable allele, tagging of genes, fine mapping of genes/QTLs and validation of outcome of linkage based mapping (Rosyara and Joshi, 2012). Association analysis is mainly based on number of meiotic recombination events happened between DNA markers and QTLs of trait of interest. When the experimental mapping populations namely; double haploids, F_2, $F_{2:3}$, recombinant inbred lines (RILs), near isogenic lines (NILs) and MAGIC population are produced by selfing for linkage analysis, the reduction in disequilibrium is not adequate as contrast to random mating population because of limited number of meiotic recombination events therefore high resolution mapping of QTLs cannot be done by using these populations. In random mating populations like germplasm lines, elite breeding lines and landraces, long-distance associations obstruct accurate positioning of the gene/QTL. In such genotypes, adequate meioses have taken place to decrease disequilibrium between fairly linked molecular markers (Sehgal et al., 2016).

In rice, there is lots of opportunity to map and tag the QTLs for different quantitative traits like yield, drought tolerance, salt tolerance, root length, panicle length and so forth through AM using available germplasm lines or landraces. Several scientists have already done AM on rice for various quantitative traits and got significant results. In this chapter we will elaborate the techniques of AM in rice and their significant outcomes for future breeding program. In this chapter we will elaborate the theoretical principles and techniques of AM in rice and their significant outcomes and applications for future breeding program.

11.2 LINKAGE ANALYSIS VERSUS LINKAGE DISEQUILIBRIUM ANALYSIS

11.2.1 LINKAGE

The inheritance of two or more genes together as a single haplotype without any substantial recombination frequency in a family or pedigree is called linkage. Two or more genes are linked due to their close proximity to each other, mostly on the genome. Genetic linkage can be defined as the tendency of two or more closely located alleles to be transmitted together from one generation to next generation. It has been now proved that genes/markers which are located closely on same chromosome may have very less chance to be separated onto different chromatids during recombination therefore it can be stated that they are linked genetically.

11.2.2 LINKAGE EQUILIBRIUM (LE)

The transmission of alleles at one locus across the generation is independent to alleles of other locus; this phenomenon is called linkage equilibrium. LE takes place when alleles can associate randomly that is genes present at one locus is completely independent of the genes at the second locus. The chance of finding one allele at one locus is independent of finding another allele at another locus. If two genes/traits/loci are in linkage equilibrium, it indicates that they are transmitted completely independently in every generation.

Those genes may be present in different chromosome or distantly located in same chromosome.

11.2.3 LINKAGE DISEQUILIBRIUM (LD)

LD is the result of physical linkage of genes. It is also called as gametic phase disequilibrium. LD can be shortly defined as the non-random association among the alleles of two or more between genetic loci. These genetic loci may be present on the same or different chromosome of the organism. When the alleles of two loci are not significantly independent to each other than LD occurs between those loci. LD can be misled by two main reasons; first is, if the two genes are unlinked then there also could be a possibility of non-random association among the alleles of these genes and second is that if two genes are linked, then this does not stand for that these genes will be present in LD condition. If two genes/loci are in LD, it means that certain alleles of each gene are inherited together. This may be due to actual genetic linkage or due to some form of functional interaction where some combinations of allele at two loci affect the viability of potential offspring that is phenotype. LD could be the outcome of recent migration, recent selection, and new mutation increased by self-pollination, inbreeding, low recombination rate, genetic isolation between lineage, population admixture, population subdivision and epistasis. Some genetic factors such as higher recombination rate, gene conversion, more out-crossing, higher mutation rate could be the reason of decay in LD (Yu and Buckler, 2006; Vinod, 2011). Mackay and Powell (2007) stated that LD decays rapidly when loci are unlinked but when the loci are closely linked then the rate of decay in LD is very slow.

11.2.4 LINKAGE BASED MAPPING AND LINKAGE DISEQUILIBRIUM BASED MAPPING (ASSOCIATION MAPPING)

AM and linkage based QTL mapping both depends upon coinheritance of functional polymorphisms and neighboring DNA markers. The main difference is that in linkage analysis, only a few meiotic recombination occurs within the families with identified ancestry, which results in comparatively low mapping resolution; whereas in LD mapping, germplasm resources or landraces are used as mapping population where several unknown historical meiotic recombination were utilized for high resolution mapping (Zhu et al., 2008). In crop plants, QTL mapping based on linkage analysis has been generally carried out by using bi-parental mapping populations which are structured populations (Fig. 11.1). LD based mapping used the communal inheritance between phenotypic trait and DNA markers for a collection of genotypes often with unseen ancestry and this common inheritance will seen continue many generations of recombination also. Basically, evolutionary and historical recombinations at the population level are exploited for the study of LD based mapping (Yu and Buckler, 2006).

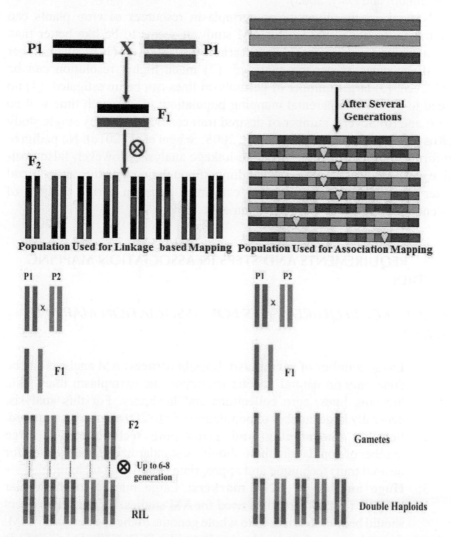

FIGURE 11.1 **(See color insert.)** Difference in population structure used for linkage analysis based mapping and linkage disequilibrium based mapping.

By using principles of LD, AM finds out the correlations between phenotypic variations and genotypic data using sophisticated computer softwares (Borba et al., 2010). As harmonizing approach, linkage analysis frequently recognizes large genomic region of desired trait with comparatively low molecular marker density whereas LD mapping gives high resolution by exploring the ancestry data on candidate genes or by exploring whole genome by large number of molecular markers (Thornsberry et al., 2001; Hirschhorn and Daly, 2005).

Natural genetic diversity or germplasm resources of crop plants can be most effectively utilized for AM study. It seems to be like better than linkage based mapping of QTLs in various ways such as, (1) greater number of allelic diversity can be analyzed; (2) much higher resolution can be achieved; (3) larger number of germplasm lines can be investigated; (4) no need to generate bi-parental mapping population; (5) research time will be less; and (6) Several number of desired trait can be mapped by single study (Rosyara and Joshi, 2012; Zhu et al., 2008; Sehgal et al., 2016). No pedigree information required for AM unlike linkage analysis. However, false positive can be increased due to mishandling of population structure and varietal inter-relationships. Therefore, fine care must be taken during analysis of phenotypic and genotypic data (Sahu et al., 2015).

11.3   REQUIREMENTS AND STEPS IN ASSOCIATION MAPPING STUDIES

11.3.1   KEY REQUIREMENTS FOR ASSOCIATION MAPPING STUDY

1. **Large number of germplasm lines/landraces:** AM analysis can be done only on natural genetic resources like germplasm lines, elite breeding lines, core collections and landraces. For this analysis, generally large number of population size (> 200 number) is required.
2. **Suitable phenotyping and genotyping technique:** For large number of population, there should be standardized phenotyping (for desired trait) technique and appropriate genotyping technique.
3. **Huge number of DNA markers:** Large number of molecular markers (> 150) should be used for AM studies. Molecular markers should be well distributed to whole genome of the species in GWAM. Several types of molecular markers like SSR, SNP, DArT, AFLP, RFLP and so forth could be used.

4. **Sophisticated molecular biology laboratory:** A well established and well equipped molecular laboratory is essential for genotyping work.
5. **Appropriate statistical software:** LD analysis requires a lot of complex statistical analysis. These analyses would be performed by statistical softwares like TASSEL, EMMA/R, ASREML, JMP Genomics, SAS, GenStat, STRUCTURE, GOLD and POWER-MAKER and so forth.
6. **Technical expert personnel:** At last but not the least, well expertise technical personnel are required for performing genotyping work and statistical analysis.

11.3.2 MAIN STEPS IN ASSOCIATION MAPPING

AM study is completed in six major steps which are explained below with the help of flowchart in Figure 11.2:

1. *Collection of plant materials to be used in study:* Collection of germplasm/landraces/natural population with wide range of genetic diversity for study.
2. *Phenotyping:* Observable traits such as morphological traits, yield, quality, resistance, and so forth in the whole genotypes should be measured properly. If possible, observations can be taken in different environments with multiple replications. The experimental trial should be grown in appropriate experimental design like augmented design, alpha lattice design and so forth.
3. *Genotyping:* Genotyping of the mapping population individuals with molecular markers like SSR, SNP, DArT, EST, and so forth. SNP and SSR markers are most effective molecular markers for population and kinship studies because they are multiallelic, highly reproducible, PCR-based and generally selectively neutral. Predominantly SNPs and SSRs markers are used to characterize population structure in rice (Garris et al., 2005; Agrama et al., 2007; Gupta et al., 2014; Sehgal et al., 2016).
4. *Measures of linkage disequilibrium:* Molecular marker data could be used to do explore the level of LD of the genome of chosen population. All these process could be performed by the help of appropriate computer software.

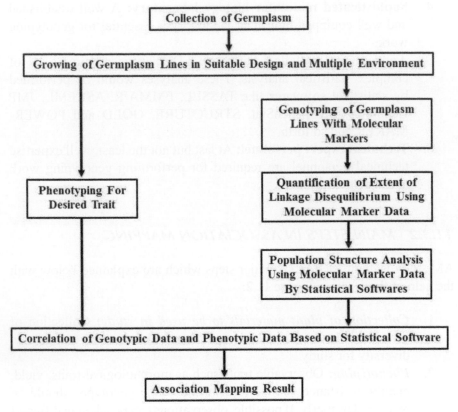

FIGURE 11.2   Flowchart of general steps of association mapping studies.

5. ***Population structure analysis:*** Population structure (level of genetic differentiation among groups within sampled population individuals) and kinship (coefficient of relatedness between pairs of each individual within a sample) would be evaluated by suitable software. Population structure can reduce the type I and type II error between DNA markers and desired phenotype during AM analysis. Population structure and ancestral relationship can be evaluated by various available computer based statistical methods such as structured association (SA) (Pritchard and Rosenberg, 1999; Falush et al., 2003), genomic control (Devlin and Roeder, 1999), mixed model approach (Yu et al., 2006) and principal component approach (Price et al., 2006; Zhu et al., 2008).

6. ***Correlation of phenotypic and genotypic data:*** After LD quantification and population structure analysis, useful information generated

to correlate the phenotypic and genotypic data by the help of suitable statistical approach. Based on this correlation, 'marker tags' can be positioned in the range of desired QTLs (Gupta et al., 2005; Sehgal et al., 2016).

11.4 TYPES OF ASSOCIATION MAPPING

Selection of AM study is depends upon the extent of LD level on the population used and objectives of the study. If LD level is higher in population then less number of DNA markers required for associating the marker-trait phenotype but this can reduces the mapping resolution. On the other hand, when reduced level of LD is present in population then it requires comparatively more number of DNA markers to scan the desired QTL/gene however this can enhance the mapping resolution. Based on the extent of LD, number of markers required and desired genome coverage, AM study can be grouped into two major divisions, namely; (1) candidate-gene based AM, and (2) GWAM.

11.4.1 CANDIDATE GENE ASSOCIATION MAPPING (CGAM)

In this approach, only selected candidate genes are targeted for mapping which has specific role in governing the desired phenotype. This technique can be done easily when some basic information about genetics of desired trait is available on the basis of previous research. There is no need to scan whole genome of organism to map the candidate gene; however it can be performed by scanning any specific chromosome or particular region of genome. An independent set of random molecular markers needs to be scored to infer genetic relationship. It is a low cost, hypothesis driven and trait specific approach but definitely miss other unknown loci. By this approach, high resolution mapping or fine mapping could be performed easily. This could also be used to find out polymorphisms (SNPs or haplotypes) linked with the trait of interest subsequent to GWAS (Gupta et al., 2014).

The scientific distance between biometrical genetics and molecular genetics can be joined by candidate gene-based AM to study complex traits (Cattivelli et al., 2008). Candidate gene studies required less number of molecular markers for analysis. Various candidate gene-based AM has been reported for various desired traits in different crops using tens to hundreds of DNA markers (Vinod, 2011). In plants, the first candidate gene-based

AM study were reported in maize for flowering time (Remington et al., 2001; Thornsberry et al., 2001), and were later followed by studies mainly involving genome-wide approach. Candidate gene-based approach can also be used in parallel with GWAM study to enhance the efficiency and accuracy of QTL identification and positioning but it takes lot of effort and cost (Gupta et al., 2014).

11.4.2   GENOME-WIDE ASSOCIATION STUDIES OR MAPPING (GWAS/GWAM)

GWAM is the comprehensive approach to systematically scan the whole genome of organism with large number of equally distributed molecular markers to detect the genetic variations for a particular trait of interest. Genome-wide prediction is also known as genomic selection (GS) which can be defined as the method of detecting individuals in an experimental population as much important as parents of the next generation of offspring using large number of evenly distributed DNA markers (Begum et al., 2015). GWAS requires high number of molecular markers to detect desired QTLs. GWAS can be used to find out all the genomic regions/QTLs/genes that may involve in expression of desired trait. GWAS does not require prior information regarding candidate genes (Risch and Merikangas 1996; Zhu et al., 2008; Braulio and Cloutier, 2012).

A number of GWAS have been carried out in crop plants to find out association of desired QTLs with molecular markers and gene tagging. However, GWAS can sometimes provide statistically significant outcome at loci unrelated to the trait of interest due to LD between multiple factors, of which only some are physically linked with the trait. These can be partially resolve by using recent models viz., multilocus mixed model (MLMM) (Segura et al., 2012) and multi-trait mixed model (MTMM) (Korte et al., 2012).

11.5   METHODS OF ASSOCIATION MAPPING ANALYSIS

The methods or models, on hand for association mapping analysis consist of the following:

1.   Single locus generalized linear model (GLM) (Reeves and Richards, 2009)

2. Single locus mixed linear model (MM or MLM) (Yu et al., 2006)
3. Multilocus mixed linear model (MLMM) (Zhang et al., 2010)

It is usually notorious that MLM and MLMM models are more informative and useful as compare to GLM. There are several approaches for handling the population structure which uses any of the above mentioned models. Brief description of these approaches is given below:

11.5.1   GENOMIC CONTROL (GC)

Delvin and Roeder (1999) firstly reported genomic control demonstrated that the effects of cryptic relatedness and population substructure on test statistics of interest are essentially constant across the genome. It is based on the principle that if population stratification affects the candidate gene, then it will also affect the unrelated null allele or loci. Genomic control follows the principle of identifying association of large number DNA markers which are not involved in controlling the trait and identified markers could be used to remove the effect of population stratification. Genomic control approach utilizes huge population-based samples and large molecular marker data to identify the trait/gene association. The GC approach exploits the fact that population substructure generates 'over-dispersion' of statistics used to assess association. Genomic control can be estimated experimentally by using large number of molecular markers data distributed though-out the genome which is different from expected null distribution. Single degree of freedom test can be the best option for genomic control (Devlin and Roeder, 1999; Reich and Goldstein, 2001).

11.5.2   STRUCTURED ASSOCIATION (SA)

This method was firstly given by Pritchard et al. (2000). Population structure can cause some spurious associations between a candidate gene/marker and trait of interest. To seek out this problem, structure association can be used. Structure association method utilizes a series of unlinked DNA markers to find out the details of population structure and the ancestry of sampled individuals. This process was performed before using genotypic data to test for associations within subpopulations. SA method is completely based on the hypothesis that association between a candidate gene/marker and the trait within a subpopulation is not due to population structure. STRUCTURE and TASSEL is the most commonly used computer program for structure association (Pritchard et al., 2000; Falush et al., 2003).

11.5.3 PRINCIPAL COMPONENT ANALYSIS (PCA)

Principal component analysis (PCA) was invented in 1901 by Karl Pearson, as an analogue of the principal axis theorem in mechanics of PCA is a common tool that has been widely used for the combined analysis of correlated phenotypes in genetic linkage and association studies. It is a simple, non-parametric method of extracting relevant information from confusing data sets. With minimal additional effort PCA provides a roadmap for how to reduce a complex data set to a lower dimension. PCA can be said as a statistical technique to alter a set of observations of possibly correlated variables into a set of values of linearly uncorrelated variables called principal components (Mackay and Powell, 2007).

11.5.4 MULTIPARENTADVANCEDGENERATIONINTERCROSS (MAGIC)

In plants community, multiparent advanced generation intercross populations were firstly described and established in *Arabidopsis* (Cavanagh et al., 2008). MAGIC population can be developed by multiple intercrossing of eight diverse genotypes. This approach has been applied for crop improvement by providing high mapping resolution and by interrogating the multiple alleles (Cavanagh et al., 2008). The consecutive rounds of recombination among the genotypes can decayed the LD which increases the accuracy of QTL/gene position. However, MAGIC population can take time and input for its development but now days these are quite beneficial for plant breeder for fine mapping of QTLs because of availability of SNP platform and advanced statistical tools for huge analysis. MAGIC lines occupy an intermediate niche between naturally occurring accessions and existing synthetic populations. MAGIC population has been developed by closely related but diverse genotypes which decayed the LD in greater level. Due to this property, MAGIC requires less number of markers to make sure the same level of LD to identify the association between phenotype and markers (Cavanagh et al., 2008; Mott and Flint, 2002; Mott et al., 2000).

11.5.5 HAPLOTYPE ANALYSIS (HA)

Haplotype can be defined as the set of strongly linked DNA markers situated on the same chromosome and have tendency to be transmitted together from one to next generation. It gives better understanding about recombination frequency within the natural diversity or germplasms and gives fair idea about range of sequence diversity. In LD mapping, haplotype analysis can be significantly used to interpret the genetic bases of complex traits in crop plants. According to Buntjer et al. (2005), use of haplotypes rather than single markers can be an effective way of improving allele or QTL detection power. According to historical relationship within the natural population, haplotype can be grouped into different classes to enhance the detection power and to estimate the phenotypic and genotypic values of unseen germplasm. Association studies based on either a single marker or haplotype have identified genetic variants and underlying genetic mechanisms of quantitative traits (Buntjer et al., 2005; Mackay and Powell, 2007).

11.5.6 TRANSMISSION DISEQUILIBRIUM TEST (TDT)

Spielman et al. (1993) first time proposed the transmission disequilibrium test (TDT) to discriminate the marker-trait associations in landraces/germplasm lines/natural populations/elite breeding lines. It is said that TDT is an application of McNemar's test. TDT can detect genetic linkage only in the presence of genetic association. This genetic association can be caused by population structure, which makes the TDT robust but it is vulnerable to enhance in false positive outcome generated by genotype error and unfair allele calling. This dilemma can be resolved by evaluating the transmission ratio for extreme phenotypes with that for control individuals (Spielman et al., 1993; Mitchell and Chakravarti, 2003).

11.6 SOFTWARES USED FOR ASSOCIATION MAPPING ANALYSIS

A list of software packages, available for AM is given below in Table 11.1. There are several free and paid softwares available for AM analysis. Most of the programmes are useful only for animal system but nowadays many softwares are available which can be easily used in plant system. Therefore, time to time new and more efficient computer softwares for AM analysis are regularly under development. STRUCTURE and TASSEL are the most

TABLE 11.1 List of Softwares Available for Association Mapping Analysis

S. No.	Name of Software	Principle	Basic information	Website
1	STRUCTURE	Genetic structure of population	Proportion of the genome of an individual of experimental population can be estimated by MCMC Bayesian analysis	http://pritch.bsd.uchicago.edu/software.html
2	TASSEL	LD and AM, stratification	Based on MLM and GLM methods, SSR markers	http://www.maizegenetics.net
3	BAPS	Genetic structure of population	Proportion of the genome of an individual of experimental population can be estimated by MCMC Bayesian analysis. BAPS software is designed to meet the increasing need for analyzing large-scale population genetics data. Also grouped the individuals into genetic clusters	http://web.abo.fi/fak/mnf/mate/jc/software/baps.html
4	EMMA	LD and population structured	Mixed model, corrects for the confounding from population structure and genetic relatedness	http://mouse.cs.ucla.edu/emma/
5	EMMAX	LD and population structured	Large-scale association mapping, corrects for the confounding from population structure and genetic relatedness, increased computational speed	http://genetics.cs.ucla.edu/emmax/
6	EIGENSOFT	PCA, LD	Uses PCA method to explain the ancestral differences between cases and controls	http://www.hsph.harvard.edu/alkes-price/software/
7	MIDAS	LD	Calculate linkage disequilibrium (LD) heat map. Inter loci and inter allelic LD variation can be estimated	http://www.genes.org.uk/software/midas/
8	SAS	Programming, AM	Generic program commonly used in data analysis; commercial	http://www.sas.com
9	PLINK	Structured AM, LD and stratification	Utilize MLM method, perform IBD and IBS analyses using SNPs	http://pngu.mgh.harvard.edu/purcell/plink/
10	GGT	Genetic analysis of LD and AM	Genetic distance can be calculated by using Jaccard similarity coefficient. Based on genetic distance, dendrograms can be prepared using Neighbor-Joining algorithm LD heatmaps and LD scatter plots are also calculated	http://www.plantbreeding.wur.nl/UK/software_ggt.html

TABLE 11.1 *(Continued)*

S. No.	Name of Software	Principle	Basic information	Website
11	SVS 7	AM, LD and population stratification	Calculate LD, analyze multiple AM for large number of markers and populations size. Estimate stratification and haplotype blocks	http://www.goldenhelix.com
12	mStruct	Genetic structure of population	Admixture model Population structure can be detected with admixing and mutations using observed DNA markers	http://www.cs.cmu.edu/~suyash/mstruct.html
13	LD heatmap	LD	LD heatmap can be estimated by using R environment software using SNPs	http://www.jstatsoft.org/v16/c03
14	LD hat 2.1	LD and recombination rate	LD heatmap can be estimated by using R environment software using SNPs. Bayesian MCMC approach for identification of hotspots using SNPs data	http://www.stats.ox.ac.uk/mcvean/ldhat.html
15	Arlequin	LD and genetic analysis	Calculate LD heatmap hierarchical analysis of genetic structure (AMOVA)	http://cmpg.unibe.ch/software/arlequin
16	Haploview	LD and haplotype study	Analyze LD and haplotype variation. Estimation of population frequency of haplotype. Testing of haplotype and single SNP	http://www.broad.mit.edu/mpg/haploview/
17	GenStat	AM, LD and population stratification	Utilize general linear model and mixed linear model for LD and AM study with SSR markers. PCA methods is also performed	http://www.vsni.co.uk/
18	GenABEL	AM	Performs GWAS for quantitative as well as binary traits	http://www.genabel.org/packages/GenABEL
19	Merlin	AM	Includes an integrated genotype inference feature for improved analysis when some genotypes are missing, does not control for population stratification of its own	http://www.sph.umich.edu/csg/abecasis/merlin/tour/assoc.html
20	Matapax	LD	GWAS is performed in R environment with EMMA and GAPIT libraries; performs all essential steps for basic GWAS, population structure, fast computation	http://matapax.mpimpgolm.mpg.de

TABLE 11.1 (Continued)

S. No.	Name of Software	Principle	Basic information	Website
21	JMP genomics	Structured AM, LD and population stratification	Candidate gene AM and Genome wide AM can be performed using SNPs Common and rare variants can be analyzed	http://www.jmp.com/software/genomics
22	ASReml	Population structure	Handle large data set, calculates population structure and pedigree-based kinship; commercial	http://www.vsni.co.uk/software/asreml
23	GenAMap	Structured AM, LD and Population Stratification	Hierarchical analysis of genetic structure AM and LD can be analyzed using SNPs	http://cogitob.ml.cmu.edu/genamap
24	FaST-LMM	LD, Population structure	Large data sets that is up to 120,000 individuals can be analyzed at a time	http://fastlmm.codeplex.com/
25	'R' software	Programming, AM	Generic, commonly used for programming	http://www.rproject.org
26	SeqFeatR	Programming, AM, Follows the frequentist and Bayesian methods for the discovery of statistical sequence-feature association	Used to identify associations between mutation patterns in biological sequences and specific selection pressures ("features")	https://seqfeatr.zmb.uni-due.de
27	The Implicitome	GWAS, Programming	Used for rationalizing gene-disease associations. Used in conjunction with experimental data resources to rationalize both known and novel associations	http://dx.doi.org/10.5061/dryad.gn219
28	GWASTools	GWAS, R/Bioconductor package. Based on R.	Used for quality control and analysis of genome-wide association studies. It brings the interactive capability and extensive statistical libraries of R to GWAS.	http://bioconductor.org/packages/GWASTools/
29	NAM 1.4.8	R package, Population stratification	Used to account prior information regarding population stratification to relax the linkage phase assumption. Allows markers to be treated as a random effect to increase the resolution power and avoid double fitting markers into the model.	https://CRAN. R-project.org/package=NAM

TABLE 11.1 *(Continued)*

S. No.	Name of Software	Principle	Basic information	Website
30	GWASpi	GWAS, mixed models	It has the ability to detect phenotype-genotype associations in the presence of population stratification and multiple levels of relatedness in genome-wide association studies	http://www.gwaspi.org/
31	GWAPP	Manhattan plots and LD plots	Enables researchers working with *Arabidopsis thaliana* to do genome-wide association mapping (GWAS) on their phenotypes	http://gwapp.gmi.oeaw.ac.at/
32	GLOGS	GWAS, Mixed model-based system	Performs a GWAS by estimating parameters of a logistic risk model based on non-genetic covariates and a polygenic effect and by evaluating associations between markers and the disease using score tests	http://www.bioinformatics.org/~stanhope/GLOGS/

Source: Braulio and Cloutier (2012); Gupta et al. (2014).

frequently used softwares for LD mapping analysis in plants. Regular updates would be done in these softwares according to need and to do analysis easier (Braulio and Cloutier, 2012). There are several purchasable and free computer softwares are available in the market for AM analyses which are listed out in Table 11.1.

11.7   ACHIEVMENTS OF ASSOCIATION MAPPING IN RICE

Rice (*Oryza sativa* L.) is the world's second most important cereal crop after maize based on grain production. Generally, Asian countries have the largest share in world rice production. China is the world's largest rice producer country while India is the second largest rice producer country. Rice is the vital staple food crop for more than half of the world's population with a total world production of 471.69 million t (Anonymous, 2016). The breeding strategy of any crop could be successful if the crop itself has large genetic diversity. Rice crop has very large genetic diversity present in farmers' cultivars, landraces, as well as germplasm lines in some established organization. AM studies in rice allows the rice breeders/researchers to utilize the natural genetic variation for identifying and positioning the new genes/ QTLs by exploiting the extent of LD (Zhu et al., 2008). This technique has been successfully applied in rice as well as in other crops also to identify marker–trait association. Some major achievements of AM in rice are given below in Table 11.2.

11.8   ADVANTAGES AND LIMITATIONS OF ASSOCIATION MAPPING

11.8.1   ADVANTAGES

1.   Facilitate high-resolution mapping/fine mapping of a QTL controlling the desired trait.
2.   It has the power to identify new and superior alleles in the genome of any organism.
3.   No need to develop mapping population. AM uses diverse germplasm lines/landraces/elite breeding lines which contains most relevant genes in segregating mode.
4.   Several QTLs can be identified and mapped at a single analysis by GWAS.

TABLE 11.2 Some Important Achievements of Association Mapping Studies in Rice Crop

No. of Genotypes Used	DNA marker used	Character studied	Number of marker–trait associations found	Computer softwares used	References
350 rice accessions	SNP	Seed germination percentage (GP) in freshly harvested seeds, and after-ripened seeds (ARS)	GP- 16 SNPs ARS- 38 SNPs	FaST-LMM	Magwa et al. (2016)
184 rice varieties	SSRs	Nitrogen use efficiency traits (NUE)	NUE-8 SSR markers	STRUCTURE, TASSEL	Liu et al. (2016)
469 Indica rice diversity	SNPs	Grain shape traits	Identified 47 SNPs within 27 significant loci for four grain shape traits. Total of 424 candidate genes were predicted	SAS 9.0, GAPIT, STRUCTURE TASSEL	Feng et al. (2016)
371 rice accessions	SNPs	Drought recovery traits	Only one significant association on chromosome 2 for drought recovery trait was found	EMMA STRUCTURE	Zaniab et al. (2015)
523 rice germplasm	SNP	Tiller number (TN), tiller angle(TA), plant height (PH), flag leaf length (FLL), flag leaf width (FLW), flag leaf length 6 and width ratio (FLLW), flag leaf angle (7FLA), panicle number (P8N), panicle type (PT), panicle length (PL), pericarp color (PC), hull color (HC)	In total, 126 and 172 significant SNPs were identified as associated with different 12 traits	STRUCTURE TASSEL SAS 9.0. SPADiGe	Lu et al. (2015)
220 rice accessions	SNP	Salinity tolerance (ST)	ST- 64 SNPs	GAPIT STRUCTURE PLINK TASSEL	Kumar et al. (2015)
63 rice landraces	SSRs	Heading date (HD) and low-temperature germinability (LTG)	HD-6 SSRs LTG-17 SSRs	FATESer	Fujino et al. (2015)

TABLE 11.2 *(Continued)*

No. of Genotypes Used	DNA marker used	Character studied	Number of marker–trait associations found	Computer softwares used	References
366 Indica rice accessions	SNP	Rice blast resistance (BR)	BR- 30 SNPs	EMMAX algorithm	Wang et al. (2014)
420 rice accessions	SNP	Cold tolerance (CT)	CT-190 SNPs	STRUCTURE TASSEL	Eizenga et al. (2014)
150 Rice landraces	SSR	Total 12 agronomic traits	76 significant trait-marker associations were detected using mixed linear model (MLM). A total of seven aforementioned trait-marker associations were verified	POWER-MARKER, STRUCTURE, SAS version 9.0, TASSEL	Zhang et al. (2014)
540 rice cultivars	SSRs	Seed vigor evaluated by root length (RL), shoot length (SL), shoot dry weight (SDW)	RL-7 markers, SL-12 markers, SDW-8 markers	STRUCTURE, Arlequin, SPAGe-Di, TASSEL	Dang et al. (2014)
20 rice accessions	SNPs	Genotype × environment interactions for agronomic traits	10 agronomic traits-23 putative loci	STRUCTURE, TASSEL	Xu et al. (2014)
167 Japonica rice accessions	SNP DArT	Root traits	19 associations were significant at P<1e-05, and 78 were significant at P<1e-04. 51 unique loci were detected;	SAS v9.2 FaST-LMM TASSEL	Courtois, et al. (2013)
347 rice accessions	SSRs	Cold tolerance	Cold tolerance- 24 markers, Yunnan (natural low temperature)-5 markers, Jilin (cold water irrigation)-19 markers	STRUCTURE, Arlequin, SPAGe-Di, NTSYSpc, TASSEL	Cui et al. (2013)

TABLE 11.2 *(Continued)*

No. of Genotypes Used	DNA marker used	Character studied	Number of marker–trait associations found	Computer softwares used	References
22 rice accessions	SSRs	Grain yield per plant, 100 grain weight, panicle length	Grain yield per plant-3 markers, 100 grain weight-1 marker, panicle length-5 markers	DARWin, STRUCTURE, TASSEL	Vanniarajan et al. (2012)
217 rice core collections	SSRs	Sheath blight (ShB) resistance	ShB resistance-10 markers	STRUCTURE, PowerMarker, MEGA 4.0, NT-SYSpc, TASSEL	Jia et al. (2012)
203 rice accessions	SSRs, InDel	Harvest index traits in temperate (Arkansas) and subtropical climates (Texas)	Harvest index traits in Arkansas-36 markers, Harvest index traits in Texas-28 markers	STRUCTURE, SPAGeDi, PDMIX800 SAS macro	Li et al. (2012)
413 rice germplasm lines	SNPs	Total 34 traits related to plant morphology, grain quality, plant development and agronomic performance	Several QTLs for different 34 traits were identified.	EIGENSOFT PLINK EMMA	Zhao et al. (2011)
242 rice accessions	SSRs	Grain yield (YLD), tiller number per plant (TILN), panicle number per plant (PANN), yield from ratooning (RYLD), amylose content (AC), head milled rice (MR)	Total of 23 SSR markers were identified as significantly associated with different grain yield and quality traits	STRUCTURE, FSTAT Genetix 4.03, SPAGeDi, TASSEL	Borba et al., (2010)
170 rice accessions	SSRs, InDel	Heading date during 2006 (HD06), heading date during 2007 (HD07), plant height in 2006 (PH06), plant height in 2007 (PH07), panicle length in 2006 (PL06), panicle length in 2007 (PL07)	HD06-3 markers, HD07-3 markers, PH06-3 markers, PH07-2 markers, PL06-2 markers, PL07-1 marker	STRUCTURE, NYSYS-pc, SPAGeDi, TASSEL	Wen et al. (2009)

TABLE 11.2 *(Continued)*

No. of Genotypes Used	DNA marker used	Character studied	Number of marker–trait associations found	Computer softwares used	References
90 rice accessions	SSRs, InDel	Single stigma exertion (SStgE), dual exertion (DStgE), Total stigma exertion (TStgE), spikelet length (SpkL), spikelet width (SpkW), spikelet length/width ratio (SpkL/W)	SStgE-4 markers, DStgE-6markers, TStgE-5markers, SpkL-4 markers, SpkW-3 markers, SpkL/W-7 markers	STRUCTURE, GenAlex, SPAGeDi, TASSEL	Yan et al. (2009)
103 rice germplasm	SSRs	Grain yield (GY), kernel length (KL), kernel width (KW), kernel length/width ratio (LWR), 1000 kernel weight (TKW)	GY-5 markers, KL-6 markers, KW-5 markers, LWR-5 markers, TKW-4 markers	STRUCTURE, Power Marker 3.23, Arlequin 3.0, TASSEL	Agrama et al. (2007)

5. Sample size and structure do not need to be as large as for linkage studies to obtain similar power of detection.
6. It can be utilized to isolate the superior individuals from a breeding population (Tian et al., 2011).
7. It has the potential to identify the epistasis between alleles at various loci and genes of minor effects.
8. AM has the potential to recognize contributory polymorphisms within a gene which are responsible for the variation between two traits (Palaisa et al., 2003).

11.8.2 LIMITATIONS

1. The main limitation arises when the desired trait under study is sturdily associated with population structure (Stich and Melchinger, 2010). When statistical methods are used to accurate population structure, the differences between subpopulations are overlooked during marker–trait association analysis. Therefore, all polymorphisms responsible for the phenotypic differences between subpopulations remain undetected; which can lower the power of AM.
2. GWAS mapping requires a large number of molecular markers for genotyping.
3. A high density of markers can only be achieved through the development of an integrated genotyping by sequencing (GBS) platform which requires lots of funds so it is a costly approach.
4. The efficiency of LD mapping can be affected by allelic frequency distribution at the functional polymorphism level. The outcomes of experimental studies suggest that a high percentage of alleles are rare. Rare alleles cannot be intended adequately because they are present only in few individuals who can cause lower map resolution (Myles et al., 2009).
5. It gives minor insight into the mechanistic basis of LD detected; therefore genomic mapping and cloning of genes based on LD may not be always easy.
6. Collection and maintenance of germplasm/landraces/elite breeding lines is a tedious job.
7. Requires sophisticated efforts and patients to obtain significant outcome.

11.9 FUTURE AND OPPORTUNITIES FOR RESEARCH

Development of new rice variety which gives good yield under stressful and unpredictable climatic conditions with having low inputs is essential for the food security of future generations. This is the most important challenge for today's rice breeders. To address these challenges, the proper understanding of genetic information/basis of crop productivity is essential for plant breeding community which can be possible by coupling plant breeding with modern DNA sequencing and molecular techniques. The above mentioned researches have demonstrated the vital potential of LD-based mapping of quantitative traits in rice germplasm lines using molecular markers. This type of LD based mapping could be a useful option to linkage mapping for the recognition of marker–phenotype associations, and show the way to utilization of marker-assisted selection in rice breeding programs. AM has more power than linkage analysis based mapping to identify alleles with weak effects that might contribute risk for common complex traits. In rice, natural alleles and alleles gained from artificially mutagenized populations provide an imperative resource for their breeding advancement. By using all accessible alleles and phenotypic data from core sets of rice lines, new genes and useful traits can be identified. Molecular tags for desirable trait could be developed using GWASs based on molecular data and phenotypic information. These data can also be used to track target traits during segregation of populations in rice breeding. The high LD predictable in rice lines and cultivars make possible the detection of marker-trait associations for applied marker assisted selection, and the mining of alleles related to important traits in germplasm. With the availability of high density maps of rice or whole genome sequence of rice, the AM tool have immense future for increasing their applications in rice breeding. AM could be helpful in identification of favorable alleles for marker aided selection and cross validation of results of linkage based mapping. It is usually performed with the objective of applying the results for genotype-based selection of superior individuals in plant breeding, or as a step toward positional cloning.

AM in rice opens a road for plant breeders to harness the new alleles/genes/QTLs of various traits which can be utilized for genetic improvement of rice cultivars and development of new rice variety. Mapped QTLs can be utilized by marker assisted selection. In future, phenotyping and genotyping approaches used for AM would be updated with high precision technique. Phenomics technique could be used for phenotyping (at least in some cases, depending upon the trait being studied) while next generation sequencing (NGS) may be used for genotyping, so that large number of sequence-based markers like SNPs, DArT, ESTs, ISBP, and CNV/PAV-based markers are used in huge quantity. For obtaining genetic information and improving any crop species, it is essential to collect large number of germplasm of that crop species. It is the first and foremost breeding activity for every plant breeder. In conclusion, AM in rice is a viable alternative to QTL mapping based on crosses between different lines.

11.10 SUMMERY AND CONCLUSION

AM or LD mapping is a method of mapping QTLs using historical meiotic recombination events performed over several generations to link phenotypes (observable characteristics) with genotypes (genetic constitution) in large germplasm populations. It is a powerful genome mapping tool in crops for high-resolution, broad allele coverage, and cost-effective gene tagging which can be utilized to evaluate the plant germplasm resources. Genetic mapping of QTLs can be performed in two main ways: (1) Linkage analysis-based mapping or conventional mapping using experimental populations or bi-parental mapping populations, and (2) LD-based mapping or "association mapping" using diverse lines from the natural populations or germplasm collections or landraces. Linkage-based QTL mapping can identify the related genes to bi-parental variations so that mapping resolution depends on the number of recombination occurred in the process of the development of mapping populations. Furthermore, construction of a suitable mapping population for study is very time consuming. AM has recently become popular for identifying and mapping QTLs with high resolution with the available collection of germplasm. It has the potential to overcome the limitations of linkage based QTL mapping. AM is based on identification of associations between phenotype and allelic frequencies using molecular markers. It gives opportunity to breeders for conducting commercial variety development and mapping process simultaneously. For phenotypes or traits that are governed by multiple genes or QTLs, diverse alleles or advantageous allele combinations should be mined by AM followed by gene tagging efforts using bi-parental crosses. Some factors must be considered to perform AM studies precisely are; large sample size, population composition, large number of molecular markers, and advanced statistical methods. In rice, there is lots of opportunity to map and tag the QTLs for different quantitative traits such as yield, drought tolerance, salt tolerance, root length, panicle length, and so forth through AM using available germplasm lines or landraces.

KEYWORDS

- Association mapping
- biparental mapping
- DNA markers
- EMMA
- gene localization
- genomics
- GWAS
- heritable phenotypic differences
- LD mapping
- marker-assisted breeding
- McNemar's test
- near isogenic lines
- phenomics
- principal component analysis
- QTL mapping
- recombinant inbred lines
- TASSEL

REFERENCES

Anonymous. Production, Supply, and Distribution Online Data Base, USDA, Foreign Agricultural Service. *Rice Outlook, RCS,* **2016** (16I), 11–12.

Agrama, H. A.; Eizenga, G. C.; Yan, W. Association Mapping of Yield and its Components in Rice Cultivars. *Mol. Breed.* **2007**, *19*, 341–356.

Asins, M. Present and Future of Quantitative Trait Locus Analysis in Plant Breeding. *Plant Breed.* **2002**, *121*, 281–291.

Balding, D. J. A Tutorial on Statistical Methods for Population Association Studies. *Nat. Genet.* **2006**, *7*, 781–791.

Begum, H.; Spindel, J. E.; Lalusin, A.; Borromeo, T.; Gregorio, G.; Hernandez, J.; Virk, P.; Vollard, B.; McCouch, S. R. Genome Wide Association Mapping for Yield and Other Agronomic Traits in an Elite Breeding Population of Tropical Rice (*Oryza sativa*). PLoS One **2015**, *10*(3), e0119873.

Borba, C.; Oliveria, D.; Brondani, R. P. V.; Breseghello, F.; Coelho, A. S. G.; Mendonca, J. A.; Rangel, P. H. N.; Brondani, C. Association Mapping for Yield and Grain Quality Traits in Rice (*Oryza sativa* L.). *Genet. Mol. Biol.* **2010**, *33*, 515–524.

Braulio, J.; Cloutier, S. Association Mapping in Plant Genomes, Genetic Diversity in Plants, Prof. Mahmut Caliskan (Ed.), InTech Publications, **2012**, 30–54.

Buntjer, J. B.; Sorensen, A. P.; Peleman, J. D. Haplotype Diversity: the Link Between Statistical and Biological Association. *Trends Plant Sci.* **2005**, *10*(10), 466–471.

Cavanagh, C.; Morell, M.; Mackay, I.; Powell, W. From Mutations to MAGIC: Resources for Gene Discovery, Validation and Delivery in Crop Plants. *Current Opinions Plant Biol.* **2008**, *11*, 215–221.

Cattivelli, L.; Rizza, F.; Badeck, F. W. Drought Tolerance Improvement in Crop Plants: An Integrated View from Breeding to Genomics. *Field Crops Res.* **2008**, *105*, 1–14.

Courtois, B.; Audebert, A.; Dardou, A.; Roques, S.; Ghneim, T. H.; Droc, G.; Frouin, J.; Rouan, L.; Goze, E.; Kilian, A.; Ahmadi, N.; Dingkuhn, M. Genome-Wide Association Mapping of Root Traits in a Japonica Rice Panel. *PLoS One* **2013**, *8*(11), e78037.

Cui, D.; Xu, C.-Y.; Tang, C.-F.; Yang, C.-G.; Yu, T.-Q.; Xin-xiang, A.; Cao, G.-L.; Xu, F.-R.; Zhang, J.-G.; Han, L.-Z. Genetic structure and association mapping of cold tolerance in improved japonica rice germplasm at the booting stage. *Euphytica*, **2013**, *193*, 369–382.

Dang, X.; Thi, T. G.; Dong, G.; Wang, H.; Edzesi, W. M.; Hong, D. Genetic Diversity and Association Mapping of Seed Vigor in Rice (*Oryza sativa* L.). *Planta* **2014**, *239*(6), 1309–1319.

Devlin, B.; Roeder, K.Genomic Control for Association Studies. *Biometrics* **1999**, *55*(4), 997–1004.

Eizenga, G. C.; Shakiba, E.; Jodari, F.; Duke, S.; Korniviel, P.; Jackson A.; Mezey, J.; McCouch, S. The Secrets of Cold Tolerance at the Seedling Stage and Heading in Rice as Revealed by Association Mapping. Int. Plant Animal Genome XXIII, San Diego, CA. 2014, 10–14 Jan 2015. https://pag.confex.com/pag/xxiii/webprogram/Paper14223.html.

Falush, D.; Stephens, M.; Pritchard, J. K. Inference of Population Structure Using Multilocus Genotype Data: Linked Loci and Correlated Allele Frequencies. *Genetics* **2003**, *164*, 1567–1587.

Feng, Y.; Lu, Q.; Zhai, R.; Zhang, M.; Xu, Q.; Yang, Y.; Wang, S.; Yuan, H.; Wang, Y.; Wei, X. Genome Wide Association Mapping for Grain Shape Traits in Indica Rice. *Planta* **2016**, *244*(4), 819–830.

Fujino, K.; Obara, M.; Shimizu, T.; Koyanagi, K. O.; Ikegaya, T. Genome-Wide Association Mapping Focusing on a Rice Population Derived from Rice Breeding Programs in a Region. *Breed. Sci.* **2015**, *65*, 403–410.

Garris, A. J.; Tai, T. H.; Coburn, J.; Kresovich, S.; McCouch, S. Genetic Structure and Diversity in *Oryza sativa* L. *Genetics* **2005**, *169*, 1631–1638.

Gupta, P. K.; Kulwal, P. L.; Jaiswal, V. Association Mapping in Crop Plants: Opportunities and Challenges. 1st Edition, *Advances in Genetics*, 2014 109–147.

Gupta, P. K.; Rastogi, S.; Kulwal, P. L. Linkage Disequilibrium and Association Studies in Higher Plants: Present Status and Future Prospects. *Plant Mol. Biol.* **2005**, *57*, 461–485.

Hirschhorn, J. N.; Daly, M. J. Genome-Wide Association Studies for Common Diseases and Complex Traits. *Nat. Rev. Genet.* **2005**, *6*, 95–108.

Jia, L.; Yan, W.; Zhu, C.; Agrama, H. A.; Jackson, A.; Li, X.; Huang, B.; Hu, B.; McClung, A.; Wu, D.; Yeater, K Allelic Analysis of Sheath Blight Resistance with Association Mapping in Rice. *PLoS One* **2012**, *7*, e32703.

Khaing, A. A.; Li, G.; Wang, X. Q.; Yoon, M. Y.; Kwon, S. W.; Lee, C. Y.; Park, B. S.; Park, Y. J. Rice Germplasm in Korea and Association Mapping. Rice–Germplasm, Genetics and Improvement. Intech publication, 2014, pp 80–103.

Korte, A.; Vilhjalmsson, B. J.; Segura, V.; Platt, A.; Long, Q.; Nordborg, M. A Mixedmodel Approach for Genome-Wide Association Studies of Correlated Traits in Structured Populations. *Nat. Genet.* **2012**, *44*, 1066–1071.

Kumar, V.; Singh, A.; Mithra, S. V. A.; Krishnamurthy, S. L.; Parida, S. K.; Jain, S.; Tiwari, K. K.; Kumar, P.; Rao, A. R.; Sharma, S. K.; Khurana, J. P.; Singh, N. K.; Mohapatra, T. Genome-Wide Association Mapping of Salinity Tolerance in Rice (*Oryza sativa* L.). *DNA Res.* **2015**, *22*(2), 133–145.

Li, X.; Yan, W.; Agrama, H. et al. Unraveling the Complex Trait of Harvest Index with Association Mapping in Rice (*Oryza sativa* L.). *PLoS One* **2012**, *7*, e29350.

Liu, Z.; Zhu, C.; Jiang, Y.; Yu, J.; Hongzhou, A.; Tang, W.; Sun, J.; Tang, J.; Chen, G.; Zhai, H.; Wang, C.; Wan, J. Association Mapping and Genetic Dissection of Nitrogen Use Efficiency Related Traits in Rice (*Oryza sativa* L.). *Funct. Integr. Genomics* **2016**, *16*(3), 323–333.

Lu, Q.; Zhang, M.; Niu, X.; Wang, S.; Xu, Q.; Feng, Y.; Deng, H.; Yuan, X.; Yu, H.; Wang, Y.; Wei, X.; Wang, C. Genetic Variation and Association Mapping for 12 Agronomic Traits in Indica Rice. *BMC Genomics* **2015**, *16*(1067), 1–17.

Mackay, I.; Powell, W. Methods for Linkage Disequilibrium Mapping in Crops. *Trends Plant Sci.* **2007**, *12*(2), 57–63.

Magwa, R. A.; Zhao, H.; Xing, Y. Genome-Wide Association Mapping Revealed a Diverse Genetic Basis of Seed Dormancy Across Subpopulations in Rice (*Oryza sativa* L.). *BMC Genet.* **2016**, *17*(28), 1–13.

Mitchell, A. A.; Chakravarti, A. Undetected Genotyping Errors Cause Apparent Overtransmission of Common Alleles in the Transmission/Disequilibrium Test. *Am. J. Hum. Genet.* **2003**, *72*, 598–610.

Mott, R.; Flint, J. Simultaneous Detection and Fine Mapping of Quantitative Trait Loci in Mice Using Heterogeneous Stocks. *Genetics* **2002**, *160*, 1609–1618.

Mott, R.; Talbot, C. J.; Turri, M. G.; Collins, A. C.; Flint, J. A Method for Fine Mapping Quantitative Trait Loci in Outbred Animal Stocks. *Proc. Natl. Acad. Sci. U. S. A.* **2000**, *97*(23), 12649–12654.

Myles, S.; Peiffer, J.; Brown, P.; Ersoz, E.; Zhang, Z.; Costich, D.; Buckler, E. Association Mapping: Critical Considerations Shift from Genotyping to Experimental Design. *Plant Cell* **2009**, *21*(8), 2194–2202.

Palaisa, K.; Morgante, M.; Tingey, S.; Rafalski, A. Contrasting effects of selection on sequence diversity and linkage disequilibrium at two phytoene synthase loci. *Plant Cell* **2003**, *15*(8), 1795–1806.

Price, A. L.; Patterson, N. J.; Plenge, R. M.; Weinblatt, M. E.; Shadick, N. A.; Reich, D. Principal Components Analysis Corrects for Stratification in Genome-Wide Association Studies. *Nat. Genet.* **2006**, *38*, 904–909.

Pritchard, J. K.; Rosenberg, N. A. Use of Unlinked Benetic Markers to Detect Population Stratification in Association Studies. *Am. J. Hum. Genet.* **1999**, *65*(1), 220–228.

Pritchard, J. K.; Stephens, M.; Rosenberg, N. A.; Donnelly, P. Association Mapping in Structured Populations. *Am. J. Hum. Genet.* **2000**, *67*(1), 170–181.

Reeves, P.; Richards, C. Accurate Inference of Subtle Population Structure (and other genetic discontinuities) Using Principal Coordinates. *PLoS One* **2009**, 4-, e4269.

Reich, D. E.; Goldstein, D. B. Detecting Association in a Case Control Study While Correcting for Population Stratification. *Genet. Epidemiol.* **2001**, *20*, 4–16.

Remington, D. L.; Thornsberry, J.; Matsuoka, Y.; Wilson, L.; Rinehart-Whitt, S.; Doebley, J. Structure of Linkage Disequilibrium and Phenotypic Associations in the Maize Genome. *Proc. Natl. Acad. Sci. U. S. A.* **2001**, *98*, 11479–11484.

Risch, N.; Merikangas, K. The Future of Genetic Studies of Complex Human Diseases. *Science* **1996**, *273*, 1516–1517.

Rosyara, U. R.; Joshi, B. K. Association Mapping for Improvement of Quantitative Traits in Plant Breeding Population. *Nepal J. Biotechnol.* **2012**, *2*(1), 72–89.

Sahu, H.; Amadabade, J.; Dhiri, N. Association Mapping: A Useful Tool. *Int. J. Plant Sci.* **2015**, *10*(1), 85–94.

Segura, V.; Vilhjalmsson, B. J.; Platt, A.; Korte, A.; Seren, U.; Long, Q. An Efficient Multi-Locus Mixed-Model Approach for Genome-Wide Association Studies in Structured Populations. *Nat. Genet.* **2012**, *44*, 825–830.

Sehgal, D.; Singh, R.; Rajpal, V. R. *Quantitative Trait Loci Mapping in Plants: Concepts and Approaches.* Molecular Breeding for Sustainable Crop Improvement. Springer International Publishing: Switzerland, **2016**, pp 31–59.

Spielman, R.; McGinnis, S.; Ewens, W. J. Transmission Test for Linkage Disequilibrium: the Insulin Gene Region and Insulin-Dependent Diabetes Mellitus (IDDM). *Am. J. Hum. Genet.* **1993**, *52*, 506–516.

Stich, B.; Melchinger, A. An Introduction to Association Mapping in Plants. *CAB Rev. Perspect. Agric. Vet. Sci. Nutr. Nat. Resour.* **2010**, *5*(39), 1–9.

Thornsberry, J. M.; Goodman, M. M.; Doebley, J.; Kresovich, S.; Nielsen, D.; Buckler, E. Dwarf Polymorphisms Associate with Variation in Flowering Time. *Nat. Genet.* **2001**, *28*, 286–289.

Tian, F.; Bradbury, P.; Brown, P.; Hung, H.; Sun, Q.; Flint-Garcia, S.; Rocheford, T.; McMullen, M.; Holland, J.; Buckler, E. Genome-Wide Association Study of Leaf Architecture in the Maize Nested Association Mapping Population. *Nat. Genet.* **2011**, *43*(2), 159–162.

Vanniarajan, C.; Vinod, K.; Pereira, A. Molecular Evaluation of Genetic Diversity and Association Studies in Rice (*Oryza sativa* L.). *J. Genet.* **2012**, *91*, 9–19.

Vinod, K. K. Association Mapping in Crop Plants. Advanced Faculty Training on "Impact of Genomics on Crop Improvement: Perceived and Achieved". TNAU, Coimbatore, **2011**.

Wang, C.; Yang, Y.; Yuan, X.; Xu, Q.; Feng, Y.; Yu, H.; Wang, Y.; Wei, X. Genome Wide Association Study of Blast Resistance in Indica Rice. *BMC Plant Biol.* **2014**, *14*(311), 1–11.

Wen, W.; Mei, H.; Feng, F.; Sibin, Y.; Huang, Z.; Wu, J.; Chen, L.; Xu, X.; Luo, L. Population Structure and Association Mapping on Chromosome 7 Using a Diverse Panel of Chinese Germplasm of Rice (*Oryza sativa* L.). *Theor. Appl. Genet.* **2009**, *119*, 459–470.

Xu, F. F.; Tang, F. F.; Shao, Y. F.; Chen, Y. L.; Tong, C.; Bao, J. S. Genotype× Environment Interactions for Agronomic Traits of Rice Revealed by Association Mapping. *Rice. Sci.* **2014**, *21*, 133–141.

Yan, W. G.; Li, Y.; Agrama, H. A.; Luo, D.; Gao, F.; Lu, X.; Ren, G. Association Mapping of Stigma and Spikelet Characteristics in Rice (*Oryza sativa* L.). *Mol. Breed.* **2009**, *24*, 277–292.

Yu, J.; Buckler, E. S. Genetic Association Mapping and Genome Organization of Maize. *Curr. Opin. Biotechnol.* **2006**, *17*, 155–160.

Yu, J.; Pressoir, G.; Briggs, W. H.; VrohBi, I.; Yamasaki, M.; Doebley, J. F.; McMullen, M. D.; Gaut, B. S.; Nielsen, D. M.; Holland, J. B.; Kresovich, S.; Buckler, E. S. A Unified Mixed

Model Method for Association Mapping that Accounts for Multiple Levels of Relatedness. *Nat. Genet.* **2006**, *38*, 203–208.

Zaniab, A. S.; Adam, H. P.; David, R. Genome Wide Association Mapping for Drought Recovery Trait in Rice (*Oryza Sativa* L.). *Int. J. Appl. Agric. Sci.* **2015**, *1*(1), 11–18.

Zhang, P.; Liu, X.; Tong, H.; Lu, Y.; Li, J. Association Mapping for Important Agronomic Traits in Core Collection of Rice (*Oryza sativa* L.) with SSR Markers. *PLoS One* **2014**, *9*(10), e111508.

Zhang, Z.; Ersoz, E.; Lai, C. Q.; Todhunter, R. J.; Tiwari, H. K.; Gore, M. A.; Bradbury, P. J.; Yu, J.; Arnett, D. K.; Ordovas, J.M.; Buckler, E. S. Mixed Linear Model Approach Adapted for Genome-Wide Association Studies. *Nat. Genet.* **2010**, *42*(4), 355–360.

Zhao, K.; Tung, C. W.; Eizenga, G. C.; Wright, M. H.; Ali, M. L.; Price, A. H.; Norton, G. J.; Islam, M. R.; Rynolds, A.; Mezey, J.; McClung, A. M.; Bustamante, C. D.; McCouch, S. R. Genome-Wide Association Mapping Reveals a Rich Genetic Architecture of Complex Traits in Oryza Sativa L. *Nat. Commun.* **2011**, *2*(467), 1–10.

Zhu, C.; Gore, M.; Buckler, E. S.; Yu, J. Status and Prospects of Association Mapping in Plants. *Plant Genome* **2008**, *1*(1), 5–20.

GLOSSARY OF TECHNICAL TERMS

2,4-D: A synthetic auxin primarily employed as a selectyse herbicide which target broad-leaved weeds but not grass and could act as a plant growth regulator.

2-Acetyl-1-Pyrroline: It is one of the volatile compounds and the key aroma compound in aromatic rice varieties responsible for the delicious popcorn-like fragrance.

5-iodoacetamidofluorescein: The thiol-reactive 5-iodoacetamidofluorescein (5-IAF) can be used to produce bioconjugates with the 5-isomer of fluorescein.

Abiotic stress: The negative effect on a plants growth and development by hindering its production or quality of yield.

Abiotic stress: The negative impact of abiotic factors on the living organisms such as plant in a specific environment called abiotic stress.

Abiotic stresses: Abiotic stress is defined as the negative impact of non-living factors on the living organisms in a specific environment.

Acid invertase: Acid invertase type of invertase occurs at pH 4.5–5.0.

Aerobic rice: In the production system, is especially developed rice varieties which are grown in well-drained, non-puddled, and non-saturated soils

Agronomic traits: The traits relating to or promoting agronomy in a plant. This includes high yield, grain weight, soil condition, disease resistance and so forth.

Aliphatic index: The aliphatic index of a protein is defined as the relative volume occupied by aliphatic side chains (alanine, valine, isoleucine, and leucine). It may be regarded as a positive factor for the increase of thermostability of globular proteins.

Allele: Alternative form of the gene found on the same position of homologous chromosome of organism.

Amino acids: The organic compound that serves as a building block for proteins having amino group and a carboxylic acid group.

Antioxidant activity: It represents the ability to inhibit the process of oxidation. And the substances that neutralize the potential will effect of free radicals are generally grouped in so called antioxidant defense system.

Antioxidant species: Substances which have antioxidant property.

Antioxidant: Any substance that prevents or inhibits the oxidation of other molecules which are damaged by highly reactive chemicals containing oxygen, so called free-radicals.

Antioxidative enzymes: The enzyme systems that catalyse reactions to neutralize free radicals and ROS. The antioxidant enzymes viz. glutathioneperoxidase, catalase, and superoxide dismutase (SOD) are metabolized oxidative toxic intermediates.

Antisense: Having a sequence of nucleotides complementary to (and hence capable of binding to) a coding (or sense) sequence, which may be either that of the strand of a DNA double helix which undergoes transcription, or that of a messenger RNA molecule.

Aromatic rice: Aromatic or scented or fragrant rice are the special rice varieties that emit delicious fragrance when cooked.

ASA-GSH cycle: It is ascorbate-glutathione cycle known as metabolic pathway that detoxifies hydrogen peroxide (H_2O_2) in which is an ROS that produced as a waste product in the metabolism. This cycle involves antioxidant metabolites viz. ascorbate, glutathione and NADPH and the enzymes that are linking with such metabolites.

*Ascorbate peroxidases***:** *Ascorbate peroxidases* (or APX) are enzymes that detoxify peroxides such as hydrogen peroxide using ascorbate as a substrate.

Ascorbic acid: Ascorbic acid is a naturally occurring organic compound with antioxidant properties. It dissolves well in water to give mildly acidic solutions. Ascorbic acid is one form of vitamin C.

Association mapping: Association mapping is a method of mapping quantitative trait loci (QTLs) using historical meiotic recombination events performed over several generations to link phenotypes with genotypes in large germplasm populations.

Auxin: Plant growth hormones are indole derivatives synthesized from tryptophan that promote phototropism, apical dominance, cell division.

Bacterial elimination: A process to kill or to remove bacteria that grow on an explant using an antibiotic solution.

Binary vector: These are shuttle vector that has the ability to replicate in multiple hosts. Used for *Agrobacterium* mediated gene transfer in plants.

Biochemistry: The branch of science concerned with the chemical and physico-chemical processes and substances which occur within living organisms.

Biological membranes: These are the form of cell membranes; often consist of a phospholipid bilayer with embedded, integral and peripheral proteins used in communication and transportation of chemicals and ions.

Bombardment: A techniques for insertion of naked DNA into explants or plant tissue using a special machine.

Breeding: A process of genetic-crossing between two selected male and female parents for the production of offspring progeny.

Broad-spectrum: Chemicals that very powerful, able to kill many types of plants or bacteria.

Broth: Bacterial medium in liquid form.

Bundle sheath: A layer of cells in plant leaves and stems that forms a sheath surrounding the vascular bundles. In C_4 plants (see C_4 pathway) the bundle sheath cells contain chloroplasts and are the site of the Calvin cycle.

Candidate-gene based association mapping: In this approach, only selected candidate genes are targeted for mapping which has specific role in controlling the desired trait of interest.

Carbohydrate metabolism: Carbohydrate metabolism denotes the various biochemical processes responsible for the formation, breakdown and interconversion of carbohydrates in living organisms. The most important carbohydrate is glucose, a simple sugar (monosaccharide) that is metabolizedby nearly all known organisms.

Carbohydrate: Any of a large group of organic compounds occurring in foods and living tissues and including sugars, starch, and cellulose. They contain hydrogen and oxygen in the same ratio as water (2:1) and typically can be broken down to release energy in the animal body.

Cas9: CRISPR associated protein 9 is an RNA-guided DNA endonuclease enzyme associated with the CRISPR adaptive immunity system in *Streptococcus pyogenes*, among other bacteria.

Catalase: An enzyme that decomposes hydrogenperoxide into oxygen and water.

Cell proliferation: Cell proliferation is the process that results in an increase of the number of cells, and is defined by the balance between cell divisions and cell loss through cell death or differentiation.

Cell signaling: Cell signaling is part of a complex system of communication that governs basic activities of cells and coordinates cell actions.

Chlorophyll "a": A molecule responsible for the green colour of many plants. This molecule absorbs sunlight and uses its energy to synthesise carbohydrates from CO_2 and water.

Chloroplast: A plastid in green plant cells which contains chlorophyll and in which photosynthesis takes place.

Co-cultivation: A condition when bacterial and plant tissues or cells are grown together.

Cold tolerance: Plant that survive and growth under lower temperature.

Comparative mapping: Comparative mapping is oOnce the two species have been mapped, the relative evolution of the two species can be compared by comparative mapping.

Competent cell: Cell that have ability to take up a foreign gene or plasmid.

CRISPR: Clustered regularly interspaced short palindromic repeats are segments of prokaryotic DNA containing short repetitions of base sequences. Each repetition is followed by short segments of "spacer DNA" from previous exposures to a bacteriophage virus or plasmid.

crRNA: CRISPRRNA encoded from the repeat CRISPR loci of bacteria and archaea.

Cytokinin: A class of plant hormones that involved in cell division, cell expansion and have also been shown to delay senescence, apical dominance.

Cytosol: The aqueous component of the cytoplasm of a cell, within which various organelles and particles are suspended.

Data mining: The performance of investigative large preexisting databases in order to generate new information.

DBD: A DNA-binding domain (DBD) is an independently folded protein domain that contains at least one structural motif that recognizes double- or single-stranded DNA.

Diamine oxidase: A copper-containing enzyme that has a high affinity for diamines such as putrescine and cadaverine.

Dicot: The dicotyledons are the flowering plants or angiosperms which are characterized by the presence of two embryogenic leaves or cotyledons.

Disulphide bond: A disulfide bond, also called an S-S bond, is a covalent bond derived from two thiol groups.

Domain: A conserved part of a given protein sequence and structure that can evolve, function, and exist independently of the rest of the protein chain.

Drought stress: Drought stress is a prolonged water deficit conditions in plants leads to cause the damage in photosystems.

Drought stress: Prolonged water deficit conditions in plants leads to cause the damage in photosystems.

Drought tolerance: Plant species that able to growth and proliferate in environment with shortage of water or water scarcity.

Ecological succession: The process by which the structure of a biological community evolves over time.

Ecology: The branch of biology that deals with the relations of organisms to one another and to their physical surroundings.

Ecosystem: Biological community of interacting organisms and their physical environment.

Electron transport chain: It is a series of compounds that transfer electrons from electron donors to electron acceptors via redox (both reduction and oxidation occurring simultaneously) reactions, and couples this electron transfer with the transfer of protons (H^+ ions) across a membrane.

Electroporation: It is a technique using an electrical pulse to increase cell permeability during gene transfer.

Encoding: Sequence of gene that represent or might be translate to the respective amino acid sequence or peptide.

Endogenous: Chemical or gene that released by the plant itself.

Environment: The physical conditions that affects the survival and growth of organism.

Enzyme: A protein molecule produced by a living organism which acts as a catalyst to bring about a specific biochemical reaction.

Eukaryotes: An organism consisting of a cell or cells in which the genetic material is DNA in the form of chromosomes contained within a distinct nucleus. Eukaryotes include all living organisms other than the eubacteria and archaea.

Exogenous: Chemicals or substances that came from outside of the cells.

Expressed sequence tags: Used to identify gene transcripts, and are instrumental in gene discovery and in gene sequence determination.

Expression: Protein or chemical that synthesis or released from a sequence of gene.

Fertilizer use efficiency: Extent of fertilisers use recovery in agriculture per crop unit.

Fructane: It is a polymer of fructose molecules, occur in foods such as agave, artichokes, asparagus, leeks, garlic, onions (including spring onions) and wheat.

Gas chromatography: Chromatography uses gas as the mobile phase to analyze or separate volatile components of mixture.

Gene expression: The procedure by which data from a gene is used in the production of a functional gene product.

Gene: Small part of DNA that transcribe to produce RNA molecule and finally express their effect.

Genetic transformation: The process by which the foreign genetic material introduced into the plant nuclear genome. The two most common methods employed for plant genetic transformation are direct (micro-projectile bombardment) and indirect (*Agrobacterium*-mediated transformation) genetic transformation.

Genome sequencing: It is used to outline out the array of DNA nucleotides, in a genome the order of As, Cs, Gs, and Ts (dNTPs) that makes an organism's genome.

Genome-wide association studies: The comprehensive approach to systematically scan the whole genome of organism with large number of equally distributed molecular markers to detect the genetic variations for particular trait of interest.

Genome: The complete set of genes or genetic material present in a cell or organism

Genomics: A discipline in genetics that applies recombinant DNA, DNA sequencing methods and bioinformatics to sequence, assemble and analyze the function and structure of genomes.

Genotype: The genetic constitution of an individual organism.

Germplasm: A set of living genetic pool such as seeds, tissues, tuber or plant that used for breeding or reproductive purposes.

Global warming: Gradual increase in the overall temperature of the earth's atmosphere generally attributed to the greenhouse effect caused by increased levels of carbon dioxide, CFCs, and other pollutants.

Glutathione reductase: It is a ubiquitous enzyme, which catalyzes the reduction of oxidized glutathione (GSSG) to glutathione (GSH). Glutathione reductase is essential for the glutathione redox cycle that maintains adequate levels of reduced cellular GSH.

Glycophyte: The plants which grow in fresh water and nutrient-rich soil are called glycophytes.

Gramineae (Poaceae): A large and nearly ubiquitous family of monocotyledonous flowering plants known as grasses. It includes the cereal grasses, bamboos and the grasses of natural grassland and cultivated lawns (turf) and pasture.

Ground water: Water held underground in the soil or in pores and crevices in rock.

Halophyte: A plant adapted to growing in saline conditions, as in a salt marsh.

Hardening: Process of subjecting the plants grown in a controlled environment into a natural growth conditions.

HDR: Homology directed repair (HDR) is a mechanism in cells to repair double strand DNA lesions. The most common form of HDR is homologous recombination.

Headspace solid-phase microextraction (HS-SPME): Solvent-free sampling technique by which the volatile compounds absorbed to a fiber coated with polymer introduced into the headspace, subsequently desorbed into a gas chromatography for analysis.

Heat stress: Heat stress often is defined as where temperatures are hot enough for sufficient time that they cause irreversible damage to plant function or development. Plants can be damaged in different ways by either high day or high night temperatures and by either high air or high soil temperatures.

Herbicide resistance: Plant that cannot be killed by herbicide.

Heterosis: It is the improved or increased function of any biological quality in a hybrid offspring.

Histochemistry: The study of identification of chemical components within tissues using histological techniques such as staining, microscopic examinations, and so forth.

HNH: The nuclease domain of the Cas9 protein that cleaves the target DNA strand complementary to the RNA guide.

Homeostasis: It is the ability to maintain a constant internal environment in response to environmental changes.

Homologous gene: A gene inherited in two species by a common ancestor. While homologous genes can be similar in sequence, similar sequences are not necessarily homologous.

Homologous recombination: A type of genetic recombination process in which nucleotide sequences are exchanged between two similar or identical molecules of DNA.

Homology: Homology is tThe existence of shared ancestry between a pair of structures, or genes, in different taxa is known as homology.

Hydroxyl radicals: Hydroxyl radicals (•OH) are the neutral form of the hydroxide ion (OH⁻). They are highly reactive (easily becoming hydroxyl groups) and consequently short-lived.

Immunity: Balanced state of having adequate biological defenses to fight infection or unwanted biological invasion.

In vitro callus: A mass of undifferentiated plant cells induced by hormone treatment in tissue culture under controlled conditions.

In vitro plant regeneration: A morphogenetic response to a stimulus that results in the entire plant generation from a single or group of cells under a controlled condition.

Infection: A process that involve the invasion of host cell by the pathogen such as *Agrobacterium*.

Infiltration: Process that chemical or solution is passing through the membrane cell or cell wall.

Instability index: Instability indexIt is a measure of proteins, used to determine whether it will be stable in a test tube.

Integration: A condition when the inserted foreign gene is correctly ligated or linked with the host genome.

Intrinsic markers: Internal signal that presence or derived from host family or itself.

Invertase: An enzyme that catalyses the hydrolysis (breakdown) of sucrose.

Ionic strength: The levels of ions in a formulated medium or culture solution.

Isobaric tag for relative and absolute quantitation: Isobaric tags for relative and absolute quantitation (iTRAQ) are an isobaric labeling method used for quantitative analysis of proteomics using tandem mass spectrometry to determine the amount of proteins from different sources in a single experiment.

Isoelectric point: Isoelectric point is the isoelectric point (pI, pH(I), IEP), is tThe pH at which a particular molecule carries no net electrical charge in the statistical mean.

Isotope-coded affinity tag: An Isotope-coded affinity tag (ICAT) is an isotopic labeling method used for quantitative proteomics by mass spectrometry that uses chemical labeling reagents.

Kinship: Coefficient of relatedness between pairs of each individual within a sample.

Landrace: Old cultivated form of a crop which are potentially adapted to local growing conditions and have wider genetic base but unimproved by modern plant breeding methods.

Linkage disequilibrium: Non-random association of alleles between genetic loci on the same or different chromosome.

Linkage equilibrium: The transmission of alleles at one locus across the generation is independent to alleles of other locus.

Linkage: Inheritance of two or more genes together as a single haplotype without any substantial recombination frequency in a family or pedigree.

Lipid peroxidation: Oxidative degradation of lipids. It is the process in which free radicals causes release of electrons from the lipids in cell membranes, resulting in cell damage. This process proceeds by a free radical chain reaction mechanism.

Mapping population: A structured population derived from an F_1, F_2 or a back-cross, which are used to construct genetic maps and to detect and locate QTL on those maps by family-based linkage analysis.

Mapping: Process used to identify the locus of a gene/QTL and the distances between genes/markers/QTLs in the genome.

Marker: Polymorphic sequence of nucleotide in the genome which can be easily identifiable and detectable.

Marker-assisted selection: It is an indirect selection process where a trait of interest is selected based on a marker linked to a trait of interest rather than on the trait itself.

Medium: Chemical formulation used for cell culture or growing of independent cell.

Membrane transport: Membrane transport is the collection of mechanisms that regulate the passage of solutes such as ions and small molecules through biological membranes, which are lipid bilayers that contain proteins embedded in them.

Metabolome: The total number of metabolites present within an organisms, cell or tissue.

Micro projectile: Insertion of gene into plant cell using a special machine.

Microarray: A deposit of DNA series on behalf of the entire set of genes of an organism, prearranged in a network prototype for use in genetic test.

Molecular biology: The branch of biology that deals with the structure and function of the macromolecules (e.g. proteins and nucleic acids) essential to life.

Molecular breeding: production of new clone using genetic engineering methods.

Monocot: The monocotyledons are the flowering plants or angiosperms which are characterized by the presence of only one embryogenic leaf or cotyledon.

Multigenic: Group of genes with similar nucleotide sequences and various functional responses.

Necrosis: It is a form of cell injury which results in the premature death of cells in living tissue by autolysis.

NHEJ: Non-homologous end joining (NHEJ) is a pathway that repairs double-strand breaks in DNA. NHEJ is referred to as "non-homologous" because the break ends are directly ligated without the need for a homologous template.

NLS: Nuclear localization signal or sequence is an amino acid sequence that 'tags' a protein for import into the cell nucleus by nuclear transport.

Non-reducing sugars: A carbohydrate that is not oxidized by a weak oxidizing agent (an oxidizing agent that oxidizes aldehydes but not alcohols, such as the Tollen's reagent) in basic aqueous solution. for example sucrose.

NUC: The nuclease lobe of Cas9 protein that carries the two nuclease domain HNH and RuvC.

Nucleic acids: Complex organic substance present in living cells, especially DNA or RNA, whose molecules consist of many nucleotides linked in a long chain.

Omics: The study of entire thing in biology ending in omics.

Orthologous gene: Orthologous are homologous genes where a gene is found in two different species, but the origin of the gene is a common ancestor.

Orthologs: Genes in different species that evolved from a common ancestral gene by speciation. Normally, orthologs retain the same function in the course of evolution.

Osmolytes: Any substance that contributes to the (regulation of) osmotic pressure in cells and tissues.

Osmoprotectants: They are small molecules that act as osmolytes and help organisms survive extreme osmotic stress. In plants, their accumulation can increase survival during stresses such as drought. Examples of compatible solutes include betaines, amino acids, and the sugar trehalose.

Osmotic adjustment: It is defined as a lowering of osmotic potential ($\psi\pi$) due to net solute accumulation in response to water stress.

Overexpression: A conceptually straightforward way that overexpression can inhibit another protein is simply to reduce the amount of that protein.

Oxidative stress: It is essentially an imbalance between the production of free radicals and the ability of the body to counteract or detoxify their harmful effects through neutralization by antioxidants.

PAM: Protospacer adjacent motif is a 2–6 base pair DNA sequence immediately following the DNA sequence targeted by the Cas9 nuclease in the CRISPR bacterial adaptive immune system.

Paralogous gene: A duplicated copy of the same gene within a species. However, they may have different sequence composition and function with course of time.

Peptide fingerprinting: An analytical technique for protein identification in which the unknown protein of interest is first cleaved into smaller peptides, whose absolute masses can be accurately measured with a mass spectrometer.

Peroxidase: An enzyme that catalyzes the oxidation of a particular substrate by hydrogen peroxide.

Peroxisomes: Cell organelle containing catalase, peroxidase, and other oxidative enzymes and other oxidative enzymes and performing essential metabolic functions, as the decomposition of fatty acids and hydrogen peroxide.

Phenome: Set of all phenotypes expressed by a cell, tissue, organ, organism, or species.

Phosphoproteome: A branch of proteomics that identifies, catalogs, and characterizes proteins containing a phosphate group as a post-translational modifications.

Photoassimilate: Any compound formed by assimilation of others under the action of light; especially such carbohydrates that are formed by photosynthes.

Photooxidation: Oxidation under the influence of radiant energy (as light).

Photorespiration: A respiratory process in many higher plants by which they take up oxygen in the light and give out some carbon dioxide, contrary to the general pattern of photosynthesis.

Photosynthesis: The process by which green plants and some other organisms use sunlight to synthesize nutrients from carbon dioxide and water. Photosynthesis in plants generally involves the green pigment chlorophyll and generates oxygen as a by-product.

Physiological pathways: A sequence of enzymatic or other reactions by which one biological material is converted to another.

Physiology: The branch of biology that deals with the study of mechanism and metabolism that are involved in growth and development of living organisms.

Pigmentation: Process of colour development or formation in cell.

Polyamine oxidase (PAOs): The PAOs found in plants catalyse the oxidation of the polyamine substrates spermidine and spermine. The reaction products are propane-1,3-diamine and 1-pyrroline or 1-(3-aminopropyl) pyrrolinium, respectively, along with hydrogen peroxide. Plant PAOs are predominantly localised in the cell wall.

Polyamines metabolism: The metabolism which is regulated by the activity of the ornithine decarboxylase (ODC) enzyme.

Polyamines: An organic compound having two or more primary amino groups $-NH_2$.

Polyols: An alcohol containing multiple hydroxyl groups. polyol" has a special meaning: food science and polymer chemistry.

Polyunsaturated fatty acid (PUFAs): Fatty acids that contain more than one double bond in their backbone. This class includes many important compounds, such as essential fatty acids and those that give drying oils their characteristic property.

Population structure: The level of genetic differentiation among groups within sampled population individuals.

Prokaryotes: Microscopic single-celled organism which has neither a distinct nucleus with a membrane nor other specialized organelles, including the bacteria and cyanobacteria.

Proline: A potent non-enzymatic antioxidant and potential inhibitor of programme cell death which is required by the microbes, animals and plants to mitigate the adverse effects of ROS.

Promoter: Promoter is a rRegion of DNA that initiates transcription of a particular gene.

Promoter: These are DNA sequences located near the transcription start sites in the upstream that initiates transcription of a particular gene.

Protein sorting: Protein sortingIt is the biological mechanism by which proteins are transported to the appropriate destinations in the cell or outside of it.

Proteome: The entire complement of proteins that is or can be expressed by a cell, tissue, or organism.

Purine catabolism: Pyrimidines are ultimately catabolized (degraded) to CO_2, H_2O, and urea. For example, cytosine can be broken down to uracil, which can be further broken down to N-carbamoyl-β-alanine, and then to beta-alanine, CO_2, and ammonia by β-ureidopropionase.

Putrescine: A foul-smelling compound produced by the breakdown of amino acids in living and dead organisms.

QTL mapping: Process of constructing linkage maps and conducting QTL analysis to identify genomic regions associated with polygenic traits is known as QTL mapping.

QTL: Genomic region that contain genes associated with particular quantitative trait.

RAMPs: Repeat Associated Mysterious Proteins (RAMPs) are similar to Cas protein and play novel role bacterial immunity by cleaving the sequences of potential viral invaders based on sequence similarity with the DNA of host organism.

Redox potential: It is a gauge of the propensity of a chemical species to obtain electrons and thus be abridged.

Reproductive phase: It is the period when an individual organism reproduces sexually.

Resistance: Plant that manage to protect them self from infection of pathogen.

Ripening period: The time after anthesis during which the fruits plant become soft textured, and accumulate soluble sugars, pigments and aroma volatiles.

RNA fingerprinting: RNA fingerprinting is a new method to screen for differences in plant litter degrading microbial communities.

RuvC: The nuclease domain of the Cas9 protein that cleaves the non-target DNA strand.

Saline soil: Non-sodic soil containing sufficient soluble salt to adversely affect the growth of most crop plants with a lower limit of electrical conductivity of the saturated extract (ECe) being 4 dS/m.

Salinity stress: Increasing the saline concentration to soil or water. It adversely affects plant growth and development.

Salinity tolerance: Explants or plants that able to growth and reproduce under extreme or high salinity levels.

Salinity: Measure of all the salts dissolved in water.

Salt stress: A condition where excessive salts in soil solution cause inhibition of organism growth or death.

Salt tolerance: The degree to which plant can withstand, without significant adverse effects, moderate or high concentrations of salt in water on its leaves or in the soil within reach of its roots.

Selection: Intensive process or practise in identification of elite clone of choice.

Semi-aquatic crop: Rice is an example of semi aquatic crop.

Semi-aquatic: Partly aquatic, growing or living in or close to water, or carrying out part of its life cycle in water.

Seminal roots: Secondary roots of a plant which support the primary root. This root system is then replaced by adventitious roots.

Sensory analysis: Organoleptic sensory analysis is one of the simple detection methods involves, smelling of the sample after physical or chemical treatment.

sgRNA: single guide RNA is a chimeric noncoding RNA that can be subdivided into three regions: a 20 nt base-pairing sequence, a 42 nt dCas9-binding hairpin and a 40 nt terminator.

Shoot regeneration: A conditions or process of shoot formation and development from unorganised tissue.

Signal transduction: Signal transduction which play a key role to converts a stimulus into a response in the cell. There are two stages in this process: A signalling molecule attaches to a receptor protein on the cell membrane. A second messenger transmits the signal into the cell, and a change takes place in the cell.

Signal transduction: The transfer or conveying the genetic message in the form of protein or product by sorting to a specific organelle.

Signaling molecule: Molecules which interact with a target cell as a ligand to cell surface receptors, and/or by entering into the cell through its membrane or by endocytosis for signaling.

Singlet oxygen: Common name of an electronically excited state of molecular oxygen which is less stable than molecular oxygen in the electronic ground state.

Site-specific mutagenesis: A molecular biology method that is used to make specific and intentional changes to the DNA sequence of a gene and any gene products.

Sodic soil: The amount of sodium held in a soil, is also called sodicity.

Somatic Embryogenesis: Somatic embryogenesis is the formation of embryo from single somatic cell.

Source-sink relationships: The integration of sugar and amino acid production in photosynthesis with sugar and amino acid utilisation in growth, storage, maintenance and production.

Spermidine: A polyamine formed from putrescine. It is found in almost all tissues in association with nucleic acids. It is found as a cation at all pH values, and is thought to help stabilize some membranes and nucleic acid structures. It is a precursor of spermine.

Spermine: A biogenic polyamine formed from spermidine. It is found in a wide variety of organisms and tissues.

Spikelet sterility: It is measurement; deal with counts of well-developed fertile spikelets in proportion to total number of spikelets on five panicles at harvest.

SSNs: Customized site-specific nucleases composed of programmable, sequence-specific DNA-binding modules linked to a non-specific DNA cleavage domain that are capable inducing double stranded breaks.

Stress-associated proteins (SAPs): Stress associated proteins (SAPs) whichThese are present in both plants and animals. It contains A20/AN1 zinc -finger domain which is responsible for abiotic stress tolerance in plants.

Structured population: A population in which individuals cannot mate with each other at random.

Subcellular localization: Subcellular localization is the cells of eukaryotic organisms are elaborately subdivided into functionally-distinct membrane-bound compartments.

Sucrose phosphate synthase: A plant enzyme involved in sucrose biosynthesis. SPS plays a major role in partitioning carbon between sucrose and starch in photosynthetic and non-photosynthetic tissues, affecting the growth and development of the plant.

Sucrose synthase: A key enzyme in plant sucrose catabolism is uniquely able to mobilize sucrose into multiple pathways involved in metabolic, structural and storage functions.

Super SAGE: Serial analysis of gene expression (SAGE) is determining the mRNA population in a desired sample by producing a snapshot of the total mRNA content.

Superoxide dismutase: An enzyme that alternately catalyzes the dismutation (or partitioning) of the superoxide (O_2^-) radical into either ordinary molecular oxygen (O_2) or hydrogen peroxide (H_2O_2). They help to break down the potentially harmful oxygen molecules in cells, which might prevent damage to tissues.

Superoxide radical: A compound that contains the superoxide anion with the chemical formula O^{-2}.

Surface sterilization: Sterilization of surface of the plant materials using mild concentration of chemicals without affecting the viability of the explant.

Susceptible: Plant that sensitive toward infection.

TALENs: Transcription activator-like effector nucleases (TALEN) are restriction enzymes made by fusing a TAL effector DNA-binding domain to a DNA cleavage domain.

TBARS (Thiobarbituric acid reactive substances): A byproduct of lipid peroxidation (i.e. as degradation products of fats) which can be detected by the TBARS assay using thiobarbituric acid as a reagent.

Temperature: conditions use during the incubation of sample.

Thermostable: cell that not affected by drastic changes of temperature.

Thylakoid: These are membrane-bound compartments inside of the chloroplasts. Thylakoids are the epicenter for photosynthetic light-reactions. They contain the chlorophyll for the plant, which is the light-collecting pigment.

Thylakoidal membranes: A membrane-bound compartments inside of the chloroplasts. Thylakoids are the epicenter for photosynthetic light-reactions. They contain the chlorophyll for the plant, which is the light-collecting pigment.

Tolerance: Plant that have capability to survive under extreme conditions.

tracrRNA: trans-activating crRNA (tracrRNA) is a small trans-encoded RNA. It was first discovered in the human pathogen Streptococcus pyogenes.

Trait: A distinct variant of a phenotypic characteristic of an organism; it may be either inherited or determined environmentally.

Transcriptional factors: It is a protein binding to a specific DNA sequence by controlling its rate of transcription

Transcriptome: The sum total of all the messenger RNA molecules expressed from the genes of an organism.

Transcriptomics: The field of transcriptomics allows for the examination of whole transcriptome changes across a variety of biological conditions.

Transformation: Modification of plant genome, where the inserted foreign gene is stably integrated with host genome and expressed.

Transgene: Gene or genetic sequence that transferred into plant tissue using genetic engineering protocols.

Transgenesis: The process of introducing an exogenous gene into a living organism so that the organism will exhibit a new property and transmit that property to its offspring.

Transgenic: An organism that contains genetic material into which DNA from an unrelated organism has been artificially introduced.

Transgenic: Explants or plant tissue that received a foreign gene or mutated.

Transpiration: The process by which moisture is carried through plants from roots to small pores on the underside of leaves, where it changes to vapor and is released to the atmosphere. Transpiration is essentially evaporation of water from plant leaves.

Transplanted rice: The rice seed is sown in one place and the seedlings after they have grown a little are transplanted to another. This is done in order to get higher yields and less weeding.

Ubiquitination: It is the process ofan addition of Uubiquitin to a substrate protein is known as ubiquitination. It can affect the proteins in several ways.- It is able to alter their cellular location, affect their activity, promote or prevent protein interactions.

Unsaturated fatty acids: They are liquid a room temperature. Unsaturated fats are derived from plants and some animals. They contain at least one double bond in their fatty acid chain.

Vegetative Phase: The period of growth between germination and flowering is known as the vegetative phase of plant development.

Visualization: observation on unstained transgenic explants by naked eyes.

Water deficit stress: Plants that receive insufficient water experience a stress called water deficit. Plants respond to soil water deficits by avoiding the occurrence of leaf water deficits (drought avoidance) or tolerating low cellular water contents (drought tolerance).

Water efficiency: The efficiency to reduce the water wastage by measuring the amount of water required for a particular purpose and the amount of water used or delivered.

Water holding capacity: It is defined as the water retained between field capacity and wilting point.

Water potential: The measure of the relative tendency of water to move from one area to another, and is commonly represented by the Greek letter Ψ (Psi).

Water use efficiency (WUE): It is refer as the ratio of water used in plant metabolism to water lost by the plant through transpiration. Water-use efficiency of productivity also known as integrated water-use efficiency, which is typically defined as the ratio of biomass produced to the rate of transpiration.

ZFNs: These are artificial restriction enzymes generated by fusing a zinc finger DNA-binding domain to a DNA-cleavage domain.

β-oxidation of fatty acids: It is the catabolic process by which fatty acid molecules are broken down in the cytosol in prokaryotes and in the mitochondria in eukaryotes to generate acetyl-CoA, which enters the citric acid cycle, and NADH and FADH2, which are co-enzyme used in electron transport chain.

INDEX

Printed and bound by CPI Group (UK) Ltd, Croydon, CR0 4YY

23/10/2024

01777703-0007